Recent Trends in Operator Theory and Applications

CONTEMPORARY MATHEMATICS

737

Recent Trends in Operator Theory and Applications

Workshop
Recent Trends in Operator Theory and Applications
May 3–5, 2018
The University of Memphis, Memphis, TN

Fernanda Botelho
Editor

AMS AMERICAN MATHEMATICAL SOCIETY

2010 *Mathematics Subject Classification*. Primary 47Bxx, 47Axx, 47L05, 46Bxx, 46Jxx, 26A15, 41Axx, 46E15, 46Lxx, 40G15.

Library of Congress Cataloging-in-Publication Data

Names: Workshop on Recent Trends in Operator Theory and Applications (2018 : Memphis, Tenn.) | Botelho, Fernanda, 1957– editor.

Title: Recent trends in operator theory and applications : Workshop on Recent Trends in Operator Theory and Applications, May 3–5, 2018, the University of Memphis, Memphis, Tennessee / Fernanda Botelho, editor.

Description: Providence, Rhode Island : American Mathematical Society, [2019] | Series: Contemporary mathematics ; volume 737 | Includes bibliographical references.

Identifiers: LCCN 2019015108 | ISBN 9781470448950 (alk. paper)

Subjects: LCSH: Operator theory–Congresses. | Functional analysis–Congresses. | AMS: Operator theory – Special classes of linear operators – Special classes of linear operators. msc | Operator theory – General theory of linear operators – General theory of linear operators. msc | Operator theory – Linear spaces and algebras of operators – Linear spaces of operators. msc | Functional analysis – Normed linear spaces and Banach spaces; Banach lattices – Normed linear spaces and Banach spaces; Banach lattices. msc | Functional analysis – Commutative Banach algebras and commutative topological algebras – Commutative Banach algebras and commutative topological algebras. msc | Linear and multilinear algebra; matrix theory – Basic linear algebra – Norms of matrices, numerical range, applications of functional analysis to matrix theory. msc | Approximations and expansions – Approximations and expansions – Approximations and expansions. msc | General topology – Maps and general types of spaces defined by maps – Function spaces. msc | Nonassociative rings and algebras – Jordan algebras (algebras, triples and pairs) – Jordan structures on Banach spaces and algebras. msc

Classification: LCC QA329 .W677 2018 | DDC 515/.724–dc23

LC record available at https://lccn.loc.gov/2019015108

Contemporary Mathematics ISSN: 0271-4132 (print); ISSN: 1098-3627 (online)

DOI: https://doi.org/10.1090/conm/737

∞ The paper used in this book is acid-free and falls within the guidelines
established to ensure permanence and durability.
Visit the AMS home page at `https://www.ams.org/`

10 9 8 7 6 5 4 3 2 1   24 23 22 21 20 19

Contents

Preface

This volume presents a collection of invited articles written by participants of the workshop on Recent Trends in Operator Theory and Applications (RTOTA 2018), held at the University of Memphis in May 2018. The contributions include both survey articles and original research papers focusing on methods and advances in operator theory.

Operator theory is an important branch of mathematics that offers a broad range of challenging and interesting research problems. Several areas of mathematics rely on techniques from operator theory, ranging from Banach space theory to partial differential equations and dynamical systems. The broad impact of this field also includes powerful tools for the development of other areas of science, including quantum computing, signal reconstruction, and approximation theory. This volume contains expository articles, co-authored by experienced and well recognized researchers and graduate students. These articles aim to introduce active fields within operator theory to early career researchers and graduate students. They provide insightful references and selection of results with articulation to modern research and advances in the area. Topics addressed by such articles include:

- Generalized numerical ranges and their application to study dilation and perturbation of operators, as well as connections to quantum error correction;
- Toeplitz operators and the existence of functions with specified zeroes in Toeplitz kernels. Their application to study linearly independent sets of reproducing kernel functions and their analogues in model spaces;
- The 2-local reflexivity problem for a set of operators on spaces of functions;
- Topics from the theory of preservers. A characterization of bijective linear maps between spaces of bounded operators that commute with the mean transformation; and
- Recent trends on the study of quotients of tensor product spaces and tensor operators.

The volume also includes research articles that present overviews of state-of-the-art techniques from operator theory with applications to recent research trends and open questions. Some of the topics addressed by these contributions are characterizations of classes of operators on JB*-triples, properties of Banach lattices and the Köthe dual space of a Banach lattice, topics from the geometry of Banach spaces, topological properties of operations, and applications of operator theory to statistical approximation.

An overall goal of all the articles is to present results accessible to the general public and, in particular, to newcomers to the topic.

The editor is most grateful to the National Science Foundation (Award DMS-1802313) for sponsorship and financial support. Special thanks go to all participants of the workshop and all contributors to this volume. Also special thanks go to many colleagues for their invaluable help with the refereeing process. The editor is also most thankful to the American Mathematical Society for making these papers widely available and by publising this volume.

The RTOTA 2018 workshop was in cooperation with the Association for Women in Mathematics (AWM) and supported their Non-Discrimination Statement.

Fernanda Botelho
Memphis, Tennessee
March 15, 2019

Contemporary Mathematics
Volume **737**, 2019
https://doi.org/10.1090/conm/737/14855

Sequence spaces on Banach lattices

SURVEY

Geraldo Botelho, Ryan Causey, and Khazhak V. Navoyan

ABSTRACT. In this survey we introduce the concept of Köthe dual for the Dedekind complete Riesz space $L_0 = L_0(X, \mu)$ of all $\mu -$ *measurable* real functions on the non-empty point set X, and then, for a Banach lattice E, we discuss the generalized sequence space $\lambda_\pi(E)$, the definition of which is based on the Köthe dual of the sequence space λ. Further, we consider the Orlicz space $\ell_\phi(E)$ as well. Finally, we introduce a useful thinning construction of a series for the purposes of further investigation.

1. Introduction

The *Köthe dual space* of the ideal A in the Dedekind complete Riesz space $L_0 = L_0(X, \mu)$ of all $\mu -$ *measurable* real functions on the non-empty point set X takes its name after G. Köthe, who was one of the first to introduce this notion. Initially this space was introduced by him for the case of sequence spaces, when X is the set of natural numbers with discrete measure, and $L_0 = L_0(X, \mu)$ in this case is the space of all real sequences [**11**].

After giving a possibly complete list of definitions and theorems in Section 2, "Preliminaries", which are necessary to read the current work, we pass in Section 3, "*Köthe dual*", to a detailed description of the *Köthe dual space* of the ideal A in the Dedekind complete Riesz space $L_0 = L_0(X, \mu)$, introduced in Zaanen's book [**11**], "Introduction to operator theory in Riesz spaces". We start with the definition of a career $c(A)$ of the ideal A which (the career), by definition, has the property that every member of A vanishes $\mu -$ *almost everywhere* on its complement. While the characteristic function $\chi_{c(A)}$ of $c(A)$ has the same property as well, there is a counterexample, showing that $\chi_{c(A)}$ is not a member of A. Nevertheless, we see that for any measurable subset Y of $c(A)$, there is an increasing sequence of measurable sets of finite measure which converge to Y, and the characteristic functions of each of these sets belong to the ideal A. Exactly this useful property is used for the further construction of the *Köthe dual space* of the ideal A. In this same Section 3, finally it is shown that an order continuous functional ϕ on the ideal A, and the corresponding to it function $t \in L_0(X, \mu)$ determine each other uniquely $\mu -$ *almost everywhere*. The set of all those $t \in L_0(X, \mu)$, corresponding to the elements $\phi \in A_n^\sim$, is a vector

2010 *Mathematics Subject Classification.* Primary 46B42, 46B45.

subspace of $L_0(X,\mu)$, and is denoted by K_A. Moreover, K_A is an ideal in $L_0(X,\mu)$, called the *Köthe dual space* of the ideal A.

In Section 4, on the basis of the sequence space λ, introduced as a subspace of $\mathbb{R}^{\mathbb{N}}$, as well as equipped with the vector lattice structure and assumed to be a Banach lattice, the Banach lattice valued sequence space $\lambda_\pi(E)$ is given, where E is a Banach lattice. Together with $\lambda_\pi(E)$ other similar Banach space and Banach lattice valued sequence spaces are introduced as well, and the norm on $\lambda_\pi(E)$ is defined, followed by very used properties of $\lambda_\pi(E)$ from [**1**–**6**]. In Section 4, the introduction and properties of the Orlicz function ϕ, Orlicz space ℓ_ϕ, Banach space valued Orlicz space $\ell_\phi(X)$ are given as well.

In Section 5, based on the proof of the Schur property for the Banach space ℓ_1 of all summable real number sequences, introduced in Megginson's book [**9**], "An introduction to Banach space theory", an interesting construction of the series used in the definition of $\lambda_\pi(E)$ is given, in a way that the "middle part" of the thinned series satisfies the property of being greater than the $\frac{3}{4}$ of the initial corresponding series. In the book of Megginson we find the series of ℓ_1 described as "thinned" without further details, which gives a good source of further thinking.

For the theory of Banach spaces the reader is refereed to [**9**], for the theory of Banach lattices to [**10**], and for the basic theory of Riesz spaces to [**11**].

2. Preliminaries

In this section we are going to give basic notions necessary to follow the rest of the material.

THEOREM 2.1. *(Radon-Nikodym theorem). If ν is a σ-additive measure on the σ-algebra Γ of subsets of the set X which is μ-absolutely continuous, then there exists a function f in the space $L_1(X,\Gamma,\mu)$ such that $\nu(A)=\int_A f d\mu$ holds for every $A\in\Gamma$.*

DEFINITION 2.2. The mapping ν from the algebra Γ into $\mathbb{R}$ is called a real finitely additive signed measure on Γ if $\nu(A_1\cup A_2)=\nu(A_1)+\nu(A_2)$ holds for all disjoint A_1 and A_2 in Γ. If Γ is a σ-algebra and $\nu(\cup_{n=1}^\infty A_n)=\sum_{n=1}^\infty \nu(A_n)$ holds for every disjoint sequence $\{A_n\}_{n=1}^\infty$ in Γ, then ν is called a σ-additive signed measure on Γ.

DEFINITION 2.3. A partially ordered set X is called Dedekind complete if every non-empty subset of X that is bounded above has a supremum. X is called Dedekind σ-complete if every non-empty finite or countable subset of X that is bounded above has a supremum. X is called a lattice, if it contains the supremum and the infimum of each of two elements.

DEFINITION 2.4. The real vector space E is called an ordered vector space if E is partially ordered in a way that the vector space structure and the order structure are compatible, that is

(i) $f\le g$ implies $f+h\le g+h$ for every $h\in E$,
(ii) $f\ge 0$ implies $\alpha f\ge 0$ for every $\alpha\ge 0$ in $\mathbb{R}$.

If, in addition, E is a lattice with respect to the partial ordering, then E is called a Riesz space or a vector lattice.

DEFINITION 2.5. The subset A of the Riesz space E is said to be solid if it follows from $f\in A$ and $|g|\le|f|$ ($g\in E$), that $g\in A$. A is called an ideal (or an order ideal to distinguish it from the algebraic ideal) if A is a solid linear subspace of E.

DEFINITION 2.6. The non-empty subset D of the Riesz space E is said to be downwards directed ($D \downarrow$) if for any two elements f and g in D there exists an element $h \in D$ such that $h \leq f \wedge g$ ($h \leq inf\{f, g\}$). If $D \downarrow$ and D has infimum f_0, then it is written $D \downarrow f_0$.

DEFINITION 2.7. Let E be a real Riesz space, equipped with a norm. The norm in E is called a Riesz norm if $|f| \leq |g|$ in E implies $\|f\| \leq \|g\|$. Any Riesz space, equipped with a Riesz norm, is called a normed Riesz space. If the normed Riesz space E is norm complete, i.e., every norm Cauchy sequence has a norm limit, then E is called a Banach lattice.

DEFINITION 2.8. The Riesz space E is said to be order separable if every set in E possessing an infimum contains a finite or countable subset having the same infimum.

DEFINITION 2.9. The Riesz space E is said to be super Dedekind complete if E is order separable and Dedekind complete, i.e., if every set in E which is bounded above has a supremum and contains a finite or countable subset having the same supremum.

DEFINITION 2.10. The normed Riesz space E is said to have order continuous norm if, for any subset $D \downarrow 0$ in E, we have $inf\{\|f\| : f \in D\} = 0$. The norm is said to be σ-order continuous if, for any sequence $f_n \downarrow 0$ in E, we have $\|f_n\| \downarrow 0$.

DEFINITION 2.11. An order interval in an ordered vector space G is a subset of G of the form $\{g : g_1 \leq g \leq g_2\}$, where g_1 and g_2 are elements of G satisfying $g_1 \leq g_2$. An operator T between ordered vector spaces E and F is called order bounded if T maps every order interval in E into an order interval in F.

DEFINITION 2.12. The linear operator T mapping the Riesz space E into the Riesz space F is said to be order continuous if for any $D \subseteq E$ such that $D \downarrow 0$ in E, we have $inf\{|Tf| : f \in D\} = 0$ in F. T is said to be σ-order continuous if, for any monotone sequence $f_n \downarrow 0$, we have $inf\{|Tf_n| : n \in \mathbb{N}\} = 0$.

3. Köthe dual

In this section we are going to describe the classical theory of the *Köthe dual space* for $L_0(X, \mu)$, defined below.

We assume that μ is a σ-finite (non-negative and σ-additive) measure in the non-empty point set X and $L_0 = L_0(X, \mu)$ is the Dedekind complete Riesz space of all $\mu-measurable$ real functions on X. In other words, μ is defined on a σ-algebra Γ the members of which are called the $\mu - measurable$ subsets of X, and the real function f on X is called $\mu-measurable$ whenever the set $\{x : x \in X,\ f(x) > \alpha\}$ is $\mu-measurable$ for every real α. It follows then that for α, β real $\{x : x \in X,\ f(x) \geq \alpha\}$, $\{x : x \in X,\ f(x) < \alpha\}$, $\{x : x \in X,\ f(x) \leq \alpha\}$, $\{x : x \in X,\ \alpha \leq f(x) \leq \beta\}$ are also $\mu - measurable$. Recall that functions differing only on a set of measure zero are identified, so that the members of $L_0(X, \mu)$ are in fact equivalence classes of measurable functions. This holds similarly for the members of the σ-algebra Γ.

Let A be an ideal in $L_0(X, \mu)$. The career of A, $c(A)$, is a measurable subset of X having the property that on its complement $n(A) = X \backslash c(A)$ every $f \in A$

vanishes $\mu - almost\ everywhere$, while $c(A)$ itself does not have any subset of positive measure on which all $f \in A$ vanish $\mu - almost\ everywhere$. Since the characteristic function $\chi_{c(A)}$ of $c(A)$ also is such that it vanishes outside $c(A)$, it is natural to ask whether $\chi_{c(A)}$ belongs to the ideal A as well. In other words, having an ideal $A \subseteq L_0(X, \mu)$, and its career $c(A)$, which, as we know, is such that all member functions of A vanish outside $c(A)$, is the characteristic function $\chi_{c(A)}$ of the set $c(A)$ necessarily among those member functions? The following counterexample shows that for an ideal $A \subseteq L_0(X, \mu)$ the characteristic function $\chi_{c(A)}$ of its career $c(A)$ maybe not a member of A [**11**].

COUNTEREXAMPLE 3.1. [**11**] *Let μ be a Lebesgue measure in $[0, 1]$ and let A be the ideal of all measurable functions f on $[0, 1]$ having the property that f vanishes on some interval $[0, \alpha_f]$, where $0 < \alpha_f \leq 1$, and α_f depends on f. Then $c(A) = [0, 1]$, and we see that $\chi_{c(A)}$ is not a member of A, because it vanishes nowhere on $[0, 1]$.*

Further, for any measurable subset Y of $c(A)$ we cannot be sure that its characteristic function χ_Y is a member of A. However, for any measurable subset Y of $c(A)$, there exists a sequence of subsets Y_n of Y with $Y_n \uparrow Y$ such that $\chi_{Y_n} \in A$ for $\forall n \in \mathbb{N}$.

THEOREM 3.2. [**11**, Theorem 29.1] *If A is an ideal in $L_0 = L_0(X, \mu)$, and Y is a measurable subset of the career $c(A)$ of A, then there exists an increasing sequence $\{Y_n\}_{n=1}^{\infty}$ of measurable sets of finite measure such that $Y_n \uparrow Y$ and $\chi_{Y_n} \in A$ for $\forall n \in \mathbb{N}$.*

PROOF. Part 1: Let us first show that if P is a measurable subset of $c(A)$ such that $\mu(P) > 0$, then P has a subset Q of positive measure such that $\chi_Q \in A$. To show this, we notice that since P is a measurable subset of $c(A)$ of positive measure, not all of $f \in A$ vanish $\mu - almost\ everywhere$ on P. Hence there exists a function $f \in A$ such that $f(x) \neq 0$ for all x in some $P_0 \subseteq P$ with $\mu(P_0) > 0$. Through replacing f by $|f|$, we may assume that $f(x) > 0$ for all $x \in P_0$. Now let us take a sequence of positive numbers $\epsilon \downarrow 0$ and $P_n = \{x : x \in P_0,\ f(x) \geq \epsilon_n\}$. We see that $\forall n \in \mathbb{N}$, $\epsilon_{n+1} \leq \epsilon_n$ implies that $P_n \subseteq P_{n+1}$. Then $P_n \uparrow P_0$ (this convergence is a result of f being positive on all points of P_0, and of the sets P_n containing more and more points of P_0 with the growth of n. The points of P_n with infinitely large index n have to be such that f on them is greater than a positive number infinitely close to zero which in limit is zero). Therefore, $\mu(P_n) \uparrow \mu(P_0) > 0$, which implies that starting from some $n = n_0$, $\mu(P_n) > 0$. Let us denote $Q = P_{n_0}$. Since for $\forall x \in Q$, the inequality $\epsilon_{n_0} \leq f(x)$ holds, and hence the equivalent to it $\epsilon_{n_0}\chi_Q(x) \leq f(x)$, $\forall x \in X$ (here also, if necessary substitute f by $|f|$, as before) holds as well, it follows that $\epsilon_{n_0}\chi_Q \leq f$. Considering that f is an element of the ideal A, we imply that $\epsilon_{n_0}\chi_Q \in A$, and A being a vector space, implies that $\chi_Q \in A$. This finishes the construction of a subset Q for P, with the required properties of having positive measure and a characteristic function belonging to A.

Part 2: Let us first assume that $\mu(Y)$ is finite. Now let

$$\alpha = sup\{\mu(Z) : Z \subseteq Y,\ \chi_Z \in A\}.$$

There exists a sequence $\{Z_n\}_{n=1}^{\infty}$ of subsets of Y such that $\chi_{Z_n} \in A$ and $\mu(Z_n) \to \alpha$, as $n \to \infty$, where it may be assumed that the sequence is increasing through, if necessary, replacing Z_n by $Z_1 \cup Z_2 \cup \cdots \cup Z_n$ (here the fact that A is an ideal is important). Then $Z := \lim_{n\to\infty} Z_n = \cup_{n=1}^{\infty} Z_n$ satisfies $\mu(Z) = \alpha$. Let us see that Z is almost equal to Y. Suppose it is not. Then the set $Y\backslash Z$ is a measurable subset of $c(A)$ having positive measure, and so, by Part 1, $Y\backslash Z$ has a subset Q of positive measure β such that $\chi_Q \in A$. Having that $\chi_{Z_n} \in A$ and $\chi_Q \in A$ it can be concluded that $\chi_{Q\cup Z_n} = \chi_Q + \chi_{Z_n} \in A$ (Q and Z_n are disjoint for each n). Thus the characteristic function of $Q \cup Z_n \subseteq Y$ is a member of A, and also

$$\mu(Q \cup Z_n) = \mu(Q) + \mu(Z_n) = \beta + \mu(Z_n) \uparrow \beta + \alpha,$$

as $n \to \infty$, so $\mu(Q \cup Z_n) > \alpha$ for n large enough. This contradicts the definition of α. Hence, Z is almost equal to Y, and so $Z_n \uparrow Y$ with $\chi_{Z_n} \in A$ for all n. This may also be expressed by saying that for any $\epsilon > 0$ there exists a set $R \subseteq Y$ such that $\chi_R \in A$ and $\mu(Y\backslash R) < \epsilon$.

Part 3: Now let us prove the general case when $\mu(Y) = \infty$. Using that the measure μ is σ-finite, we can write $Y = \cup_{n=1}^{\infty} Y_n^{\sim}$ with $\mu(Y_n^{\sim})$ finite for all n. We may also assume that $Y_n^{\sim}$ is increasing as n increases, so $Y_n^{\sim} \uparrow Y$. According to the last statement of Part 2, each $Y_n^{\sim}$ has a subset Y_n such that $\chi_{Y_n} \in A$ and $\mu(Y_n^{\sim}\backslash Y_n) < \frac{1}{n}$. One more time, it may be assumed that Y_n increases with n, so $Y_\infty := \lim_{n\to\infty} Y_n = \cup_{n=1}^{\infty} Y_n$ exists and we have $Y_n^{\sim}\backslash Y_\infty \subseteq Y_n^{\sim}\backslash Y_n$, which implies that $\mu(Y_n^{\sim}\backslash Y_\infty) < \frac{1}{n}$ for $n \in \mathbb{N}$. But

$$Y_n^{\sim}\backslash Y_\infty \uparrow Y\backslash Y_\infty, \text{ so } \mu(Y\backslash Y_\infty) = \lim_{n\to\infty} (\mu(Y_n^{\sim}\backslash Y_\infty)) \leq \lim_{n\to\infty} \frac{1}{n} = 0.$$

This shows that $Y = Y_\infty$ almost everywhere, and that except for a set of measure zero, $Y_n \uparrow Y$. Therefore, the constructed sequence $\{Y_n\}_{n=1}^{\infty}$ is the desired one. □

This theorem is important since it helps to precisely obtain the space $A_n^{\sim}$ of all order continuous linear functionals on a given ideal A in $L_0 = L_0(X, \mu)$. The remark after [**11**, Theorem 1.4] shows that $L_0(X, \mu)$ is super Dedekind complete, so in particular $L_0(X, \mu)$ is order separable. It follows that every ideal A in $L_0(X, \mu)$ is order separable, which means that every downwards directed set D in A satisfying $D \downarrow 0$ contains a sequence $\{f_n\}_{n=1}^{\infty}$ satisfying $f_n \to 0$. Consequently, we see that an order bounded linear functional on A is order continuous if and only if it is σ-order continuous.

THEOREM 3.3. [**11**, Theorem 29.2] *If A is an ideal in $L_0 = L_0(X, \mu)$ and the order bounded linear functional ϕ on A is order continuous, i.e., if $\phi \in A_n^{\sim}$, then there exists a real μ- measurable function t on X such that*

(i) *for every $f \in A$ the product tf is μ-summable over X,*
(ii) *$\phi(f) = \int_X tf d\mu$ holds for every $f \in A$.*

The function t thus corresponding to ϕ is uniquely determined μ − almost everywhere on the career $c(A)$ of A. If ϕ is positive, then $t(x) \geq 0$ for μ − almost every $x \in c(A)$.

Conversely, if t is a real μ − measurable function on X possessing the property that tf is μ − summable over X for every $f \in A$, then the linear functional ϕ on A, defined by

$$\phi(f) = \int_X tf d\mu \text{ for every } f \in A, \tag{3.1}$$

is order bounded and order continuous. Trivially, ϕ is uniquely determined by t. If $t(x) \geq 0$ for $\mu -$ almost every $x \in c(A)$, then ϕ is positive.

PROOF. Let us assume first that $\phi \in A_n^\sim$. Then ϕ^+ and ϕ^- are positive members of $A_n^\sim$, so it is sufficient to show that each of ϕ^+ and ϕ^- has the desired properties, then it will follow that the desired properties also hold for $\phi = \phi^+ - \phi^-$. So, let us take $0 \leq \phi \in A_n^\sim$ and prove that there exists a $\mu -$ *measurable* function t, non-negative on $c(A)$ and such that (i) and (ii) hold. Using Theorem 3.2, a sequence $\{Y_n\}_{n=1}^\infty$ of measurable subsets of $c(A)$ is chosen, such that each Y_n has a finite measure, $Y_n \uparrow c(A)$ and $\chi_{Y_n} \in A$ for $\forall n \in \mathbb{N}$. Therefore, if Z is a $\mu -$ *measurable* subset of some Y_n, we have $\chi_Z \leq \chi_{Y_n} \in A$, and since A is an ideal, $\chi_Z \in A$. So $\phi(\chi_Z)$ is well-defined. $\phi(\chi_Z)$ is briefly denoted by $\nu(Z)$ and let us look at the subsets of one fixed Y_n, say for $n = n_0$. Given such a subset $Z \subseteq Y_{n_0}$, we apply Theorem 3.2 one more time to choose $Z_k \uparrow Z$ as $k \to \infty$, and since ϕ is positive and order continuous, it follows that $\phi(\chi_{Z_k}) \uparrow \phi(\chi_Z)$. In other words, $\nu(Z_k) \uparrow \nu(Z)$, which shows that ν is a finite, non-negative, σ-additive measure on the σ-algebra of $\mu -$ *measurable* subsets of Y_{n_0}. Furthermore, the measure ν is $\mu -$ *absolutely continuous*, i.e., $\mu(Z) = 0$ implies that $\nu(Z) = 0$. Hence, by the Radon-Nikodym theorem, there exists a non-negative, $\mu -$ *measurable* function t on Y_{n_0} such that $\nu(Z) = \int_Z t d\mu$ holds for every $\mu -$ *measurable* subset Z of Y_{n_0}. The function t is $\mu -$ *almost* uniquely determined on Y_{n_0} ([**11**, Exercise 28.4]). By varying n_0, the function t can now be extended $\mu -$ *almost* uniquely to $\cup_{n=1}^\infty Y_n = c(A)$ such that $\nu(Z) = \int_Z t d\mu$ holds for every Z satisfying $Z \subseteq Y_n$ for some n. Setting $t(x) = 0$ for all $x \in X \backslash c(A)$, the following holds:

$$\phi(\chi_Z) = \nu(Z) = \int_X t\chi_Z d\mu$$

for every Z of this particular type. Hence, if s is now a step function $s = \sum_{k=1}^p a_k \chi_{Z_k}$ with all Z_k of this type, it is evident that $\phi(s) = \int_X ts d\mu$. Finally, let $0 \leq f \in A$ be given. It is well-known that there exists a sequence $\{s_n^\sim\}_{n=1}^\infty$ of $\mu -$ *measurable* step functions such that $0 \leq s_n^\sim \uparrow f$. Then, for each n, $s_n = s_n^\sim \chi_{Y_n}$ is a step function of the particular type we need and $0 \leq s_n \uparrow f$ still holds. Hence $\phi(s_n) = \int_X ts_n d\mu$ for all n, by what was proved above. Furthermore, $\phi(s_n) \uparrow \phi(f)$, since ϕ is positive and order continuous, and

$$\int_X ts_n d\mu \uparrow \int_X tf d\mu$$

by the well-known theorem on integration of monotone sequences. It follows, as desired, that $\phi(f) = \int_X tf d\mu$. Finally, for any arbitrary $f \in A$, we have

$$\phi(f) = \phi(f^+) - \phi(f^-) = \int_X tf^+ d\mu - \int_X tf^- d\mu = \int_X tf d\mu.$$

Conversely, assume that the real $\mu -$ *measurable* function t on X has the property that tf is $\mu -$ *summable* over X for every $f \in A$. Then $|tf|$ is $\mu -$ *summable* and so $t^+ f$ and $t^- f$ are $\mu -$ *summable*. It is evident that ϕ, as defined in (3.1), is a linear functional on A and

$$\phi(f) = \int_X tf d\mu = \int_X t^+ f d\mu - \int_X t^- f d\mu = \phi_1(f) - \phi_2(f)$$

for every $f \in A$, so ϕ is the difference of the positive linear functionals ϕ_1 and ϕ_2. This shows already that ϕ is order bounded. Furthermore, $f_n \downarrow 0$ in A implies

$\phi_1(f_n) = \int_X t^+ f_n d\mu \downarrow 0$, so ϕ_1 is order continuous. Similarly, ϕ_2 is order continuous. It follows that ϕ is order continuous. □

As before, let A be an ideal in $L_0 = L_0(X, \mu)$ with carrier $c(A)$. Let us assume that $c(A) = X$. As already shown above, to any $\phi \in A_n^\sim$, there corresponds a function $t \in L_0$ such that $\phi(f) = \int_X tf d\mu$ for every $f \in A$. Conversely, if the function $t \in L_0$ has the property that tf is $\mu -$ *summable* over X for every $f \in A$, then the linear functional ϕ on A, defined by $\phi(f) = \int_X tf d\mu$, satisfies $\phi \in A_n^\sim$. The functional $\phi \in A_n^\sim$ and the corresponding $t \in L_0$ determine each other uniquely $\mu -$ *almost everywhere*, as far as it concerns t. The set of all $t \in L_0$ thus corresponding to elements $\phi \in A_n^\sim$ is obviously a linear subspace of L_0 which is denoted by K_A. Let us see that K_A is an ideal in L_0. If $t \in K_A$, then tf is summable over X for every $f \in A$, i.e., $|tf|$ is summable for every $f \in A$, and so $|t|f$ is summable for every $f \in A$. This shows that $|t| \in K_A$. Furthermore, it is evident that if $0 \le t_2 \le t_1$, and $t_1 \in K_A$, then $t_2 \in K_A$. In particular, $|t| \in K_A$ implies that t^+ and t^- are members of K_A, and so $t = t^+ - t^- \in K_A$. To summarize, it is now proved that $t \in K_A$ if and only if $|t| \in K_A$ and $0 \le t_2 \le t_1 \in K_A$ implies that $t_2 \in K_A$. Hence, K_A is an ideal in L_0. The ideal K_A is sometimes called the *Köthe dual space* of A, named after G. Köthe, who was one of the first to introduce this notion, first for the case of sequence spaces, i.e., when X is the set of natural numbers with discrete measure, so that $L_0 = L_0(X, \mu)$ in this case is the space of all real sequences [**11**].

4. Sequence spaces

For Banach spaces E and F, $\mathcal{L}(E; F)$ will denote the space of bounded linear operators from E to F. In case that E and F are Banach lattices, $\mathcal{L}^r(E; F)$ will denote the space of regular linear operators from E to F with its regular norm $\|\cdot\|_r$.

For a Banach space H, H^* will denote its topological dual and B_H will denote its closed unit ball. For a vector lattice G, the G-valued sequence space $G^{\mathbb{N}}$ is a vector lattice with the following lattice operations

$$\bar{x} \wedge \bar{y} = (x_i \wedge y_i)_i, \quad \bar{x} \vee \bar{y} = (x_i \vee y_i)_i,$$

$$\bar{x} = (x_i)_i, \quad \bar{y} = (y_i)_i \in G^{\mathbb{N}}.$$

For each $n \in \mathbb{N}$ and each $\bar{x} = (x_i)_i \in G^{\mathbb{N}}$, let

$$\bar{x}(\ge n) = (0, \cdots, 0, x_n, x_{n+1}, \cdots), \quad \bar{x}(n) = (0, \cdots, 0, x_n, 0, \cdots).$$

G^+ will denote the positive cone of G and for each $x \in G$, x^+ and x^- will denote the positive part and the negative part of x, respectively.

Let λ be a sequence space, i.e., a subspace of $\mathbb{R}^{\mathbb{N}}$. The Köthe dual of λ is defined by

$$\lambda' = \Big\{(b_i)_i \in \mathbb{R}^{\mathbb{N}} : \sum_{i=1}^{\infty} |a_i b_i| < +\infty, \ \forall (a_i)_i \in \lambda\Big\}.$$

λ is called Köthe perfect if $\lambda'' = \lambda$. In addition, if λ is an order continuous Banach lattice, then $\lambda' = \lambda^*$. Thus λ' is also a Banach lattice and for each $a = (a_i)_i \in \lambda$

and each $b = (b_i)_i \in \lambda'$,

$$\|a\|_\lambda = sup\Big\{|\sum_{i=1}^{\infty} a_i b_i| : (b_i)_i \in B_{\lambda'}\Big\},$$

and

$$\|b\|_{\lambda'} = \|b\|_{\lambda^*} = sup\Big\{|\sum_{i=1}^{\infty} a_i b_i| : (a_i)_i \in B_\lambda\Big\}.$$

Let λ'_0 denote the closed subspace of λ' consisting of all such sequences of λ' whose tails converge to 0, i.e.,

$$\lambda'_0 = \Big\{a = (a_i)_i \in \lambda' : \lim_n \|a(\geq n)\|_{\lambda'} = 0\Big\}.$$

Then $(\lambda'_0)^* = (\lambda'_0)' = \lambda''$.

LEMMA 4.1. [**2**, Lemma 3.4], [**6**, Lemma 4], [**3**, Lemma 1] *If λ is an order continuous Banach lattice, then a bounded subset B of λ is relatively compact if and only if*

$$\limsup_n \Big\{\|a(\geq n)\|_\lambda : a = (a_i)_i \in B\Big\} = 0.$$

From now on it will be assumed that λ is a Köthe perfect ($\lambda'' = \lambda$) and order continuous Banach sequence lattice with $\|e_i\|_\lambda = 1$ for all $i \in \mathbb{N}$, where $e_i's$ are standard unit vectors in the sequence space λ. Let G be a Banach lattice as before.

For a Banach space H, let

$$\lambda_\omega(H) = \lambda_{weak}(H) = \Big\{\bar{x} = (x_i)_i \in H^{\mathbb{N}} : (x^*(x_i))_i \in \lambda,\ \forall x^* \in H^*\Big\}$$

and

$$\|\bar{x}\|_{\lambda_\omega(H)} = sup\Big\{\|(x^*(x_i))_i\|_\lambda : x^* \in B_{H^*}\Big\},\quad \forall \bar{x} = (x_i)_i \in \lambda_\omega(H).$$

Then $\lambda_\omega(H)$ is a Banach space.

Let $\lambda_{\omega,0}(H)$ denote the closed subspace of $\lambda_\omega(H)$ consisting of all such elements of $\lambda_\omega(H)$ whose tails converge to 0, i.e.,

$$\lambda_{\omega,0}(H) = \Big\{\bar{x} \in \lambda_\omega(H) : \lim_n \|\bar{x}(\geq n)\|_{\lambda_\omega(H)} = 0\Big\}.$$

REMARK 4.2. We notice that the Banach space $\ell_1^\omega(H)$ of weakly 1-summable H-valued sequences with norm $\|(x_j)_{j=1}^\infty\|_{\omega,1} = \sup_{\phi \in B_{H^*}} \|(\phi(x_j))_{j=1}^\infty\|_1$ is a particular case of $\lambda_\omega(H)$. And $\ell_1^u(H) = \Big\{(x_j)_{j=1}^\infty \in \ell_1^\omega(H) : \lim_k \|(x_j)_{j=k}^\infty\|_{\omega,1} = 0\Big\}$, the Banach space of unconditionally 1-summable sequences, with the norm inherited from $\ell_1^\omega(H)$ is a particular case of $\lambda_{\omega,0}(H)$.

For each $\bar{x} = (x_i)_i \in \lambda_\omega(H)$, each $a = (a_i)_i \in \lambda'_0$, and each $n \in \mathbb{N}$,

$$\|\sum_{i=n}^{\infty} a_i x_i\|_H = sup\Big\{|\sum_{i=n}^{\infty} a_i x^*(x_i)| : x^* \in B_{H^*}\Big\} \leq$$

$$\leq sup\Big\{\|a(\geq n)\|_{\lambda'} \cdot \|(x^*(x_i))_{i=n}^\infty\|_\lambda : x^* \in B_{H^*}\Big\} \leq$$

$$\leq \|a(\geq n)\|_{\lambda'} \cdot \|\bar{x}\|_{\lambda_\omega(H)} \to 0, \; as \; n \to \infty.[\mathbf{3}] \tag{4.1}$$

Since the above is true for each $n \in \mathbb{N}$, for $n = 1$ it has the following form:

$$\|\sum_{i=1}^{\infty} a_i x_i\|_H = sup\Big\{|\sum_{i=1}^{\infty} a_i x^*(x_i)| : x^* \in B_{H^*}\Big\} \leq$$

$$\leq \|(a_i)_i\|_{\lambda'} \cdot \|(x_i)_i\|_{\lambda_\omega(H)}.$$

Taking supremum when $(a_i)_i \in B_{\lambda'}$, the following is true:

$$sup\Big\{\|\sum_{i=1}^{\infty} a_i x_i\|_H : (a_i)_i \in B_{\lambda'}\Big\} =$$

$$= sup\Big\{|\sum_{i=1}^{\infty} a_i x^*(x_i)| : x^* \in B_{H^*}, (a_i)_i \in B_{\lambda'}\Big\} \leq$$

$$\leq sup\Big\{\|(a_i)_i\|_{\lambda'} \cdot \|(x_i)_i\|_{\lambda_\omega(H)} : (a_i)_i \in B_{\lambda'}\Big\}.$$

Therefore,

$$sup\Big\{\|\sum_{i=1}^{\infty} a_i x_i\|_H : (a_i)_i \in B_{\lambda'}\Big\} =$$

$$= sup\Big\{\|(x^*(x_i))_i\|_{\lambda} : x^* \in B_{H^*}\Big\} \leq$$

$$\leq \|x^*(x_i)\|_{\lambda_\omega(H)}.$$

Middle part of the inequalities above is $\|(x_i)_i\|_{\lambda_\omega(H)}$, hence

$$sup\Big\{\|\sum_{i=1}^{\infty} a_i x_i\|_H : (a_i)_i \in B_{\lambda'}\Big\} =$$

$$= \|x^*(x_i)\|_{\lambda_\omega(H)} \leq \|x^*(x_i)\|_{\lambda_\omega(H)}.$$

By (4.1), and the explanations following (4.1), the following lemma holds:

LEMMA 4.3. [**3**, Lemma 2] $\bar{x} = (x_i)_i \in \lambda_\omega(H)$ *if and only if for each* $(a_i)_i \in \lambda'_0$, *the series* $\sum_i a_i x_i$ *converges in H. In this case, we have:*

$$\|\bar{x}\|_{\lambda_\omega(H)} = sup\Big\{\|\sum_{i=1}^{\infty} a_i x_i\|_H : (a_i)_i \in B_{\lambda'_0}\Big\}.$$

For each $\bar{x} = (x_i)_i \in \lambda_\omega(H)$, define a linear operator $T_{\bar{x}}$ from λ'_0 to H by $T_{\bar{x}}(a) = \sum_{i=1}^{\infty} a_i x_i$, $\forall a = (a_i)_i \in \lambda'_0$.

PROPOSITION 4.4. [**3**, Proposition 3] Let H be a Banach space. Then $\lambda_\omega(H)$ is isometrically isomorphic to $\mathcal{L}(\lambda'_0, H)$ under the mapping $\bar{x} \to T_{\bar{x}}$. Moreover, $T_{\bar{x}}$ is compact if and only if $\bar{x} \in \lambda_{\omega,0}(H)$.

For a Banach lattice G, let

$$\lambda_\epsilon(G) = \Big\{\bar{x} = (x_i)_i \in G^{\mathbb{N}} : (x^*(|x_i|))_i \in \lambda,\ \forall x^* \in G^{*+}\Big\},$$

and

$$\|\bar{x}\|_{\lambda_\epsilon(G)} = sup\Big\{\|(x^*(|x_i|))_i\|_\lambda : x^* \in B_{G^{*+}}\Big\},\ \forall \bar{x} = (x_i)_i \in \lambda_\epsilon(G).$$

Then $\lambda_\epsilon(G)$ is a Banach lattice (see [**2**]). Let $\lambda_{\epsilon,0}(G)$ denote the closed sublattice of $\lambda_\epsilon(G)$, consisting of all such elements of $\lambda_\epsilon(G)$ whose tails converge to 0, i.e.,

$$\lambda_{\epsilon,0}(G) = \Big\{\bar{x} \in \lambda_\epsilon(G) : \lim_n \|\bar{x}(\geq n)\|_{\lambda_\epsilon(G)} = 0\Big\}.$$

Then $\lambda_{\epsilon,0}(G)$ is an order ideal of $\lambda_\epsilon(G)$. The following lemma is straightforward.

LEMMA 4.5. [**3**, Lemma 4] *Let G be a Banach lattice. Then $\lambda_\epsilon(G) \subseteq \lambda_\omega(G)$ and for each $\bar{x} \in \lambda_\epsilon(G)$, $\|\bar{x}\|_{\lambda_\epsilon(G)} \geq \|\bar{x}\|_{\lambda_\omega(G)}$. In addition, if $\bar{x} \geq 0$, $\bar{x} \in \lambda_\epsilon(G) \iff \bar{x} \in \lambda_\omega(G)$. In this case, $\|\bar{x}\|_{\lambda_\epsilon(G)} = \|\bar{x}\|_{\lambda_\omega(G)}$.*

PROPOSITION 4.6. [**3**, Proposition 5] Let G be a Dedekind complete Banach lattice. Then $\lambda_\epsilon(G)$ is isometrically isomorphic and lattice homomorphic to $\mathcal{L}^r(\lambda_0', G)$ under the mapping $\bar{x} \to T_{\bar{x}}$. Moreover, $T_{\bar{x}} \in \mathcal{K}^r(\lambda_0', G)$ if and only if $\bar{x} \in \lambda_{\epsilon,0}(G)$, where $\mathcal{K}^r(\lambda_0', G)$ will denote the linear span of positive compact operators from λ_0' to G.

For a Banach lattice G, define

$$\lambda_\omega'(G^*) = \lambda_{weak}'(G^*) = \Big\{\bar{x}^* = (x_i^*)_i \in G^{*^{\mathbb{N}}} : (x_i^*(x))_i \in \lambda',\ \forall x \in G\Big\}$$

and

$$\|\bar{x}^*\|_{\lambda_\omega'(G^*)} = sup\Big\{\|(x_i^*(x))_i\|_{\lambda'} : x \in B_G\Big\},\ \forall \bar{x}^* = (x_i^*)_i \in \lambda_\omega'(G^*).$$

Then $\lambda_\omega'(G^*)$ is a Banach space (it may not be a Banach lattice).

Define

$$\lambda_s(G) = \lambda_{strong}(G) = \Big\{\bar{x} = (x_i)_i \in G^{\mathbb{N}} : \sum_{i=1}^{\infty} |x_i^*(x_i)| < +\infty,\ \forall (x_i^*)_i \in \lambda_\omega'(G^*)\Big\}$$

and

$$\|\bar{x}\|_{\lambda_s(G)} = sup\Big\{|\sum_{i=1}^{\infty} x_i^*(x_i)| : (x_i^*)_i \in B_{\lambda_\omega'(G^*)}\Big\},\ \forall \bar{x} = (x_i)_i \in \lambda_s(G).$$

Then $\lambda_s(G)$ is a Banach space (it may not be a Banach lattice).

REMARK 4.7. We notice that the Banach space of cohen strongly 1-summable sequences $\ell_1\langle G\rangle = \Big\{(x_j)_{j=1}^\infty \in G^{\mathbb{N}} : \|(x_j)_{j=1}^\infty\|_{\ell_1\langle G\rangle} := sup\{\|(\phi_j(x_j))_{j=1}^\infty\|_1 : (\phi_j)_{j=1}^\infty \in B_{\ell_\infty^\omega(G^*)}\} < +\infty\Big\}$, is a particular case of $\lambda_s(G)$.

Define

$$\lambda_\epsilon'(G^*) = \Big\{\bar{x}^* = (x_i^*)_i \in G^{*^{\mathbb{N}}} : (|x_i^*|(x))_i \in \lambda',\ \forall x \in G^+\Big\}$$

$$(Compare\ \lambda_\epsilon(G) = \Big\{\bar{x} = (x_i)_i \in G^{\mathbb{N}} : (x^*(|x_i|))_i \in \lambda,\ \forall x^* \in G^{*+}\Big\}.)$$

and

$$\|\bar{x}^*\|_{\lambda'_\epsilon(G^*)} = sup\Big\{\|(|x_i^*|(x))_i\|_{\lambda'} : x \in B_{G+}\Big\},\ \forall \bar{x}^* = (x_i^*)_i \in \lambda'_\epsilon(G^*).$$

Then $\lambda'_\epsilon(G^*)$ is a Banach lattice.

Define

$$\lambda_\pi(G) = \Big\{\bar{x} = (x_i)_i \in G^{\mathbb{N}} : \sum_{i=1}^{\infty} x_i^*(|x_i|) < +\infty,\ \forall (x_i^*)_i \in \lambda'_\epsilon(G^*)^+\Big\}$$

and

$$\|\bar{x}\|_{\lambda_\pi(G)} = sup\Big\{\sum_{i=1}^{\infty} x_i^*(|x_i|) : (x_i^*)_i \in B_{\lambda'_\epsilon(G^*)+}\Big\},\ \forall \bar{x} = (x_i)_i \in \lambda_\pi(G).$$

Then $\lambda_\pi(G)$ is a Banach lattice.

Let $\lambda_{\pi,0}(G)$ denote the closed sublattice of $\lambda_\pi(G)$ consisting of all such elements of $\lambda_\pi(G)$ whose tails converge to 0, that is,

$$\lambda_{\pi,0}(G) = \Big\{\bar{x} \in \lambda_\pi(G) : \underset{n}{lim}\|\bar{x}(\geq n)\|_{\lambda_\pi(G)} = 0\Big\}.$$

Then $\lambda_{\pi,0}(G)$ is an ideal of $\lambda_\pi(G)$.

PROPOSITION 4.8. [**5**, Proposition 1] Let G and λ be as defined above. If λ is σ-order continuous, then $\lambda_{\pi,0}(G) = \lambda_\pi(G)$.

For Archimedean Riesz spaces E and F, let $E\bar{\otimes}F$ denote the Riesz space tensor product of E and F, with the positive cone C_p defined as follows.

$$C_p = \Big\{\sum_{k=1}^{n} x_k \otimes y_k : n \in \mathbb{N}, x_k \in E^+, y_k \in F^+\Big\}.$$

If, in addition, E and F are Banach lattices, then the positive projective tensor norm on $E\bar{\otimes}F$ is defined by

$$\|u\|_{|\pi|} = inf\{\sum_{k=1}^{n} \|x_k\| \cdot \|y_k\| : x_k \in E^+, y_k \in F^+,\ |u| \leq \sum_{k=1}^{n} x_k \otimes y_k\}$$

for every $u \in E\bar{\otimes}F$. Let $E\hat{\otimes}_{|\pi|}F$ denote the completion of $E\bar{\otimes}F$ with respect to $\|\cdot\|_{|\pi|}$. Then $E\hat{\otimes}_{|\pi|}F$ is a Banach lattice, called the positive projective tensor product.

PROPOSITION 4.9. [**5**, Proposition 2][**2**, Theorem 7.3] If λ is σ-order continuous then $\lambda\hat{\otimes}_{|\pi|}G$ is isometrically isomorphic and lattice homomorphic to $\lambda_{\pi,0}(G)$.

Define $\lambda(G) = \{\bar{x} = (x_i)_i \in G^{\mathbb{N}} : (\|x_i\|)_i \in \lambda\}$ and

$$\|\bar{x}\|_{\lambda(G)} = \|(\|x_i\|)_i\|_\lambda.$$

Then $\lambda(G)$ is a Banach lattice. If λ is σ-order continuous, it follows from [**2**, Lemma 3.4] that for every $a = (a_i)_i \in \lambda$ we have $\underset{n}{lim}\|a(\geq n)\|_\lambda = 0$ and hence, for every $\bar{x} \in \lambda(G)$, $\underset{n}{lim}\|\bar{x}(\geq n)\|_{\lambda(G)} = 0$.

PROPOSITION 4.10. [**5**, Proposition 3] Let λ be σ-order continuous. Then a subset B of $\lambda(G)$ is relatively compact if and only if for each $i \in \mathbb{N}$, the set $\{x_i : \bar{x} = (x_i)_i \in B\}$ is a relatively compact subset of G, and

$$\underset{n}{limsup}\{\|\bar{x}(\geq n)\|_{\lambda(G)} : \bar{x} \in B\} = 0.$$

An Orlicz function ϕ is a continuous, nondecreasing and convex function, which maps $[0,\infty)$ to $[0,\infty)$ and is zero only at $t = 0$, with $\underset{t\to\infty}{lim}\,\phi(t) = \infty$. Every Orlicz function ϕ has a right derivative p and

$$\phi(t) = \int_0^t p(u)du, \; t \geq 0.$$

If p satisfies $p(0) = 0$ and $\underset{t\to\infty}{lim}\,p(t) = \infty$ (after making these restrictions we exclude only the case $\ell_\phi = \ell_1$ when $\phi(t)$ is equivalent to t), then the right inverse q of p,

$$q(s) = sup\{t : p(t) \leq s\}, \; s \geq 0,$$

is a right continuous nondecreasing function such that q is zero at zero and $q(s) > 0$ whenever $s > 0$. Define

$$\phi^*(s) = \int_0^s q(u)du, \; s \geq 0.$$

This ϕ^* is also an Orlicz function, and q is its right derivative. ϕ^* is called the function complementary to ϕ. ϕ has its complementary function if its right derivative p satisfies $p(0) = 0$ and $\underset{t\to\infty}{lim}\,p(t) = \infty$.

DEFINITION 4.11. [**4**] An Orlicz function ϕ is said to satisfy the Δ_2 condition at zero, if there exist $K > 0$ and $t_0 > 0$ such that $\phi(2t) \leq K\phi(t)$ for every $0 < t \leq t_0$.

When we have the definition for the Δ_2 condition, let us look at the definition of the Orlicz sequence space ℓ_ϕ, and notice its similarity to the definition of the Köthe dual λ', given above and appearing in [**3**, **5**].

$$\ell_\phi = \Big\{(a_i)_{i=1}^\infty \in \mathbb{R}^{\mathbb{N}} : \sum_{i=1}^\infty \phi(|\lambda a_i|) < \infty \; for\ some\ \lambda > 0.\Big\}.$$

On the contrary to the general case of λ', the norm for each element $(a_i)_{i=1}^\infty$ of ℓ_ϕ has a given form of

$$\|(a_i)_{i=1}^\infty\|_{\ell_\phi} = inf\Big\{\lambda > 0 : \sum_{i=1}^\infty \phi\Big(\frac{|a_i|}{\lambda}\Big) \leq 1\Big\}.$$

With this norm, ℓ_ϕ is a Banach space.

Next comes the X-valued Orlicz sequence space $\ell_\phi(X)$, defined by

$$\ell_\phi(X) = \Big\{(x_i)_{i=1}^\infty \in X^{\mathbb{N}} : \sum_{i=1}^\infty \phi(\|\lambda x_i\|) < \infty \; for\ some\ \lambda > 0.\Big\}.$$

For each element $(x_i)_{i=1}^\infty$ of $\ell_\phi(X)$, the $\ell_\phi(X)$ norm is given by:

$$\|(x_i)_{i=1}^\infty\|_{\ell_\phi(X)} = inf\Big\{\lambda > 0 : \sum_{i=1}^\infty \phi\Big(\frac{\|x_i\|}{\lambda}\Big) \leq 1\Big\}.$$

Under this norm, $\ell_\phi(X)$ is a Banach space.

COROLLARY 4.12. [**5**, Corollary 4] Let λ be σ- order continuous. Then a subset B of λ is relatively compact if and only if for each $i \in \mathbb{N}$, the set $\{a_i : a = (a_i)_i \in B\}$ is a bounded subset of $\mathbb{R}$, and

$$\underset{n}{limsup}\{\|a(\geq n)\|_\lambda : a = (a_i)_i \in B\} = 0.$$

REMARK 4.13. An Orlicz function ϕ satisfies the Δ_2- condition if and only if for each $(x_i)_{i=1}^\infty \in \ell_\phi(X)$, $\underset{n}{lim}\|(x_i)_{i=n}^\infty\|_{\ell_\phi(X)} = 0$ [**4**]. This statement allows us to think that the Δ_2- condition plays the role of σ-order continuity of the sequence space λ in only one direction, since in [**5**] we have, that if λ is σ-order continuous then $\lambda_{\pi,0}(X) = \lambda_\pi(X)$, where

$$\lambda_{\pi,0}(X) = \Big\{(x_i)_{i=1}^\infty \in \lambda_\pi(X) : \underset{n}{lim}\|(x_i)_{i=n}^\infty\|_{\lambda_\pi(X)} = 0\Big\},$$

and a Banach lattice Y is called σ-order continuous if $0 \leq x_n \downarrow 0$ in Y when $x_n \to 0$ in Y.

With the proposition that follows the compact subsets of $\ell_\phi(X)$ are characterized, and actually this is the proposition the proof of which contains a useful technique, possibly applicable for the general sequence space case also.

PROPOSITION 4.14. [**4**, Proposition 1] Let ϕ be an Orlicz function satisfying the Δ_2 condition. Then a subset B of $\ell_\phi(X)$ is relatively compact if and only if for each $i \in \mathbb{N}$, the set $\{x_i : (x_i)_{i=1}^\infty \in B\}$ is a relatively compact subset of X, and

$$\limsup\{\|(x_i)_{i=n}^\infty\|_{\ell_\phi(X)} : (x_i)_{i=1}^\infty \in B\} = 0. \tag{4.2}$$

PROOF. Suppose that B is a relatively compact subset of $\ell_\phi(X)$. It can be shown that $\{x_i : (x_i)_{i=1}^\infty \in B\}$ is a relatively compact subset of X for each $i \in \mathbb{N}$. Next assume that (4.2) does not hold. Note that $\underset{n}{lim}\|(x_i)_{i=n}^\infty\|_{\ell_\phi(X)} = 0$ for each $(x_i)_{i=1}^\infty \in \ell_\phi(X)$ since ϕ satisfies the Δ_2 condition.

We want to show that there exists an $\epsilon_0 > 0$, $(x_i^{(k)})_{i=1}^\infty \in B$ for each $k \in \mathbb{N}$, and a subsequence $n_1 < m_1 < n_2 < m_2 < \dots$ such that

$$\Big\|(x_i^{(k)})_{i=n_k}^\infty\Big\|_{\ell_\phi(X)} \geq \epsilon_0,\ k = 1, 2, \dots,$$

$$\Big\|(x_i^{(k)})_{i=m}^\infty\Big\|_{\ell_\phi(X)} \leq \frac{\epsilon_0}{2},\ m > m_k,\ k = 1, 2, \dots.$$

Since condition (4.2) *is not satisfied, there exists* $\epsilon_0 > 0$ *such that for* $\forall N \in \mathbb{N}$, $\exists n \geq N$ *with*

$$sup\Big\{\|(x_i)_{i=n}^\infty\|_{\ell_\phi(X)} : (x_i)_{i=1}^\infty \in B\Big\} \geq \epsilon_0.$$

Having the assumption above, let us take $N = 1$, *and hence* $\exists n_1 \geq 1$ *such that*

$$sup\Big\{\|(x_i)_{i=n_1}^\infty\|_{\ell_\phi(X)} : (x_i)_{i=1}^\infty \in B\Big\} \geq \epsilon_0.$$

For this n_1, $\exists (x_i^{(1)})_{i=1}^\infty \in B$ *with*

$$\|(x_i^{(1)})_{i=n_1}^\infty\|_{\ell_\phi(X)} \geq \epsilon_0.$$

When n_1 *is already chosen, we consider that* $\underset{n}{lim}\|(x_i)_{i=n}^\infty\|_{\ell_\phi(X)} = 0$ *holds for any* $(x_i)_{i=1}^\infty \in \ell_\phi(X)$, *and in particular* $\underset{n}{lim}\|(x_i^{(1)})_{i=n}^\infty\|_{\ell_\phi(X)} = 0$, *and hence, for* $\frac{\epsilon_0}{2}$, *there exists* $N_1 \in \mathbb{N}$ *such that for* $\forall m \geq N_1$

$$\|(x_i^{(1)})_{i=m}^{\infty}\|_{\ell_\phi(X)} \leq \frac{\epsilon_0}{2}.$$

Take m_1*, such that* $m_1 > max\{N_1, n_1\}$*. Then for* $\forall m \geq m_1$*,*

$$\|(x_i^{(1)})_{i=m}^{\infty}\|_{\ell_\phi(X)} \leq \frac{\epsilon_0}{2}.$$

Now we return to the assumption that (4.2) *does not hold, and choose the arbitrary* N *to be the already constructed* m_1*. This means that there exists* $n_2 > m_1$ *such that*

$$sup\Big\{\|(x_i)_{i=n_2}^{\infty}\|_{\ell_\phi(X)} : (x_i)_{i=1}^{\infty} \in B\Big\} \geq \epsilon_0.$$

For n_2*,* $\exists (x_i^{(2)})_{i=1}^{\infty} \in B$ *such that*

$$\|(x_i^{(2)})_{i=n_2}^{\infty}\|_{\ell_\phi(X)} \geq \epsilon_0.$$

Then, since $\lim_n \|(x_i^{(2)})_{i=n}^{\infty}\|_{\ell_\phi(X)} = 0$*, we choose* $N_2 \in \mathbb{N}$ *such that for* $\forall m \geq N_2$*,*

$$\|(x_i^{(2)})_{i=m}^{\infty}\|_{\ell_\phi(X)} \leq \frac{\epsilon_0}{2}.$$

Then we take $m_2 > max\{N_2, n_2\}$*, so for* $\forall m \geq m_2$*,*

$$\|(x_i^{(2)})_{i=m}^{\infty}\|_{\ell_\phi(X)} \leq \frac{\epsilon_0}{2}.$$

By induction, we can consider the construction of $n_1 < m_1 < n_2 < m_2 < \ldots$ *complete.*

For each $k, j \in \mathbb{N}$ with $k > j$ (and hence $n_k \geq n_j$),

$$\|(x_i^{(k)})_{i=1}^{\infty} - (x_i^{(j)})_{i=1}^{\infty}\|_{\ell_\phi(X)} \geq$$

$$\|(x_i^{(k)})_{i=n_k}^{\infty} - (x_i^{(j)})_{i=n_k}^{\infty}\|_{\ell_\phi(X)} \geq$$

$$\|(x_i^{(k)})_{i=n_k}^{\infty}\|_{\ell_\phi(X)} - \|(x_i^{(j)})_{i=n_k}^{\infty}\|_{\ell_\phi(X)} \geq$$

$$\epsilon_0 - \frac{\epsilon_0}{2} = \frac{\epsilon_0}{2}.$$

Thus the sequence $(x_i^{(k)})_{i=1}^{\infty}$ in B cannot have any limit point in $\ell_\phi(X)$, which shows that B is not a relatively compact subset of $\ell_\phi(X)$. This contradiction shows that (4.2) holds.

On the other hand, suppose that each $\{x_i : (x_i)_{i=1}^{\infty} \in B\}$ is a relatively compact subset of X and (4.2) holds. Take a sequence $((x_i^{(m)})_{i=1}^{\infty})_{m=1}^{\infty}$ in B. Below we show that by the diagonal method, there exists a subsequence $((x_i^{(m_k)})_{i=1}^{\infty})_{k=1}^{\infty}$ of $((x_i^{(m)})_{i=1}^{\infty})_{m=1}^{\infty}$ such that

$$\lim_k x_i^{(m_k)} \tag{4.3}$$

exists in X for each $i \in \mathbb{N}$.

Figure 4.1 (Diagonal method)

$$\begin{array}{cccccc} x_1^{(1)} & x_1^{(2)} & \cdots\cdots\cdots & x_1^{(n)} & \cdots\cdots\cdots \\ x_2^{(1)} & x_2^{(2)} & \cdots\cdots\cdots & x_2^{(n)} & \cdots\cdots\cdots \\ x_3^{(1)} & x_3^{(2)} & \cdots\cdots\cdots & x_3^{(n)} & \cdots\cdots\cdots \\ x_4^{(1)} & x_4^{(2)} & \cdots\cdots\cdots & x_4^{(n)} & \cdots\cdots\cdots \\ \cdots & \cdots & \cdots\cdots\cdots & \cdots & \cdots\cdots\cdots \end{array}$$

In Figure 4.1, the ith row is a subset of $\{x_i : (x_i)_{i=1}^\infty \in B\}$, which is given to be relatively compact. In the first row, choose a convergent subsequence $(x_1^{(m_k^1)})_{k=1}^\infty$. Then look at $(x_2^{(m_k^1)})_{k=1}^\infty$, and choose a convergent subsequence of $(x_2^{(m_k^1)})_{k=1}^\infty$, denoted by $(x_2^{(m_k^2)})_{k=1}^\infty$. By continuing this process at the ith step we choose a convergent subsequence of $(x_i^{(m_k^{i-1})})_{k-1}^\infty$, denoted by $(x_i^{(m_k^i)})_{k=1}^\infty$. By induction, the required subsequence is constructed if we denote $((x_i^{(m_k^i)})_{i=1}^\infty)_{k=1}^\infty$ by $((x_i^{(m_k)})_{i=1}^\infty)_{k=1}^\infty$.

For each $\epsilon > 0$, there exists, by (4.2), an $n_0 \in \mathbb{N}$ such that

$$\sup\{\|(x_i)_{i=n_0+1}^\infty\|_{\ell_\phi(X)} : (x_i)_{i=1}^\infty \in B\} < \frac{\epsilon}{4},$$

that is

$$\|(x_i)_{i=n_0+1}^\infty\|_{\ell_\phi(X)} < \frac{\epsilon}{4},\ \forall (x_i)_{i=1}^\infty \in B.$$

By (4.3), there exists a $k_0 \in \mathbb{N}$ such that for each $k, j \in \mathbb{N}$ with $k, j > k_0$,

$$\left\|x_i^{(m_k)} - x_i^{(m_j)}\right\|_X < \frac{\epsilon}{2n_0},\ i = 1, 2, \ldots, n_0.$$

Thus for each $k, j \in \mathbb{N}$ with $k, j > k_0$,

$$\begin{aligned} &\left\|(x_i^{(m_k)})_{i=1}^\infty - (x_i^{(m_j)})_{i=1}^\infty\right\|_{\ell_\phi(X)} \leq \\ &\sum_{i=1}^{n_0} \left\|x_i^{(m_k)} - x_i^{(m_j)}\right\|_X + \\ &\left\|(x_i^{(m_k)})_{i=n_0+1}^\infty\right\|_{\ell_\phi(X)} + \\ &\left\|(x_i^{(m_j)})_{i=n_0+1}^\infty\right\|_{\ell_\phi(X)} < \\ &\frac{\epsilon}{2} + \frac{\epsilon}{4} + \frac{\epsilon}{4} = \epsilon. \end{aligned}$$

Therefore, $((x_i^{(m_k)})_{i=1}^\infty)_{k=1}^\infty$ is a Cauchy sequence in $\ell_\phi(X)$ and hence a convergent sequence in $\ell_\phi(X)$. This shows that B is a relatively compact subset of $\ell_\phi(X)$. $\square$

When $\mathbb{R}$ is taken instead of X, the following corollary follows, and in this case $\ell_\phi(X)$ is simply ℓ_ϕ.

COROLLARY 4.15. [**4**, Corollary 2] Let ϕ be an Orlicz function satisfying the Δ_2 condition. Then a subset B of ℓ_ϕ is relatively compact if and only if for each $i \in \mathbb{N}$, the set $\{a_i : (a_i)_{i=1}^\infty \in B\}$ is a bounded subset of $\mathbb{R}$, and

$$\lim_n sup\{\|(a_i)_{i=n}^\infty\|_{\ell_\phi} : (a_i)_{i=1}^\infty \in B\} = 0.$$

Before introducing [**4**, Theorem 3], we need to define what it means for a Banach space to have the type I-Λ-complete continuity property (I-Λ-CCP).

Let G be a compact metrizable Abelian group, and let $B(G)$ be the σ-algebra of Borel subsets of G. Let λ be a normalized Haar measure on G, and let Γ be the dual group of G. For a real or complex Banach space X, we denote by $L_1(G, X)$ the Banach space of all equivalence classes of λ-Bochner integrable functions on G with values in X.

If μ is a countably additive X-valued measure on $B(G)$, we say that it is of bounded variation if $sup \sum_{A\in\pi} \|\mu(A)\| < \infty$, where the supremum is taken over all finite measurable partitions of G.

The measure μ is said to be of bounded average range if there is a positive constant c so that $\|\mu(A)\| \le c\lambda(A)$ for every $A \in B(G)$.

$M_1(G, X)$ denotes the space of all X-valued measures on $B(G)$ that are of bounded variation.

$M_\infty(G, X)$ denotes the space of all X-valued measures on $B(G)$ that are of bounded average range.

For $\mu \in M_1(G, X)$, define the Fourier coefficient of μ at γ by

$$\hat{\mu}(\gamma) = \int_G \bar{\gamma}(t) d\mu(t).$$

Let Λ be a subset of Γ. A measure μ in $M_1(G, X)$ is called a Λ - measure if $\hat{\mu}(\gamma) = 0$ for all $\gamma \notin \Lambda$.

DEFINITION 4.16. [**4**] A Banach space X is said to have the type I-Λ-complete continuity property (I - Λ - CCP) if every Λ - measure μ in $M_\infty(G, X)$ has a relatively compact range.

DEFINITION 4.17. [**4**] A Banach space X is said to have the type II-Λ-complete continuity property (II - Λ - CCP) if every Λ - measure μ in $M_1(G, X)$ has a relatively compact range.

THEOREM 4.18. [**4**, Theorem 3] *Let ϕ be an Orlicz function satisfying the Δ_2-condition. Let G be a compact metrizable Abelian group, Γ the dual group of G and Λ a subset of Γ. Then $\ell_\phi(X)$ has II - Λ - CCP if and only if both ℓ_ϕ and X have II - Λ - CCP.*

PROOF. We need only to show that if ℓ_ϕ and X have II - Λ - CCP then $\ell_\phi(X)$ has II - Λ - CCP.

Let $\mu : B(G) \to \ell_\phi(X)$ be a λ-continuous, Λ-measure of bounded variation. We want to show that $\{\mu(E) : E \in B(G)\}$, the range of μ, is a relatively compact subset of $\ell_\phi(X)$.

For each $i \in \mathbb{N}$, define

$$\mu_i : B(G) \to X, \quad E \to \mu(E)_i,$$

where $\mu(E)_i$ is the *ith* coordinate of $\mu(E)$. Note that for each $E \in B(G)$,

$$\|\mu_i(E)\|_X \leq \|\mu(E)\|_{\ell_\phi(X)}.$$

Each μ_i is also a λ - continuous Λ - measure of bounded variation. Since X has II - Λ - CCP,

$$\{\mu(E)_i = \mu_i(E) : E \in B(G)\} \tag{4.4}$$

is a relatively compact subset of X for each $i \in \mathbb{N}$.

Define

$$\tilde{\mu} : B(G) \to \ell_\phi, \quad E \to (\|\mu(E)_i\|_X)_i.$$

Note that, for each $E \in B(G)$,

$$\|\tilde{\mu}(E)\|_{\ell_\phi} = inf\{\lambda > 0 : \sum_{i=1}^{\infty} \phi\Big(\frac{\|\mu(E)_i\|_X}{\lambda}\Big) \leq 1\} = \|\mu(E)\|_{\ell_\phi(X)}.$$

Thus $\tilde{\mu}$ is a λ - continuous Λ - measure of bounded variation. Since ℓ_ϕ has II - Λ - CCP, $\{\tilde{\mu}(E) : E \in B(G)\}$ is a relatively compact subset of ℓ_ϕ. By Corollary 4.15,

$$\lim_n sup\Big\{\Big\|(0,\ldots,0,\|\mu(E)_n\|_X, \|\mu(E)_{n+1}\|_X,\ldots)\Big\|_{\ell_\phi} : E \in B(G)\Big\} = 0.$$

That is,

$$\lim_n sup\Big\{\Big\|(0,\ldots,0,\mu(E)_n, \mu(E)_{n+1},\ldots)\Big\|_{\ell_\phi(X)} : E \in B(G)\Big\} = 0. \tag{4.5}$$

It follows from (4.4), (4.5) and Proposition 4.14 that the set $\{\mu(E) : E \in B(G)\}$ is a relatively compact subset of $\ell_\phi(X)$.

□

What comes now is of a valuable interest, since the spaces defined below are quite similar to general sequence spaces.

For a Banach lattice X and an Orlicz function ϕ, define

$$\ell_\phi^\epsilon(X) = \Big\{(x_i)_{i=1}^\infty \in X^{\mathbb{N}} : \Big(x^*(|x_i|)\Big)_{i=1}^\infty \in \ell_\phi, for\ all\ x^* \in {X^*}^+\Big\}$$

and

$$\ell_\phi^\pi(X) = \Big\{(x_i)_{i=1}^\infty \in X^{\mathbb{N}} : \sum_{i=1}^{\infty} x_i^*(|x_i|) < \infty, for\ all\ (x_i^*)_{i=1}^\infty \in \ell_{\phi^*}^\epsilon(X^*)^+\Big\},$$

with the corresponding norms

$$\|(x_i^*)_{i=1}^\infty\|_{\ell_\phi^\epsilon(X)} = sup\Big\{\|(x^*(|x_i|))_{i=1}^\infty\|_{\ell_\phi} : x^* \in \mathcal{B}_{X^{*+}}\Big\},\ \forall (x_i^*)_{i=1}^\infty \in \ell_\phi^\epsilon(X),$$

and

$$\|(x_i)_{i=1}^\infty\|_{\ell_\phi^\pi(X)} = sup\Big\{\sum_{i=1}^{\infty} x_i^*(|x_i|) : (x_i^*)_{i=1}^\infty \in \mathcal{B}_{\ell_{\phi^*}^\epsilon(X^*)^+}\Big\},\ \forall (x_i)_{i=1}^\infty \in \ell_\phi^\pi(X).$$

Then $\ell_\phi^\pi(X)$ and $\ell_\phi^\epsilon(X)$ are Banach lattices under the corresponding norms.

In Proposition 4.19 a sequential representation of the Fremlin projective tensor product [**7**] of ℓ_ϕ and X is given.

PROPOSITION 4.19. [**4**, Proposition] Let X be a Banach lattice and ϕ an Orlicz function that has its complementary function. If ϕ satisfies the Δ_2 - condition, then $\ell_\phi \hat{\otimes}_F X$ is isometrically lattice isomorphic to $\ell_\phi^\pi(X)$.

A Banach space X is said to semi-embed into a Banach space Y if there is a one-to-one continuous linear operator from X to Y such that the image of the closed unit ball of X is closed in Y. A Banach space property $\mathcal{P}$ is said (i) to be separably determined if a Banach space X has $\mathcal{P}$ whenever every separable closed subspace of X has $\mathcal{P}$, and (ii) to be separably semi- embeddably stable if a separable Banach space X has $\mathcal{P}$ whenever X semi-embeds into a Banach space Y with $\mathcal{P}$. Conditions (i) and (ii) imply that if a Banach space X has $\mathcal{P}$, then so does every closed subspace of X. Conditions (i) and (ii) also imply that if X and Y are isomorphic Banach spaces and Y has $\mathcal{P}$, then X has $\mathcal{P}$.

THEOREM 4.20. [**4**, Theorem 5] *Let X be a Banach lattice, and let ϕ be an Orlicz function that has its complementary function and satisfies the Δ_2-condition. Let $\mathcal{P}$ be a Banach space property such that $\mathcal{P}$ is separably determined and separably semi-embeddably stable. If $\ell_\phi(X)$ has $\mathcal{P}$, then $\ell_\phi \hat{\otimes}_F X$ also has $\mathcal{P}$.*

5. Thinning sequences in $\lambda_\pi(X)$

In this section we give a detailed explanation and clarification of the adapted thinning process mentioned in Megginson's book [**9**], on page 219 in Example 2.5.24.

Let us take a sequence $\left(\bar{x}^{(n)}\right) = \left((x_i^{(n)})_i\right)$ of positive elements of $\lambda_\pi(X)$, which is weakly convergent to zero. May we find conditions under which $\|\left(x_i^{(n)}\right)\|_{\lambda_\pi(X)} \to 0$, as $n \to \infty$? Let us see what conclusions we may achieve if we assume that $\|\left(x_i^{(n)}\right)\|_{\lambda_\pi(X)}$ does not tend to zero.

For the sequence $\left(\bar{x}^{(n)}\right) = \left((x_i^{(n)})_i\right)$ to be weakly null means that $\lim\limits_n \sum_{i=1}^\infty x_i^*(|x_i^{(n)}|) = 0$ for $\forall \bar{x}^* = (x_i^*)_i \in \lambda_\epsilon^{'}(X^*)$. It may be as well assumed that $x_i^{(n)} = |x_i^{(n)}|$.

Figure 5.1

$x_1^*(\lvert x_1^{(1)}\rvert)$	$x_1^*(\lvert x_1^{(2)}\rvert)$	....	$x_1^*(\lvert x_1^{(n)}\rvert)$	....
$x_2^*(\lvert x_2^{(1)}\rvert)$	$x_2^*(\lvert x_2^{(2)}\rvert)$	....	$x_2^*(\lvert x_2^{(n)}\rvert)$	....
$x_3^*(\lvert x_3^{(1)}\rvert)$	$x_3^*(\lvert x_3^{(2)}\rvert)$	....	$x_3^*(\lvert x_3^{(n)}\rvert)$	....
$x_4^*(\lvert x_4^{(1)}\rvert)$	$x_4^*(\lvert x_4^{(2)}\rvert)$	....	$x_4^*(\lvert x_4^{(n)}\rvert)$	....
...	...	....	...	....
...	...	....	...	....
(i)	(ii)	....	(nth)	....
$\in \ell_1$	$\in \ell_1$	....	$\in \ell_1$	....

REMARK 5.1. The sequence of elements of ℓ_1, introduced in Figure 4.1, is convergent to zero in ℓ_1, because $\left((x_i^{(n)})_{i=1}^\infty\right)_{n=1}^\infty$ is chosen to be weakly convergent to zero. Let us first suppose that we have the initial series (not the thinned one).

For a sequence in ℓ_1, to be convergent to zero, it means that for $\forall\epsilon > 0$, there exists $N \in \mathbb{N}$ such that for $\forall n \geq N$,

$$\|(x_i^*(x_i^{(n)}))_i\|_{\ell_1} = \sum_{i=1}^{\infty} x_i^*(x_i^{(n)}) < \epsilon.$$

Below, when we will thin this sequence, we will keep all the columns (that is, the elements of the sequence) in their places but we will remove some number of elements from each of the members of the sequence. For the resulting series this inequality will still hold, because its $n-th$ element will differ from the $n-th$ element of the initial series by only not having the first, let us say r (r_n, depending on each n) elements. Hence the ℓ_1 norm of the $n-th$ element of the sequence of ℓ_1, resulted after thinning (or throwing away first r elements), will be a formal positive series less than the series, corresponding to the initial sequence (the ℓ_1 norm of the $n-th$ element of the initial series in ℓ_1), which is less than ϵ. Hence the thinned series will also be convergent to zero in ℓ_1.

We want to find conditions under which

$$\mathop{lim}_n \sup\Big\{\sum_{i=1}^{\infty} x_i^*(|x_i^{(n)}|) : (x_i^*)_i \in \mathcal{B}_{\lambda_\epsilon'(X^*)^+}\Big\} = 0,$$

which is what $\|\cdot\|_{\lambda_\pi(X)}$-convergence means by the definition.

Let us notice one more time that since

$$\left(\bar{x}^{(n)}\right) = \left((x_i^{(n)})_i\right) \subseteq \lambda_\pi(X),$$

it follows that each of $(i), (ii), \ldots, (nth), \ldots$, is an element of the space ℓ_1 of summable sequences.

Moreover, we have to stress that not only

$$\sum_{i=1}^{\infty} x_i^*(|x_i^{(n)}|) < \infty$$

for each fixed $\bar{x}^* = (x_i^*)_i \in \lambda_\epsilon'(X^*)$ and each fixed $n \in \mathbb{N}$, but also

$$\sup\Big\{\sum_{i=1}^{\infty} x_i^*(|x_i^{(n)}|) : \ \forall (x_i^*)_i \in \mathcal{B}_{\lambda_\epsilon'(X^*)^+}\Big\} < \infty,$$

for each fixed $n \in \mathbb{N}$.

The last relation follows because $\lambda_\pi(X)$-norm of $\left(x_i^{(n)}\right)_i$ is finite for each fixed $n \in \mathbb{N}$. Besides, the weak nullity of $\left(\bar{x}^{(n)}\right) = \left((x_i^{(n)})_i\right)$, that is, the fact that

$$\mathop{lim}_n \sum_{i=1}^{\infty} x_i^*(|x_i^{(n)}|) = 0,$$

for $\forall \bar{x}^* = (x_i^*)_i \in \lambda_\epsilon'(X^*)$, implies that for each fixed $\bar{x}^* = (x_i^*)_i \in \lambda_\epsilon'(X^*)$, $\left((x_i^*(|x_i^{(n)}|))_i\right)$ is a norm convergent to zero sequence of l_1.

$$\left(\mathop{lim}_n \sum_{i=1}^{\infty} x_i^*(|x_i^{(n)}|) = 0 \Longleftrightarrow \mathop{lim}_n \Big\|\left(x_i^*(|x_i^{(n)}|)\right)_i\Big\|_{\ell_1} = 0\right).$$

Another observation is that by Proposition 1 of [**5**], since λ is σ-order continuous, it follows that $\lambda_{\pi,0}(X) = \lambda_\pi(X)$, which means that for each fixed $n \in \mathbb{N}$

$$\mathop{lim}_{k\to\infty} \sup\Big\{\sum_{i=k}^{\infty} x_i^*(|x_i^{(n)}|) : \ (x_i^*)_i \in \mathcal{B}_{\lambda'_\epsilon(X^*)^+}\Big\} = 0$$

("sup" tail converges to zero).

After making the few observations above, let us return to our assumption that $\|\left(x_i^{(n)}\right)\|_{\lambda_\pi(X)}$ does not converge to zero as n converges to infinity. May this assumption eventually lead us to a contradiction?

From this assumption it follows that there is a subsequence (n_j) of $\mathbb{N}$ and a positive scalar t such that

$$\sup\Big\{t\cdot\sum_{i=1}^{\infty} x_i^*(|x_i^{(n_j)}|) : \ (x_i^*)_i \in \mathcal{B}_{\lambda'_\epsilon(X^*)^+}\Big\} \geq 1.$$

By thinning $t\cdot\sum_{i=1}^{\infty} x_i^*(|x_i^{(n_j)}|)$ if necessary, it may be assumed that there is a sequence (t_{n_j}) of nonnegative integers such that $0 = t_{n_1} < t_{n_2} < \ldots$ and

$$t\cdot\sum_{i=t_{n_j}+1}^{t_{n_{j+1}}} x_i^*(|x_i^{(n_j)}|) > \frac{3}{4}\cdot t\sum_{i=1}^{\infty} x_i^*(|x_i^{(n_j)}|).$$

Let us describe more in detail the process of constructing such a sequence. We take $t_{n_1} = 0$. Having n_1 fixed at this step, we take $t_{n_2} > t_{n_1} = 0$ such that

$$t\cdot\sum_{i=1}^{t_{n_2}} x_i^*(|x_i^{(n_1)}|) > \frac{3}{4}\cdot t\sum_{i=1}^{\infty} x_i^*(|x_i^{(n_1)}|)$$

(tail converges to zero).

Now let us look at $t\cdot\sum_{i=1}^{t_{n_2}} x_i^*(|x_i^{(n_2)}|)$.

If

$$t\cdot\sum_{i=1}^{t_{n_2}} x_i^*(|x_i^{(n_2)}|) \leq \frac{1}{8}\cdot t\sum_{i=1}^{\infty} x_i^*(|x_i^{(n_2)}|),$$

then we choose $t_{n_3} > t_{n_2}$ to be such that also

$$t\cdot\sum_{i=t_{n_3}+1}^{\infty} x_i^*(|x_i^{(n_2)}|) \leq \frac{1}{8}\cdot t\sum_{i=1}^{\infty} x_i^*(|x_i^{(n_2)}|).$$

Then

$$t\cdot\sum_{i=t_{n_2}+1}^{t_{n_3}} x_i^*(|x_i^{(n_2)}|) \geq \frac{3}{4}\cdot t\sum_{i=1}^{\infty} x_i^*(|x_i^{(n_2)}|).$$

In general, we would have been lucky to have

$$t\cdot\sum_{i=1}^{t_{n_2}} x_i^*(|x_i^{(n_2)}|) \leq \frac{1}{8}\cdot t\sum_{i=1}^{\infty} x_i^*(|x_i^{(n_2)}|),$$

so we have to consider the case when

$$t \cdot \sum_{i=1}^{t_{n_2}} x_i^*(|x_i^{(n_2)}|) > \frac{1}{8} \cdot t \sum_{i=1}^{\infty} x_i^*(|x_i^{(n_2)}|).$$

In this case, we throw away the first t_{n_2} members of

$$\left(x_i^*(|x_i^{(n_2)}|)\right)_i,$$

renumber it, so that in the sum below $x_1^*(|x_1^{(n_2)}|)$ was previously $x_{t_{n_2}+1}^*(|x_{t_{n_2}+1}^{(n_2)}|)$, further $x_2^*(|x_2^{(n_2)}|)$ was previously $x_{t_{n_2}+2}^*(|x_{t_{n_2}+2}^{(n_2)}|)$, next $x_3^*(|x_3^{(n_2)}|)$ was previously $x_{t_{n_2}+3}^*(|x_{t_{n_2}+3}^{(n_2)}|)$, etc., and again look at

$$t \cdot \sum_{i=1}^{t_{n_2}} x_i^*(|x_i^{(n_2)}|)$$

(this last sum is under the new numbering).

If this last sum is $\leq$ than

$$\frac{1}{8} \sum_{i=1}^{\infty} x_i^*(|x_i^{(n_2)}|),$$

then we are satisfied, and if not we repeat the step, by throwing it away.

(The sum from 1 to ∞ is the initial sum, when nothing has been thrown away from it yet).

At most after 8 steps, after having thrown away $7 \cdot t_{n_2}$ members of $\left(x_i^*(|x_i^{(n_2)}|)\right)_i$, and each time renumbering it, we get that

$$t \cdot \sum_{i=1}^{t_{n_2}} x_i^*(|x_i^{(n_2)}|) \leq \frac{1}{8} \cdot t \sum_{i=1}^{\infty} x_i^*(|x_i^{(n_2)}|),$$

where the series in the right corresponds to the initial numbering (when nothing was thrown away).

Next, we choose $t_{n_3} > t_{n_2}$ such that

$$t \cdot \sum_{i=t_{n_3}+1}^{\infty} x_i^*(|x_i^{(n_2)}|) \leq \frac{1}{8} \cdot t \sum_{i=1}^{\infty} x_i^*(|x_i^{(n_2)}|),$$

(where on the right is still the initial series, with none of its members thrown away yet).

Finally we have that

$$t \cdot \sum_{i=t_{n_2}+1}^{t_{n_3}} x_i^*(|x_i^{(n_2)}|) \geq \frac{3}{4} \cdot t \sum_{i=1}^{\infty} x_i^*(|x_i^{(n_2)}|),$$

where on the right may as well be taken the thinned series (its $7 \cdot t_{n_2}$ many members removed).

By induction, we may consider the required sequence (t_{n_j}), $0 = t_{n_1} < t_{n_2} < \dots$, be constructed.

Now we will show the same "thinning" construction for supremums.

Take $t_{n_1} = 0$. Here n_1 is fixed, and we already stated that for each fixed n

$$\lim_{k\to\infty} \sup\left\{\sum_{i=k}^{\infty} x_i^*(|x_i^{(n)}|) : (x_i^*)_i \in \mathcal{B}_{\lambda'_\epsilon(X^*)^+}\right\} = 0,$$

which is the supremum tail convergence to zero. Therefore, we can choose $t_{n_2} > t_{n_1} = 0$ such that

$$\sup\left\{t \cdot \sum_{i=1}^{t_{n_2}} x_i^*(|x_i^{(n_1)}|) : (x_i^*)_i \in \mathcal{B}_{\lambda'_\epsilon(X^*)^+}\right\}$$
$$\geq \frac{3}{4} \cdot \sup\left\{t \sum_{i=1}^{\infty} x_i^*(|x_i^{(n_1)}|) : (x_i^*)_i \in \mathcal{B}_{\lambda'_\epsilon(X^*)^+}\right\}.$$

Next, if

$$\sup\left\{t \cdot \sum_{i=1}^{t_{n_2}} x_i^*(|x_i^{(n_2)}|) : (x_i^*)_i \in \mathcal{B}_{\lambda'_\epsilon(X^*)^+}\right\}$$
$$\leq \frac{1}{8} \cdot \sup\left\{t \sum_{i=1}^{\infty} x_i^*(|x_i^{(n_2)}|) : (x_i^*)_i \in \mathcal{B}_{\lambda'_\epsilon(X^*)^+}\right\},$$

then we choose $t_{n_3} > t_{n_2}$ in such a way that

$$\sup\left\{t \cdot \sum_{i=t_{n_3}+1}^{\infty} x_i^*(|x_i^{(n_2)}|) : (x_i^*)_i \in \mathcal{B}_{\lambda'_\epsilon(X^*)^+}\right\}$$
$$\leq \frac{1}{8} \cdot \sup\left\{t \sum_{i=1}^{\infty} x_i^*(|x_i^{(n_2)}|) : (x_i^*)_i \in \mathcal{B}_{\lambda'_\epsilon(X^*)^+}\right\}.$$

If we could have that

$$\sup\left\{t \cdot \sum_{i=1}^{\infty} x_i^*(|x_i^{(n_2)}|) : (x_i^*)_i \in \mathcal{B}_{\lambda'_\epsilon(X^*)^+}\right\}$$
$$= \sup\left\{t \sum_{i=1}^{t_{n_2}} x_i^*(|x_i^{(n_2)}|) : (x_i^*)_i \in \mathcal{B}_{\lambda'_\epsilon(X^*)^+}\right\} +$$

$$+ \sup\left\{t \cdot \sum_{i=t_{n_2}+1}^{t_{n_3}} x_i^*(|x_i^{(n_2)}|) : (x_i^*)_i \in \mathcal{B}_{\lambda'_\epsilon(X^*)^+}\right\}$$
$$+ \sup\left\{t \sum_{i=t_{n_3}+1}^{\infty} x_i^*(|x_i^{(n_2)}|) : (x_i^*)_i \in \mathcal{B}_{\lambda'_\epsilon(X^*)^+}\right\},$$

then it would follow that

$$\sup\left\{t\cdot\sum_{i=t_{n_2}+1}^{t_{n_3}} x_i^*(|x_i^{(n_2)}|) : (x_i^*)_i \in \mathcal{B}_{\lambda'_\epsilon(X^*)^+}\right\}$$
$$\geq \frac{3}{4}\cdot\sup\left\{t\sum_{i=1}^{\infty} x_i^*(|x_i^{(n_2)}|) : (x_i^*)_i \in \mathcal{B}_{\lambda'_\epsilon(X^*)^+}\right\}.$$

Next, if

$$\sup\left\{t\cdot\sum_{i=1}^{t_{n_2}} x_i^*(|x_i^{(n_2)}|) : (x_i^*)_i \in \mathcal{B}_{\lambda'_\epsilon(X^*)^+}\right\}$$
$$> \frac{1}{8}\cdot\sup\left\{t\sum_{i=1}^{\infty} x_i^*(|x_i^{(n_2)}|) : (x_i^*)_i \in \mathcal{B}_{\lambda'_\epsilon(X^*)^+}\right\},$$

then we throw away at most $7\cdot t_{n_2}$ many terms of

$$\left(x_i^*(|x_i^{(n_2)}|)\right)_i,$$

and after renumbering it, get that

$$\sup\left\{t\cdot\sum_{i=1}^{t_{n_2}} x_i^*(|x_i^{(n_2)}|) : (x_i^*)_i \in \mathcal{B}_{\lambda'_\epsilon(X^*)^+}\right\}$$
$$\leq \frac{1}{8}\cdot\sup\left\{t\sum_{i=1}^{\infty} x_i^*(|x_i^{(n_2)}|) : (x_i^*)_i \in \mathcal{B}_{\lambda'_\epsilon(X^*)^+}\right\},$$

where the supremum in the right is taken for the initial series, with no members thrown away. Then we choose $t_{n_3} > t_{n_2}$ with

$$\sup\left\{t\cdot\sum_{i=t_{n_3}+1}^{\infty} x_i^*(|x_i^{(n_2)}|) : (x_i^*)_i \in \mathcal{B}_{\lambda'_\epsilon(X^*)^+}\right\}$$
$$\leq \frac{1}{8}\cdot\sup\left\{t\sum_{i=1}^{\infty} x_i^*(|x_i^{(n_2)}|) : (x_i^*)_i \in \mathcal{B}_{\lambda'_\epsilon(X^*)^+}\right\},$$

where the supremum of the series in the right is still for the initial series, with no terms removed.

Finally we get that

$$\sup\left\{t\cdot\sum_{i=t_{n_2}+1}^{t_{n_3}} x_i^*(|x_i^{(n_2)}|) : (x_i^*)_i \in \mathcal{B}_{\lambda'_\epsilon(X^*)^+}\right\}$$
$$\geq \frac{3}{4}\cdot\sup\left\{t\sum_{i=1}^{\infty} x_i^*(|x_i^{(n_2)}|) : (x_i^*)_i \in \mathcal{B}_{\lambda'_\epsilon(X^*)^+}\right\},$$

where the supremum in the right may also be taken for the final renumbered series, with at most $7\cdot t_{n_2}$ many terms removed.

By induction, the construction is finalized.

References

[1] G. Botelho and J. R. Campos, *On the transformation of vector-valued sequences by linear and multilinear operators*, Monatsh. Math. **183** (2017), no. 3, 415–435, DOI 10.1007/s00605-016-0963-4. MR3662075

[2] Q. Bu and G. Buskes, *Schauder decompositions and the Fremlin projective tensor product of Banach lattices*, J. Math. Anal. Appl. **355** (2009), no. 1, 335–351, DOI 10.1016/j.jmaa.2009.01.061. MR2514471

[3] Q. Bu, Y. Li, and X. Xue, *Some properties of the space of regular operators on atomic Banach lattices*, Collect. Math. **62** (2011), no. 2, 131–137, DOI 10.1007/s13348-010-0007-7. MR2792516

[4] Q. Bu, D. Ji, and X. Xue, *Complete continuity properties for the Fremlin projective tensor product of Orlicz sequence spaces and Banach lattices*, Rocky Mountain J. Math. **40** (2010), no. 6, 1797–1808, DOI 10.1216/RMJ-2010-40-6-1797. MR2764221

[5] D. Ji, Y. Li, and Q. Bu, *The complete continuity properties for the positive projective tensor product of atomic Banach lattices*, Positivity **17** (2013), no. 1, 17–25, DOI 10.1007/s11117-011-0141-9. MR3027643

[6] N. Dunford and J. T. Schwartz, *Linear operators. Part I*, Wiley Classics Library, John Wiley & Sons, Inc., New York, 1988. General theory; With the assistance of William G. Bade and Robert G. Bartle; Reprint of the 1958 original; A Wiley-Interscience Publication. MR1009162

[7] D. H. Fremlin, *Tensor products of Archimedean vector lattices*, Amer. J. Math. **94** (1972), 777–798, DOI 10.2307/2373758. MR0312203

[8] D. H. Fremlin, *Tensor products of Banach lattices*, Math. Ann. **211** (1974), 87–106, DOI 10.1007/BF01344164. MR0367620

[9] R. E. Megginson, *An introduction to Banach space theory*, Graduate Texts in Mathematics, vol. 183, Springer-Verlag, New York, 1998. MR1650235

[10] P. Meyer-Nieberg, *Banach lattices*, Universitext, Springer-Verlag, Berlin, 1991. MR1128093

[11] A. C. Zaanen, *Introduction to operator theory in Riesz spaces*, Springer-Verlag, Berlin, 1997. MR1631533

Email address: Botelho@ufu.br

Email address: CauseyRM@miamioh.edu

Email address: Kh_Navoyan@ufu.br

Contemporary Mathematics
Volume **737**, 2019
https://doi.org/10.1090/conm/737/14856

Generalized numerical ranges, dilation, and quantum error correction

Sara Botelho-Andrade and Chi-Kwong Li

ABSTRACT. A survey is given to some recent results on how generalized numerical ranges relate to the study dilation and perturbation of operators. The connection to the study quantum error correction, and unital completely positive linear maps from a Calkin algebra to a matrix space is also discussed.

1. Introduction

Let $\mathcal{B}(\mathcal{H})$ be the set of bounded linear operators on a Hilbert space $\mathcal{H}$ equipped with the inner product $\langle x, y\rangle$. If $\mathcal{H}$ has dimension n, then it is identified as $\mathbb{C}^n$ with the usual inner product $\langle x, y\rangle = y^*x$, and $\mathcal{B}(\mathcal{H})$ is identified as M_n, the algebra of $n \times n$ complex matrices. The study of quadratic forms and their applications appear in many areas of mathematics and other branches of sciences. One such form is the numerical range (a.k.a. the field of values), defined as follows.

DEFINITION 1.1. The *numerical range* of an operator $A \in \mathcal{B}(\mathcal{H})$ is the set

$$W(A) = \{\langle Ax, x\rangle : x \in \mathcal{B}(\mathcal{H}) \text{ and } \langle x, x\rangle = 1\}$$

EXAMPLE 1.2. Here are some simple examples, which will appear again in our subsequent discussion.

- If $A = \begin{pmatrix} 1 & 0 \\ 0 & 0 \end{pmatrix}$, then $W(A)$ is the line segment joining 0 and 1.
- If $A = \begin{pmatrix} 0 & 2 \\ 0 & 0 \end{pmatrix}$, then $W(A) = \{\mu \in \mathbb{C} : |\mu| \le 1\}$, the unit disk centered at the origin.
- If $A = \operatorname{diag}(a_1, a_2, a_3)$, then $W(A)$ is the triangular disk with vertices a_1, a_2, a_3.

Informally, the numerical range of an operator A can be viewed as a "picture" of A, and every point $\langle Ax, x\rangle$ in $W(A)$ can be viewed as a "pixel" of the picture. The "picture" can provide useful information of the operator. In fact, the numerical range of an operator A can be used to deduce algebraic or analytic properties, locate

2010 *Mathematics Subject Classification.* Primary 15A60, 47A12, 47A13, 47A20.

the spectrum $\sigma(A)$, and obtain norm bounds of A. The information can then be used to study useful properties such as the invertibility, stability, and convergence of the sequence $\{A^m : m = 1, \dots\}$ of the operator.

In this paper, we survey some results concerning the use of numerical range and generalized numerical ranges to study dilation and compression of operators. The connection of these results to the study quantum error correction, and unital completely positive linear maps from a Calkin algebra to a matrix space will be discussed.

We first present some basic results in Section 2. In Section 3 we describe some ideas on how one can use the inclusion relation of $W(B) \subseteq W(A)$ to ensure that B has a dilation of the $I \otimes A$. Section 4 concerns the joint numerical ranges of several operators and the joint dilation problem. In Sections 5 and 6, we discuss different kinds of generalized numerical ranges arising in the study of quantum error correction codes. The non-emptyness of such general numerical ranges associated with the error operators of a noisy quantum channel will ensure the existence of different types of quantum error correction codes. In such a case, the element in the generalized numerical range will be useful for the construction of a quantum error correction code for the given channel. It turns out that the results and insights developed in the study of quantum error correction is useful in the study of joint essential matricial ranges, and also the images of unital completely positive linear maps from the Calkin algebra associated with a Hilbert space to matrix spaces. These results will be described in Section 7.

For most results we will present the statements without proofs. Nevertheless, we will present three short proofs for the convexity of the numerical range. Also, we give several short new proofs for a few selected results that are different from those in the literature.

2. Basic results

We begin with some results which can be readily deduced from the definition.

PROPOSITION 2.1. *Let $A = H + iG \in \mathcal{B}(\mathcal{H})$, where $H = H^*$ and $G = G^*$.*

a) *$W(aH + ibG) = \{ah + ibg : h + ig \in W(A)\}$ for any $a, b \in \mathbb{R}$.*
b) *$W(aA + bI) = aW(A) + b$ for any $a, b \in \mathbb{C}$.*
c) *$W(A) = W(A^T)$ and $W(A^*) = \overline{W(A)} = \{\bar{\mu} : \mu \in W(A)\}$.*
d) *$W(X^*AX) \subseteq W(A)$ for any subspace $\mathcal{K}$ of $\mathcal{H}$ and $X : \mathcal{K} \to \mathcal{H}$ satisfies $X^*X = I_{\mathcal{K}}$. The equality holds if $\mathcal{K} = \mathcal{H}$ and X is unitary.*

A fundamental and useful result on the numerical range is the celebrated Töplitz-Hausdorff theorem [**18, 33**], proved 100 years ago, asserting that the numerical range of an operator is always convex. There have been many different proofs of this result; see [**4**]. Here we present three short proofs.

Proof 1. Let $a = \langle Ax, x\rangle$ and $b = \langle Ay, y\rangle$ be two different elements in $W(A)$. By Proposition 2.1, we may replace A by $\frac{1}{b-a}(A-aI)$ and assume that $(a,b) = (0,1)$. We will show that $[a,b] \subseteq W(A)$. Because $\langle Ax, x\rangle \neq \langle Ay, y\rangle$, the vectors x and y are linearly independent. Consider the family of unit vectors $z(t) = \frac{(1-t)x+te^{i\theta}y}{\|(1-t)x+te^{i\theta}y\|}$, where $\theta \in [0, 2\pi)$ satisfies $\langle Ax, e^{i\theta}y\rangle \geq 0$. Then by our choice of θ,

$$t \mapsto \mu(t) = \langle Az(t), z(t)\rangle = (1-t)^2\langle Ax, x\rangle + 2t(1-t)\langle Ax, e^{i\theta}y\rangle + t^2\langle Ay, y\rangle,$$

is a continuous real-valued function on $[0,1]$ with $\mu(0) = 0$ and $\mu(1) = 1$. Hence, $[0,1] \subseteq \{\mu(t) : t \in [0,1]\} \subseteq W(A)$. □

Proof 2. Let $a = \langle Ax, x\rangle$ and $b = \langle Ay, y\rangle$ be two different elements in $W(A)$. Suppose $\mathcal{K} = \operatorname{span}\{x, y\} \subseteq \mathcal{H}$, and $X : \mathcal{K} \to \mathcal{H}$ satisfying $X^*X = I_{\mathcal{K}}$. We may identify $B = X^*AX \in M_2$ and $\operatorname{span}\{x,y\} = \mathbb{C}^2$. Then

$$W(B) = \{u^*Bu : u \in \mathbb{C}^2, u^*u = 1\} = \{\operatorname{tr}(Buu^*) : u^*u = 1\}$$

can be viewed as the image of the "sphere"

$$\{uu^* : u \in \mathbb{C}^2, u^*u = 1\} = \left\{\frac{1}{2}\begin{pmatrix} 1+a & b-ic \\ b+ic & 1-a\end{pmatrix} : a,b,c \in \mathbb{R},\ a^2+b^2+c^2 = 1\right\}$$

in $\mathbb{R}^3$ under the real linear map $X \mapsto \operatorname{tr}(BX) \in \mathbb{C} \equiv \mathbb{R}^2$. Thus, $W(B)$ is an elliptical disk containing the two points $a = \operatorname{tr}(Bxx^*)$ and $b = \operatorname{tr}(Byy^*)$. By Proposition 2.1, $W(B) \subseteq W(A)$. The result follows. □

Proof 3. We refine the second part of Proof 2, and show that for $B \in M_2$ with eigenvalues $\lambda_1, \lambda_2 \in \mathbb{C}$ and $b = \sqrt{\operatorname{tr}(BB^* - |\lambda_1|^2 - |\lambda_2|^2}$, $W(B)$ is an elliptical disk with foci λ_1, λ_2 and length of minor axis b. Replacing B by $B - \frac{\operatorname{tr} B}{2}I_2$, we may assume that $\operatorname{tr}(B) = 0$ and $B = \begin{pmatrix}\lambda & b \\ 0 & -\lambda\end{pmatrix}$. If $b = 0$, then

$$W(B) = \{|u_1|^2\lambda - |u_2|^2\lambda : u_1, u_2 \in \mathbb{C}, |u_1|^2 + |u_2|^2 = 1\}$$

is a line segment joining λ and $-\lambda$. Suppose $b > 0$. If $\lambda = 0$, then

$$W(B) = \{bu_2\bar{u}_1 : u_1, u_2 \in \mathbb{C}, |u_1|^2 + |u_2|^2 = 1\}$$

with diameter b. If $\lambda \neq 0$, we may further replace B by D^*BD/λ, where $D = \operatorname{diag}(1, e^{i\theta})$ satisfies $e^{i\theta}b/\lambda = c > 0$. Let $\gamma = \sqrt{(2/c)^2 + 1}$, and $B = H + iG$ with $H = H^*$, $G = G^*$. Then $\hat{B} = H + i\gamma G$ is rank one nilpotent and is unitarily similar to $\begin{pmatrix} 0 & \sqrt{4+c^2} \\ 0 & 0\end{pmatrix}$ so that $W(\hat{B}) = \{\mu \in \mathbb{C} : |\mu| \leq \sqrt{1+(c/2)^2}\}$. Since $W(\hat{B}) = W(H + i\gamma G) = \{h + i\gamma g : h + ig \in W(B)\}$ by Proposition 2.1, $W(B)$ is an elliptical disk with major axis $[-\sqrt{1+(c/2)^2}, \sqrt{1+(c/2)^2}]$ and minor axis $\{ir : r \in [-c/2, c/2]\}$. □

Note that Proof 1 is quite standard, and used in many textbooks, e.g., see [**35**]. Proof 2 is based on [**14**] and Proof 3 is based on [**23**]. We summarize the above results into the following.

THEOREM 2.2. *The numerical range of $A \in \mathcal{B}(\mathcal{H})$ is convex. In particular, if $A \in M_2$ has eigenvalues λ_1, λ_2, then $W(A)$ is an elliptical disk with foci λ_1, λ_2, and minor axis with length $\sqrt{\operatorname{tr} A^*A - |\lambda_1|^2 - |\lambda_2|^2}$. Consequently, if $A = \begin{pmatrix} \lambda_1 & b \\ 0 & \lambda_2 \end{pmatrix}$, then the minor axis of the elliptical disk $W(A)$ has length $|b|$.*

Next, we list some results showing that there is an interesting interplay between the algebraic and analytic properties of $A \in \mathcal{B}(\mathcal{H})$ and the geometrical properties of $W(A)$. We use $\operatorname{conv} S$ and $\operatorname{cl}(S)$ to denote the convex hull and closure of the set $S \subseteq \mathbb{C}$.

PROPOSITION 2.3. *Let $A \in \mathcal{B}(\mathcal{H})$. Then*

a) *$A = \mu I$ if and only if $W(A) = \{\mu\}$*
b) *$A = A^*$ if and only if $W(A) \subseteq \mathbb{R}$.*
c) *A is positive semi-definite if and only if $W(A) \subseteq [0, \infty)$*
d) *A is unitary if and only if $W(A)$ and $W(A^{-1})$ lie in the disk.*
e) $\operatorname{conv} \sigma(A) \subseteq \operatorname{cl}(W(A))$*; the set equality holds if A is normal.*
f) *If $\mathcal{H} = \mathcal{H}_1 \oplus \mathcal{H}_2$ and $A_1 \oplus A_2 \in \mathcal{B}(\mathcal{H}_1) \oplus \mathcal{B}(\mathcal{H}_2)$, then*
$$W(A) = \operatorname{conv}\{W(A_1) \cup W(A_2)\}.$$
g) *$W(I \otimes A) = W(A)$.*

The proof of (a) – (c) and (e) – (f) can be verified readily. The proof of (d) is more tricky, especially for the infinite dimensional case. One may see [**2**] for details.

3. Dilation and numerical range inclusion

A useful technique in studying an operator T is to dilate T to a "larger" operator $A \in \mathcal{B}(\mathcal{H})$ with "nice" structure so that one can obtain information about T using the properties of A. Formally, we have the following definition.

DEFINITION 3.1. Let $T \in \mathcal{B}(\mathcal{K})$ and $A \in \mathcal{B}(\mathcal{H})$. We say that A is a *dilation* of T, equivalently, T is a *compression* of A, if $\mathcal{K}$ can be embedded in $\mathcal{H}$, and A has operator matrix $\begin{pmatrix} T & \star \\ \star & \star \end{pmatrix}$ with respect to an orthonormal basis using the vectors in $\mathcal{K}$ and $\mathcal{K}^\perp$.

The following example illustrates how one can dilate an operator to one with nice structure.

EXAMPLE 3.2. Every contraction $T \in \mathcal{B}(\mathcal{K})$, i.e., $T \in \mathcal{B}(\mathcal{K})$ with $\|T\| \le 1$, admits a unitary dilation of the form
$$A = \begin{pmatrix} T & \sqrt{I - TT^*} \\ \sqrt{I - TT^*} & -T^* \end{pmatrix} \in \mathcal{B}(\mathcal{K} \oplus \mathcal{K}).$$

It turns out that the numerical range can be used in studying dilation. Evidently, $A \in \mathcal{B}(\mathcal{H})$ is a dilation of $T \in \mathcal{B}(\mathcal{K})$ with $\mathcal{K} \subseteq \mathcal{H}$ if there exists $X : \mathcal{K} \to \mathcal{H}$ with $X^*X = I_{\mathcal{K}}$ such that $X^*AX = T$. By Proposition 2.1 (d), we have $W(T) \subseteq W(A)$. But the converse may not hold, i.e., $W(T) \subseteq W(A)$ does not ensure that A is a dilation of T as shown in the following.

EXAMPLE 3.3. Let $A = \begin{pmatrix} 0 & 2 \\ 0 & 0 \end{pmatrix} \in M_2$ and $T = 0_3 \in M_3$, then

$$W(T) = \{0\} \subset \{\mu \in \mathbb{C} : |\mu| \leq 1\} = W(A),$$

but A is not a dilation of T as the dimension of A is too low.

By Proposition 2.3 (g), $W(I \otimes A) = W(A)$. This inspires the following.

PROBLEM 3.4. Identify "good" operators $A \in \mathcal{B}(\mathcal{H})$ such that $T \in \mathcal{B}(\mathcal{K})$ has a dilation of the form $I \otimes A$ whenever $W(T) \subseteq W(A)$.

The following theorem was obtained in [**31**]; see also [**32**].

THEOREM 3.5. *Let $A \in M_3$ be a normal matrix with eigenvalues $a_1, a_2, a_3 \in \mathbb{C}$. Then $T \in \mathcal{B}(\mathcal{K})$ satisfies $W(T) \subseteq W(A) = \operatorname{conv}\{a_1, a_2, a_3\}$ if and only if T has a dilation of the form $I \otimes A$.*

Note that in applying the above theorem, one does not need to fix the matrix A in advance. For a given operator T, one may choose any triangle with vertices a_1, a_2, a_3 such that $W(T)$ lies inside the triangle. Then T will admit a dilation of the form $I \otimes \operatorname{diag}(a_1, a_2, a_3)$.

Here we give a new short proof for Theorem 3.5 using the following observation, which can be extended to prove some later results in our discussion.

LEMMA 3.6. *Let $A = H + iG$ with $(H, G) = (H^*, G^*)$. Suppose $a_1, a_2, b_1, b_2, c_1, c_2$ are real numbers such that $\begin{pmatrix} a_1 & b_1 \\ a_2 & b_2 \end{pmatrix}$ is invertible, and*

$$\tilde{A} = (a_1 H + b_1 G + c_1 I) + i(a_2 H + b_2 G + c_2 I).$$

Then $T = T_1 + iT_2$ with $(T_1, T_2) = (T_1^*, T_2^*)$ has a dilation of the form $I \otimes A$ if and only if $\tilde{T} = (a_1 T_1 + b_1 T_2 + c_1 I) + i(a_2 T_1 + b_2 T_2 + c_2 I)$ has a dilation of the form $I \otimes \tilde{A}$.

Proof of Theorem 3.5. If $T \in \mathcal{B}(\mathcal{K})$ is a compression of an operator of the form $I \otimes A$, then $W(T) \subseteq W(A)$. To prove the converse, we consider three cases. If $a_1 = a_2 = a_3$, then $W(A) = \{a_1\}$ and $W(T) \subseteq W(A)$ implies that $T = a_1 I_{\mathcal{K}}$. The result follows.

Suppose A is not a scalar matrix, and a_1, a_2, a_3 are collinear so that $W(A)$ is a line segment, say, with end points a_1, a_3. By Lemma 3.6, we can replace A by $\frac{1}{a_1 - a_3}(A - a_3 I)$ and assume that $(a_1, a_2, a_3) = (0, r, 1)$ for some $r \in [0, 1]$. If

$W(T) \subseteq W(A)$ then T is a positive operator with $T \le I_{\mathcal{K}}$. Then, T is a compression of the operator

$$\begin{pmatrix} \sqrt{T} \\ \sqrt{I-T} \end{pmatrix} \begin{pmatrix} \sqrt{T} & \sqrt{I-T} \end{pmatrix} = \begin{pmatrix} T & \sqrt{T-T^2} \\ \sqrt{T-T^2} & I-T \end{pmatrix},$$

which is unitarily similar to $I_{\mathcal{K}} \otimes \operatorname{diag}(0,1)$. So, T is a compression of a matrix of the form $I_{\mathcal{K}} \otimes \operatorname{diag}(0,r,1)$.

Finally, suppose a_1, a_2, a_3 are three non-collinear points in $\mathbb{C}$. We may replace A by $\frac{1}{a_2-a_1}(A - a_1 I)$ and assume that $(a_1, a_2) = (1, 0)$. Up to unitary similarity, we may assume that $A = \operatorname{diag}(0, 1, r+is) = H + iG$ with $(H,G) = (H^*, G^*)$. We may further replace A by $(H - \frac{r}{s}G) + i\frac{1}{s}G$, and assume that $A = \operatorname{diag}(0,1,i)$. If $T = T_1 + iT_2$ with $(T_1, T_2) = (T_1^*, T_2^*)$ satisfies $W(T) \subseteq W(A)$, then for any unit vector x, the point $\langle Tx, x\rangle$ lies inside the triangle with vertices $0, 1, i$. Hence,

$$\langle T_1 x, x\rangle \ge 0, \quad \langle T_2 x, x\rangle \ge 0, \quad \langle (T_1+T_2)x, x\rangle \le 1 \quad \text{for all } x \in \mathcal{K} \text{ with } \|x\| = 1.$$

Thus, T_1, T_2 are positive operators such that $T_1 + T_2 \le I_{\mathcal{K}}$. Let X be such that $X = [\sqrt{I_{\mathcal{K}} - T_1 - T_2}\ \sqrt{T_1}\ \sqrt{T_2}]$ satisfies $X^*X = I_{\mathcal{K}}$ and $T = X^*(A \otimes I_{\mathcal{K}})X$. Hence T is a compression of an operator of the form $I_{\mathcal{K}} \otimes A$. □

The following result was obtained in [**11**] extending a result in [**1**] (see also [**3**]) corresponding to the special case when $A = \begin{pmatrix} 0 & 2 \\ 0 & 0 \end{pmatrix}$.

THEOREM 3.7. *Let $A \in M_2$ so that $W(A)$ is the elliptical disk with eigenvalues a_1, a_2 as foci and minor axis of length $b = \sqrt{\operatorname{tr} A^*A - |a_1|^2 - |a_2|^2}$. Then $T \in \mathcal{B}(\mathcal{H})$ satisfies $W(T) \subseteq W(A)$ if and only if T has a dilation of the form $I \otimes A$.*

In Theorem 3.7, if $A \in M_2$ is normal, then one can use the proof of Theorem 3.5 to get the conclusion. If A is not normal, one can use Lemma 3.6 to reduce the problem to the case treated in [**1**] (and also [**3**]) as follows. Replace A by $aA+bI$ and assume that $W(A)$ is a standard ellipse with major axis equal to $[-1,1]$ and minor axis $\{ri : r \in [-b,b]\}$. Then we can further replace A by $\tilde{A} = \frac{1}{2}(A+A^*) + \frac{1}{2b}(A-A^*)$ so that $W(\tilde{A})$ is the unit disk centered at origin.

Note that in the application of Theorem 3.7, one needs not specify the matrix A in advance. For a given operator T, one may consider an ellipse $\mathcal{E}$ such that $W(T) \subseteq \mathcal{E}$. One can then construct $A = \begin{pmatrix} a_1 & b \\ 0 & a_2 \end{pmatrix} \in M_2$, where a_1, a_2 are the foci of $\mathcal{E}$ and b is the length of the minor axis of $\mathcal{E}$. Then T will admit a dilation of the form $I \otimes A$.

Theorems 3.5 and 3.7 were further extended in [**12**] to the following.

THEOREM 3.8. *Let $A \in M_3$ have a reducing eigenvalue so that A is unitarily similar to $[\alpha] \oplus A_1$, with $A_1 \in M_2$, so that $W(A)$ is the convex hull of α and the elliptical disk $W(A_1)$. Then $T \in \mathcal{B}(\mathcal{H})$ satisfies $W(T) \subseteq W(A)$ if and only if T has a dilation of the form $I \otimes A$.*

Theorem 3.8 may fail for general matrices $A \in M_3$ or normal matrices $A \in M_4$ as shown in the following.

EXAMPLE 3.9. (a) Let $A = \begin{pmatrix} 0 & 1 & 0 \\ 0 & 0 & 1 \\ 0 & 0 & 0 \end{pmatrix}$ and $T = \begin{pmatrix} 0 & \sqrt{2} \\ 0 & 0 \end{pmatrix}$, then

$$W(A) = W(T) = \{\mu \in \mathbb{C} : |\mu| \leq 1/\sqrt{2}\}.$$

However $\|T\| = \sqrt{2} > 1 = \|A\|$, therefore T has no dilation of the form $I \otimes A$.

(b) Let T be as in the previous example and $A = \operatorname{diag}(1, i, -1, -i)$, then note $W(T) \subseteq \operatorname{conv}\{1, i, -1, -i\} = W(A)$. As in the previous example, $\|A\| = 1$ therefore T has no dilation of the form $I \otimes A$.

In connection to Example 3.9 (b), we have the following; see [**11**, Theorem 2.5].

THEOREM 3.10. *Let $A = \operatorname{diag}(1, i, -1, -i)$. Then $T \in B(H)$ has a dilation of the form $T \otimes A$ if and only if $W(\tilde{T})$ lies inside the unit disk, where*

$$\tilde{T} = \begin{pmatrix} 0 & T + T^* \\ i(T^* - T) & 0 \end{pmatrix}.$$

It is interesting to note that the proof of Theorem 3.8 in [**12**] relies on results of completely positive linear maps and the following theorem, which is the key to affirm a conjecture of Halmos [**17**] that we will state as Corollary 3.12.

THEOREM 3.11. *Suppose $T \in \mathcal{B}(\mathcal{H})$ is a contraction with*

$$W(T) \subseteq S = \{\mu : |\mu| \leq 1, \mu + \overline{\mu} \leq r\},$$

then T has a unitary dilation $A \in \mathcal{B}(\mathcal{H} \oplus \mathcal{H})$ with $W(A) \subseteq S$.

Denote by $\mathrm{cl}(\mathcal{R})$ the closure of a set $\mathcal{R} \subseteq \mathbb{C}$.

COROLLARY 3.12. Let $T \in \mathcal{B}(\mathcal{H})$ be a contraction. Then

$$\mathrm{cl}(W(T)) = \cap\{\mathrm{cl}(W(U)) : U \in \mathcal{B}(\mathcal{H} \oplus \mathcal{H}) \text{ is a unitary dilation of } T\}.$$

There are many open problems concerning dilation and numerical range inclusion. We list a few in the following.

1. Determine $A \in \mathcal{B}(\mathcal{H})$ such that an operator $B \in \mathcal{B}(\mathcal{K})$ has a dilation of the form $I \otimes A$ whenever $W(B) \subseteq W(A)$.
2. Determine $B \in \mathcal{B}(\mathcal{K})$ such that B has a dilation of the form $I \otimes A$ for an operator $A \in \mathcal{B}(\mathcal{H})$ whenever $W(B) \subseteq W(A)$.
3. One may also consider a special region $\mathcal{R}$ in $\mathbb{C}$ such that $W(B) \subseteq \mathcal{R}$ will ensure that B has a dilation of the form $I \otimes A$ for some $A \in \mathcal{B}(\mathcal{H})$ with simple structure.

In connection to Problem 1, by Theorem 3.8 if $A \in M_3$ has a reducing eigenvalue, then for any $T \in \mathcal{B}(\mathcal{H})$ satisfying $W(T) \subseteq W(A)$ will ensure that T has dilation of the form $I \otimes A$. In a forthcoming paper, C.K. Li and Y.T. Poon show that if a matrix $A \in M_3$ is such that the boundary of $W(A)$ contains a line segment, then any operator T satisfying $W(T) \subseteq W(A)$ will have a dilation of the form $I \otimes A$. This will further extend Theorem 3.8. For Problem 2, it is clear that all normal operators B satisfy the said property. It would be nice to determine whether the converse is true. For Problem 3, it was shown in [**11**] that if $\mathcal{R}$ is a trapezoidal region in $\mathbb{R}^2 \equiv \mathbb{C}$ and $B \in M_n$ satisfies $W(B) \subseteq \mathcal{R}$, then B has a dilation of the form $A_1 \oplus \cdots \oplus A_n$ with $A_1, \ldots, A_n \in M_2$.

4. Joint numerical ranges and joint dilation

DEFINITION 4.1. For $A_1, ..., A_k \in \mathcal{B}(\mathcal{H})$, define their *joint numerical range* by

$$W(A_1, ..., A_k) = \{(\langle A_1 x, x\rangle, ..., \langle A_k x, x\rangle) : x \in \mathcal{H} \text{ and } \langle x, x\rangle = 1\}$$

Identifying $\mathbb{C}$ with $\mathbb{R}^2$, we have $W(A) = W(A_1, A_2)$ if $A = A_1 + iA_2$ with $(A_1, A_2) = (A_1^*, A_2^*)$. So we can focus on $A_1, \ldots, A_m$ lying in $\mathcal{S}(\mathcal{H})$, the real linear space of self-adjoint operators in $\mathcal{B}(\mathcal{H})$.

A natural property to consider is the convexity of the joint numerical range. The following result was obtained in [**25**]; see also [**4**].

THEOREM 4.2. *Let* $\mathbf{A} = (A_1, \ldots, A_m) \in \mathcal{S}(\mathcal{H})^m$.

a) *If the span of* $\{I, A_1, \ldots, A_m\}$ *has dimension not larger than* 3, *then* $W(\mathbf{A})$ *is convex.*

b) *If* $\dim \mathcal{H} \geq 3$ *and the span of* $\{I, A_1, \ldots, A_m\}$ *has dimension 4, then* $W(\mathbf{A})$ *is convex.*

c) *Let* $B_1 = \begin{pmatrix} 0 & 1 \\ 1 & 0 \end{pmatrix}$, $B_2 = \begin{pmatrix} 0 & i \\ -i & 0 \end{pmatrix}$, $B_3 = \begin{pmatrix} 1 & 0 \\ 0 & -1 \end{pmatrix}$. *Then*

$$W(B_1, B_2, B_3) = \{(b_1, b_2, b_3) : b_1, b_2, b_3 \in \mathbb{R}, b_1^2 + b_2^2 + b_3^2 = 1\}$$

is not convex.

d) *If* $\{I, A_1, A_2, A_3\}$ *are linearly independent, then there is a rank-2 orthogonal projection* $A_0 \in \mathcal{B}(\mathcal{H})$ *such that* $W(A_0, A_1, A_2, A_3)$ *is not convex.*

One can extend the result by Mirman to the joint numerical range setting. Suppose $W(B_1, B_2, B_3)$ has interior points and lies inside a simplex $\mathcal{S}$ in $\mathbb{R}^3$ with vertices

$$v_1 = \begin{pmatrix} a_1 \\ a_2 \\ a_3 \\ a_4 \end{pmatrix} \quad v_2 = \begin{pmatrix} b_1 \\ b_2 \\ b_3 \\ b_4 \end{pmatrix} \quad v_3 = \begin{pmatrix} c_1 \\ c_2 \\ c_3 \\ c_4 \end{pmatrix} \quad v_4 = \begin{pmatrix} d_1 \\ d_2 \\ d_3 \\ d_4 \end{pmatrix}$$

Then (B_1, B_2, B_3) has a joint dilation (D_1, D_2, D_3) with $D_j = I \otimes \mathrm{diag}\,(a_j, b_j, c_j, d_j)$ for $j = 1, 2, 3$. In other words, there exists a unitary U such that

$$U^* D_j U = \begin{pmatrix} B_j & \star \\ \star & \star \end{pmatrix} \qquad \text{for } j = 1, 2, 3.$$

More generally, we have the following result proved in [**5**].

THEOREM 4.3. *Let* $\mathbf{B} = (B_1, \ldots, B_m) \in \mathcal{S}(\mathcal{H})^m$ *be such that* $W(\mathbf{B})$ *has non-empty interior in* $\mathbb{R}^m$. *That is,* $\{I, B_1, \ldots, B_m\}$ *is linearly independent. Suppose* $\mathcal{S} \subseteq \mathbb{R}^m$ *is a simplex with vertices*

$$v_1 = \begin{pmatrix} v_{1,1} \\ \vdots \\ v_{1,m} \end{pmatrix}, \ldots, v_{m+1} = \begin{pmatrix} v_{m+1,1} \\ \vdots \\ v_{m+1,m} \end{pmatrix} \in \mathbb{R}^m.$$

Then $W(B_1, \ldots, B_m) \subseteq S$ *if and only if* $B_1, \ldots, B_m$ *has a joint dilation to the diagonal operators*

$$I_N \otimes D_j \quad \textit{with} \quad D_j = \mathrm{diag}\,(v_{1j}, \ldots, v_{m+1,j}) \in M_{m+1}, \quad \textit{for } j = 1, \ldots, m.$$

One can extend the idea of Lemma 3.6 and give a short proof for Theorem 4.3.

Proof of Theorem 4.3. We first reduce the problem to the special case for S to the standard simplex with vertices $0, e_1, \ldots, e_m$, where $\{e_1, \ldots, e_m\}$ is the standard basis for $\mathbb{R}^{1\times m}$. To this end, consider the inveritble affine map $f : \mathbb{R}^{1\times m} \to \mathbb{R}^{1\times m}$ defined by

$$(b_1, \ldots, b_m) \mapsto (b_1, \ldots, b_m)R + v,$$

where $R \in M_m$ is a real invertible matrix and $v = (v_1, \ldots, v_m) \in \mathbb{R}^{1\times m}$. One may extend the affine map to $f : \mathcal{B}(\mathcal{H})^m \to \mathcal{B}(\mathcal{H})^m$ defined by

$$(B_1, \ldots, B_m) \mapsto (B_1, \ldots, B_m)(R \otimes I) + (v_1 I, \ldots, v_m I).$$

Then the conclusion of the theorem holds for $(\mathbf{B}, S)$ if and only if it holds for $(f(\mathbf{B}), f(S))$. Thus, one may apply a suitable invertible affine map to transform S to the standard simplex, and prove the result for this special case.

Now, suppose $\mathcal{S}$ is the standard simplex. Then $W(\mathbf{B}) \subseteq \mathcal{S}$ if and only if $B_1, \ldots, B_m$ are positive operators such that $B_1 + \cdots + B_m \leq I$. Let X be such that $X^* = [\sqrt{B_0} \ \cdots \ \sqrt{B_m}]$, where $B_0 = I - (B_1 + \cdots + B_m)$. Then $X^*X = I$ and $X^*(A_j \otimes I)X = B_j$ for $A_j = A_{j0} \oplus \cdots \oplus A_{jm}$, where $A_{jj} = I_{\mathcal{H}}$ and $A_{j\ell} = 0_{\mathcal{H}}$ otherwise. □

Similar to the remark after Theorem 3.5, instead of fixing a simplex in advance, one may choose any simplex $\mathcal{S}$ such that $W(\mathcal{B}) \subseteq \mathcal{S}$ and use the vertices of $\mathcal{S}$ to get an m-tuple of diagonal matrices $(D_1, \ldots, D_m)$ such that $(B_1, \ldots, B_m)$ has a joint dilation of the form $(I \otimes D_1, \ldots, I \otimes D_m)$. Thus, we have the following.

COROLLARY 4.4. *Let* $\boldsymbol{A} \in \mathcal{S}(\mathcal{H})^m$. *Then the closure of* $\mathrm{conv}\,(W(\boldsymbol{A}))$ *equals the intersection of* $W(D_1, \ldots, D_n)$, *where* $D_1, \ldots, D_m \in \mathcal{S}(\mathcal{H})$ *are mutually commuting operators such that* $\mathbf{D}$ *is a joint dilation of* $\boldsymbol{A}$.

In [**5**], the authors use Theorem 4.3 (or the above corollary) to define a norm on $\mathbf{A} = (A_1, \ldots, A_m) \in \mathcal{B}(\mathcal{H})^m$ by

$$\|\mathbf{A}\| = \inf\{\|(\tilde{A}_1, \ldots, \tilde{A}_m)\| : \tilde{\mathbf{A}} \in \mathcal{D}(\mathbf{A})\},$$

where $\mathcal{D}(\mathbf{A})$ consists of $(\tilde{A}_1, \ldots, \tilde{A}_m)$ such that $\{\tilde{A}_1, \ldots, \tilde{A}_m\}$ is a set of mutually commuting normal operators and there is X satisfying $X^*X = I$, $X^*\tilde{A}_jX = A_j$ for $j = 1, \ldots, m$. Such a norm is invariant under any permutation of the components of $\mathbf{A}$, and the change of any of the component A_j to A_j^t, A_j^*, etc.

Recall that an operator system of $\mathcal{B}(\mathcal{H})$ is a subspace spanned by some self-adjoint operators and the identity operator. Let $\mathcal{A} \subseteq \mathcal{B}(\mathcal{H})$ and $\mathcal{B} \subseteq \mathcal{B}(\mathcal{K})$ be operators systems. A map $\phi : \mathcal{A} \to \mathcal{B}$ is positive if $\Phi(A) \in \mathcal{B}$ is positive semidefinite whenever $A \in \mathcal{A}$ is positive semidefinite. For a positive integer k, the map ϕ is k-positive if $(\phi(A_{ij})) \in M_k(\mathcal{B})$ is positive whenever $(A_{ij}) \in M_k(\mathcal{A})$ is positive. If ϕ is k-positive for all positive integers k, then ϕ is completely positive. The following results connect the notion of unital positive maps and unital completely positive maps with the inclusion relation of numerical ranges and joint dilation of operators; see [**11**].

THEOREM 4.5. *Let $B_1, \ldots, B_m \in \mathcal{S}(\mathcal{H})$ and $A_1, \ldots, A_m \in M_n$ be Hermitian matrices. Consider the map $\phi : M_n \to \mathcal{B}(\mathcal{H})$ defined by*

$$\phi(\mu_0 I + \mu_1 A_1 + \ldots + \mu_m A_m) = \mu_0 I + \mu_1 B_1 + \ldots + \mu_m B_m \quad \text{for any } \mu_0, \ldots, \mu_m \in \mathbb{C},$$

on $\operatorname{span}\{I, A_1, \ldots, A_m\}$. *Then*

- *ϕ is a positive linear map if and only if*

$$W(B_1, \ldots, B_m) \subseteq \operatorname{conv} W(A_1, \ldots, A_m).$$

- *ϕ is a completely positive (linear) map if and only if*

$$(B_1, \ldots, B_m) \text{ has joint dilation } (I \otimes A_1, \ldots, I \otimes A_m).$$

5. Quantum Channels and Higher Rank Numerical Ranges

In the mathematical setting, quantum states are density operators, i.e. positive semidefinite operators of trace 1. Quantum channels and quantum operations are trace preserving completely positive maps. In the finite dimensional case, a quantum channel Φ transforming quantum states in M_n to quantum states in M_m admits the operator sum representation:

$$\Phi(X) = F_1XF_1^* + \ldots + F_rXF_r^*$$

for some $m \times n$ and F_i satisfying $\sum_{j=1}^r F_j^*F_j = I_n$; see [**6**] and [**20**]. The matrices $F_1, \ldots, F_r$ are known as the Choi-Kraus operators or the error operators of the channel Φ.

We say that a quantum channel $\Phi : \mathcal{B}(\mathcal{H}) \to \mathcal{B}(\mathcal{H})$ has a quantum error code V, which is a subspace of $\mathcal{H}$, provided that there is a quantum channel $\Psi : M_n \to M_n$ satisfying $\Psi \circ \Phi(X) = X$ where $P_VXP_V = X$, where P_V is the orthogonal projection of $\mathcal{H}$ onto the coding subspace V. If $\mathcal{H} = \mathbb{C}^n$ and Φ has error operators $F_1, \ldots, F_r$, it is shown in [**19**] that the search of subspace V and P_V reduces to the search of P_V satisfying $PF_i^*F_jF = f_{ij}P$ for all $1 \le i, j \le r$. In this connection, researchers consider the *rank p-numerical range* of $\mathbf{A} = (A_1, ..., A_m) \in M_n^m$ by

$$\Lambda_p(\mathbf{A}) = \{(a_1, ..., a_m) : X^*A_jX = a_jI_p \text{ for some } X \in \mathcal{V}_p\}$$

where $\mathcal{V}_p$ is the set of linear maps $X : \mathbb{C}^p \longrightarrow \mathcal{H}$ satisfying $X^*X = I_p$; see [**7**–**10**, **27**, **34**]. Note that the quantum channel Φ has a quantum error correction code of dimension p if and only if $\Lambda_p(\mathbf{A}) \neq \emptyset$ with $\mathbf{A} = (F_1^*F_1, F_1^*F_2, \ldots, F_r^*F_r) \in M_n^{r^2}$. Also, observe that $(a_1, \ldots, a_m) \in \Lambda_p(\mathbf{A})$ if and only if there is a unitary $U = [X, \tilde{X}]$ such that

$$U^*A_jU = \begin{pmatrix} a_jI_p & \star \\ \star & \star \end{pmatrix}, \qquad j = 1, \ldots, m.$$

Using the higher rank numerical ranges, one can change the problem of searching for an error correct code for a quantum channel Φ to the problem of studying the non-empty-ness of the set $\Lambda_p(\mathbf{A}) \subseteq \mathbb{C}^m$, which is closely related to the *joint unitary orbit* of $\mathbf{A}$:

$$\mathcal{U}(\mathbf{A}) = \{(U^*A_1U, \ldots, U^*A_mU) : U \in M_n,\ U^*U = I_n\}.$$

Therefore, one can apply algebraic, analytic, and geometrical techniques to study the problem.

If $m = 1$ and $A_1 = A_1^*$ has eigenvalues $\lambda_1, \ldots, \lambda_n$, and $p \le (n+1)/2$, then $\Lambda_p(A_1) = [\lambda_{n-p+1}, \lambda_p]$. To see this, assume that $\{x_1, \ldots, x_n\}$ is a set of orthonormal eigenvectors of A such that $Ax_j = \lambda_j x_j$ for $j = 1, \ldots, n$. Then for any $\mu \in [\lambda_{n-p+1}, \lambda_p]$ there is a unit vector $y_j \in \text{span}\{x_{n-p+1}, x_p\}$ such that $y_j^*Ay_j = \mu$. Let $Y = [y_1 \ldots y_p]$. Then $Y^*Y = I_p$ and $Y^*AY = \mu I_p$. Conversely, if Y is $n \times p$ such that $Y^*Y = I_p$ and $Y^*AY = \mu I_p$, then by the interlacing inequalities, see [**15**], we see that $\lambda_p \ge \mu \ge \lambda_{n-p+1}$. However, $\Lambda_p(A_1)$ may be empty if $p > (n+1)/2$.

It is non-trivial to determine $\Lambda_p(A)$ even if A is a normal matrix. In [**9**], the authors conjectured the following result, which was confirmed in [**27**].

THEOREM 5.1. *Suppose $A \in M_n$ is a normal matrix with eigenvalues $\lambda_1, \ldots, \lambda_n$. Let $1 \le p \le n$. Then*

$$\Lambda_p(A) = \bigcap_{1 \le j_1 < \cdots < j_{n-p+1} \le n} \text{conv}\,\{\lambda_{j_1}, \ldots, \lambda_{n-p+1}\}.$$

The authors in [**9**] also conjectured that for $A \in M_n$, the set $\Lambda_p(A)$ is convex, and they reduced the problem to the existence of the solution of certain matrix equation. This is confirmed in [**34**]. In [**29**], the authors used the theory of canonical form of matrices under $*$-congruence to give a description of the set $\Lambda_p(A)$ as the intersection of half spaces in $\mathbb{C}$. It then follows that $\Lambda_p(A)$ is convex. Here is the statement of the result.

THEOREM 5.2. *Suppose* $A = H + iG \in M_n$ *with* $(H, G) = (H^*, G^*)$, *and* $1 \le p \le n$. *Then*

$$\Lambda_p(A) = \bigcap_{\theta\in[0,2\pi)} \{h + ig : \cos\theta h + \sin\theta g \le \lambda_1(\cos\theta H + \sin\theta G)\},$$

where $\lambda_1(K)$ *is the largest eigenvalue of the Hermitian matrix* K. *Consequently,* $\Lambda_p(A)$ *is a compact convex set.*

The study of $\Lambda_p(A_1, \ldots, A_m)$ for $m \ge 2$ is more intricate. Similar to the study of the joint numerical range, if we write $A_j = H_j + iG_j$ with $(H_j, G_j) = (H_j^*, G_j^*)$ for $j = 1, \ldots, m$, then $\Lambda_j(A_1, \ldots, A_m) \subseteq \mathbb{C}^m$ and can be identified with $\Lambda_j(H_1, G_1, \ldots, H_m, G_m) \subseteq \mathbb{R}^{2m}$. So, we can focus on $\Lambda_p(\mathbf{A})$ for $\mathbf{A} = (A_1, \ldots, A_m) \in \mathcal{S}(\mathcal{H})^m$. Now, suppose $R \in M_m$ is a real invertible matrix and $\tilde{\mathbf{A}} = (\tilde{A}_1, \ldots, \tilde{A}_m)$ with $(\tilde{A}_1 \cdots \tilde{A}_m) = (A_1 \cdots A_m)(R \otimes I_n)$, Then $\tilde{\mathbf{a}} \in \Lambda_p(\tilde{\mathbf{A}})$ if and only if $\tilde{\mathbf{a}} = \mathbf{a}R$ with $\mathbf{a} \in \Lambda_p(A)$. Hence, we can choose a suitable $R \in M_m$ such that $\tilde{\mathbf{A}} = (\tilde{A}_1, \ldots, \tilde{A}_k, 0, \ldots, 0)$, where $\{\tilde{A}_1, \ldots, \tilde{A}_k\}$ is linearly independent. Then $\Lambda_p(\mathbf{A})$ is completely determined by $\Lambda_p(\tilde{A}_1, \ldots, \tilde{A}_k)$. Furthermore, if $I \in \operatorname{span}\{\tilde{A}_1, \ldots, \tilde{A}_k\}$, which is the case in the study of quantum error correction, then we may assume that $\tilde{A}_k = I_n$ so that every $\tilde{\mathbf{a}} \in \Lambda_p(\tilde{A}_1, \ldots, \tilde{A}_k)$ will have last entry equal to 1. So, we need only consider $\Lambda_p(\tilde{A}_1, \ldots, \tilde{A}_{k-1})$. In the following we shall always assume that $\{A_1, \ldots, A_m\}$ is linearly independent and $I \notin \operatorname{span}\{A_1, \ldots, A_m\}$.

Return to the problem concerning the non-empty-ness of $\Lambda_p(A_1, \ldots, A_m)$ related to the study of quantum error correction. We have the following; see [**26**].

THEOREM 5.3. *Let* $\boldsymbol{A} = (A_1, \ldots, A_m) \in \mathcal{S}(\mathcal{H})^m$. *Then* $\Lambda_p(A_1, \ldots, A_m)$ *is non-empty if* $\dim \mathcal{H} \ge (p-1)(m+1)^2$.

It is a challenging problem to find minimum dimension of $\mathcal{H}$ that ensure $\Lambda_p(\mathbf{A}) \ne \emptyset$ for $\mathbf{A} = (A_1, \ldots, A_m) \in \mathcal{S}(\mathcal{H})^m$. In case the set is non-empty, one may ask for other properties of the set $\Lambda_p(\mathbf{A})$. The following results were also obtained in [**26**].

THEOREM 5.4. *Let* $\boldsymbol{A} = (A_1, \ldots, A_m) \in \mathcal{S}(\mathcal{H})^m$. *Suppose* $\mathcal{H}_1$ *is a subspace of* $\mathcal{H}$ *of dimension* r, *where* $1 \le r < p \le \dim \mathcal{H}$, *and* $X : \mathcal{H}_1^\perp \to \mathcal{H}$ *satisfying* $X^*X = I_{\mathcal{H}_1^\perp}$. *Then*

$$\Lambda_p(\boldsymbol{A}) \subseteq \Lambda_{p-r}(X^*A_1X, \ldots, X^*A_mX).$$

From this theorem, we can obtain the following.

THEOREM 5.5. *Let* $\boldsymbol{A} = (A_1, \ldots, A_m) \in \mathcal{S}(\mathcal{H})^m$. *Suppose* $\mathbf{K} = (K_1, \ldots, K_m) \in \mathcal{S}(\mathcal{H})^m$ *is an* m*-tuple of finite rank operators such that* $\operatorname{rank}(K_1^2 + \cdots + K_m^2) = r$, *with* $1 \le r < p$. *Then*

$$\Lambda_p(\boldsymbol{A}) \subseteq \Lambda_{p-r}(\boldsymbol{A} + \mathbf{K}).$$

The above two theorems provide information about $\Lambda_p(\tilde{\mathbf{A}})$ if $\tilde{\mathbf{A}}$ is a (joint) compression of $\mathbf{A}$ or perturbation of $\mathbf{A}$. These results will be extended to a more general version of numerical ranges arising in more sophisticated quantum error correction schemes.

In general, $\Lambda_p(A_1,\ldots,A_m)$ is not convex if $m>2$. It is interesting to determine the conditions on $(A_1,\ldots,A_m)$ so that the set is convex. Note that if $\Lambda_p(A_1,\ldots,A_m)$ is convex, one can derive efficient algorithms to find its elements (if they exist) and construct quantum error correction codes accordingly.

6. The (p,q)-matricial ranges

DEFINITION 6.1. For $\mathbf{A}=(A_1,...,A_m)\in\mathcal{B}(\mathcal{H})^m$, the *joint q-matricial range* of $\mathbf{A}$ is given by

$$W(q:\mathbf{A})=\{(X^*A_1X,...,X^*A_mX):X\in\mathcal{V}_q\},$$

where $\mathcal{V}_q$ denote the set of operators $X:\mathbb{C}^p\to\mathcal{H}$ satisfying $X^*X=I_q$.

In case, $m=1$, we simply write $W(q:A_1)$. Researchers have used the set $W(q:A_1)$ to study operators $A_1\in\mathcal{B}(\mathcal{H})$. For example, two compact operators $A_1,B_1\in\mathcal{B}(\mathcal{H})$ are unitarily similar if and only if $W(q:A_1)=W(q:B_1)$ for all $q=1,2,\ldots$; see **[13]**. One may see the survey **[16]** and its references for more interesting results. However, not many geometrical results have been obtained from $W(q:\mathbf{A})$. Again, for the geometrical properties of the set $W(q:\mathbf{A})$, one can use the Hermitian decomposition of $A_j=H_j+iG_j$ for $j=1,\ldots,m.$ and focus on the study of $W(q:\mathbf{A})$ for $A=(A_1,\ldots,A_m)\in\mathcal{S}(\mathcal{H})^m$. Moreover, we can assume that $\{A_1,\ldots,A_m\}$ is linearly independent and I is not in the linear span. When $m=1$, the following was proved in **[30]**.

THEOREM 6.2. *Let $A\in M_n$ be a Hermitian matrix with eigenvalues $a_1\geq...\geq a_n$. Then the set $W(q:\mathbf{A})$ consists of Hermitian matrices $B\in M_q$ with eigenvalues $b_1\geq\cdots\geq b_q$ satisfying $a_j\geq b_j\geq a_{n-q+j}$ for $j=1,...,q$. Consequently, the set $W(q:\mathbf{A})$ is convex if and only if $a_1=a_q$ and $a_{n-q+1}=a_n$.*

By the above theorem, we see that $W(q:\mathbf{A})$ may not be convex even if $\mathbf{A}=(A_1)$ with $A_1=A_1^*$. In general, it is difficult to determine the structure of $W(q:\mathbf{A})$. Even if $\mathbf{A}=(A_1,A_2)$ for two commuting Hermitian matrices $A_1,A_2\in M_n$ so that $W(q:(A_1,A_2))\equiv W(q:A)$ for the normal matrix $A=A_1+iA_2$, it is quite difficult to determine the set $W(q:A)$ of all $q\times q$ principal submatrices of U^*AU, where $U\in M_n$ is unitary. This may be a reason why the study of $W(q:\mathbf{A})$ has not been very active in the last two decades.

In the pursuit of "better" quantum error correction codes, researchers proposed the following concept covering both $\Lambda_p(\mathbf{A})$ and $W(q:\mathbf{A})$; see **[21]** and **[28]**.

DEFINITION 6.3. The *(p,q)-matricial range* of $\mathbf{A}=(A_1,...,A_m)\in\mathcal{B}(\mathcal{H})^m$, denoted by $\Lambda_{p,q}(\mathbf{A})$, is the set of m-tupe Hermitian matrices $(B_1,...,B_m)\in M_q^m$

such that

$$X^*A_jX = I_p \otimes B_j = \begin{pmatrix} B_j & & \\ & \ddots & \\ & & B_j \end{pmatrix} \qquad \text{for } j = 1, ..., m$$

for some $X \in \mathcal{V}_{pq}$.

An equivalent formulation is to say there is a unitary $U = [X, \tilde{X}]$ such that

$$\begin{pmatrix} I_p \otimes B_j & \star \\ \star & \star \end{pmatrix}.$$

Observe that joint numerical range, joint rank p-numerical range and q-matricial range are all particular cases of (p, q)-*matricial range.*

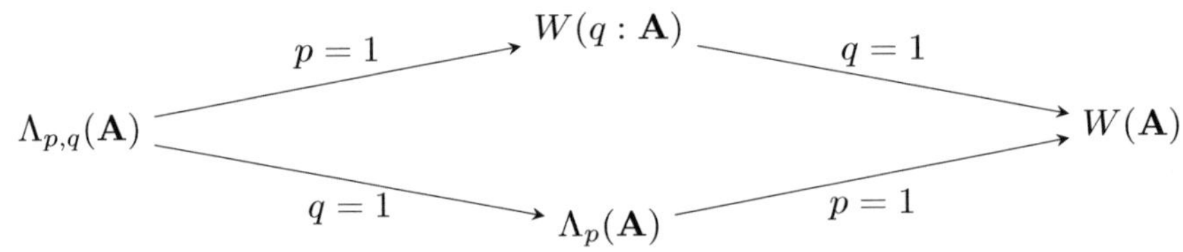

There is no general convexity result for $\Lambda_{p,q}(\mathbf{A})$. In [**22**] it was shown that if $\dim \mathcal{H}$ is sufficiently large, then $\Lambda_{p,q}(\mathbf{A})$ is star-shaped and non-empty. To state the theorem we first recall the definition of star-shaped.

Definition 6.4. A set $S \subseteq \mathbb{R}^N$ is *star-shaped* if there is a star center, $v_0 \in S$, such that the line segment $[v_0, v] = \{tv_0 + (1-t)v : t \in [0,1]\}$ lies entirely in the set S for any $v \in S$.

The following result was proved in [**22**].

Theorem 6.5. *Let $\boldsymbol{A} = (A_1, \ldots, A_m) \in \mathcal{S}(\mathcal{H})^m$, and p, q be positive integers. Then*

(a) *If $\dim(\mathcal{H}) \geq (pq-1)(m+1)^2$ then $\Lambda_{pq}(\boldsymbol{A})$ and $\Lambda_{p,q}(\boldsymbol{A})$ are non-empty.*
(b) *If $\dim(\mathcal{H}) \geq (N-1)(m+1)^2$ for $N = pq(m+2)$, then every element in $\operatorname{conv}\{(a_1 I_q, ..., a_m I_q) : (a_1, ..., a_m) \in \Lambda_N(\boldsymbol{A})\}$ is a star center of $\Lambda_{p,q}(\boldsymbol{A})$.*
(c) *For any r with $1 \leq qr < p \leq \dim \mathcal{H}$, if $\boldsymbol{K} = (K_1, \ldots, K_m) \in \mathcal{S}(\mathcal{H})^m$ is such that $\operatorname{rank}(K_1^2 + \ldots + K_m^2) \leq r$, then $\Lambda_{p,q}(\boldsymbol{A}) \subseteq \Lambda_{p-qr,q}(\boldsymbol{A} + \boldsymbol{K})$.*
(d) *Suppose $\mathcal{H}_1$ is a subspace of $\mathcal{H}$ of dimension r with $1 \leq qr < p \leq \dim \mathcal{H}$, and $X : \mathcal{H}_1^\perp \to \mathcal{H}$ satisfies $X^*X = I_{\mathcal{H}_1^\perp}$. Then*

$$\Lambda_{p,q}(\boldsymbol{A}) \subseteq \Lambda_{p-qr,q}(X^*A_1X, \ldots, X^*A_mX).$$

When $q = 1$, Theorem 6.5 reduces to the results in Section 5. By Theorem 6.5 and the results in Section 5, if $\dim \mathcal{H}$ is high, then one can always find a (basic type or more sophisticated type of) quantum error correction code for a noisy quantum channel. If the dimension is much higher, then one can get a convex subset of higher rank numerical range so that there will be more choices for quantum error correction codes, and one may be able to select one with additional nice properties. Moreover, if the quantum channel is under perturbation so that the operator system associated with the error operators changed from the $\mathrm{span}\{I, A_1, \ldots, A_m\}$ to $\mathrm{span}\{I, A_1 + K_1, \ldots, A_m + K_m\}$ with low rank operators $K_1, \ldots, K_m$, one can still get a quantum error correction with reduced dimension based on those of the original channel. Similarly, if one considers the compression of $\mathrm{span}\{I, A_1, \ldots, A_m\}$ to $\mathrm{span}\{I, X^*A_1X, \ldots, X^*A_mX\}$, the quantum error correction codes for the original system will still be useful for the compressed system.

When $p = 1$, Theorem 6.5 reduces to the following.

THEOREM 6.6. *Let* $\mathbf{A} = (A_1, \ldots, A_m) \subset \mathcal{S}(\mathcal{H})^m$. *The set*

$$W(q : \mathbf{A}) = \{(X^*A_1X, \ldots, X^*A_mX) : X \in \mathcal{V}_q\}$$

is star-shaped if $\dim \mathcal{H} \geq (N-1)(m+1)^2$ *with* $N = q(m+2)$.

In particular, the set $\Lambda_N(\mathbf{A})$ *is non-empty, and* $(a_1 I_1, \ldots, a_m I_q)$ *is a star center whenever* $(a_1, \ldots, a_m) \in \mathrm{conv}\, \Lambda_N(\mathbf{A})$.

7. The joint essential numerical range

If $\dim \mathcal{H} = \infty$, we consider the following generalization of $\Lambda_q(\mathbf{A})$ and $\Lambda_{p,q}(\mathbf{A})$. Let

$$\Lambda_\infty(\mathbf{A}) = \cap_{r \in \mathbb{N}} \Lambda_r(\mathbf{A}) \subseteq \mathbb{R}^m,$$

i.e., $(a_1, \ldots, a_m) \in \Lambda_\infty(\mathbf{A})$ if there is $X \in \mathcal{V}_\infty$, where $\mathcal{V}_\infty$ consists of $X : \mathcal{K} \to \mathcal{H}$ for an infinite dimensional closed subspace of $\mathcal{H}$ satisfying $X^*X = I_\infty$, and $X^*A_jX = a_j I_\infty$ for $j = 1, \ldots m$;

$$\Lambda_{\infty,q}(\mathbf{A}) = \cap_{r \in \mathbb{N}} \Lambda_{r,q}(\mathbf{A}) \subseteq M_q^m,$$

i.e., $(B_1, \ldots, B_m) \in \Lambda_{\infty,q}(\mathbf{A})$ if there is $X \in \mathcal{V}_\infty$ such that $X^*A_jX = I_\infty \otimes B_j$ for $j = 1, \ldots m$.

THEOREM 7.1. **[22]** *Let* $\mathbf{A} = (A_1, \ldots, A_m) \in \mathcal{B}(\mathcal{H})$, *where* $\dim \mathcal{H} = \infty$, *then the set* $\Lambda_{\infty,q}(\mathbf{A})$ *is convex.*

The statement above does not guarantee that $\Lambda_{\infty,q}(\mathbf{A})$ is non-empty. For instance, if $A = \mathrm{diag}\,(1, 1/2, 1/3, \ldots)$ then $\Lambda_{\infty,q}(A) = \emptyset$ for any integer $q > 0$.

DEFINITION 7.2. Let $\mathcal{K}(\mathcal{H})$ denote the set of compact operators in $\mathcal{B}(\mathcal{H})$. Define the *essential* (p,q)*-matricial range* of $\mathbf{A} \in \mathcal{B}(\mathcal{H})^m$ by

$$\Lambda_{p,q}^{ess}(\mathbf{A}) = \cap\{\mathrm{cl}(\Lambda_{p,q}(\mathbf{A} + \mathbf{K})) : \mathbf{K} \in \mathcal{K}(\mathcal{H})^m\}.$$

This definition says that $\mathbf{B} = (B_1, \ldots, B_m) \in \Lambda_{p,q}^{ess}(\mathbf{A})$ if for any $\mathbf{K} \in \mathcal{K}(\mathcal{H})^m$ then $\mathbf{B} \in \mathrm{cl}(\Lambda_{p,q}(\mathbf{A}+\mathbf{K}))$. Note that if $p = 1$, we get the essential q-matricial range defined as

$$W_{ess}(q : \mathbf{A}) = \cap\{\mathrm{cl}(W(q : \mathbf{A} + \mathbf{K})) : \mathbf{K} \in \mathcal{K}(\mathcal{H})^m\}.$$

That is, $\mathbf{B} \in W_{ess}(q : \mathbf{A})$ if $\mathbf{B} \in \text{cl}(W(q : \mathbf{A} + \mathbf{K}))$ for every $\mathbf{K} \in \mathcal{K}(\mathcal{H})^m$. The following is obtained in [**22**].

THEOREM 7.3. *Let* $\boldsymbol{A} = (A_1, \ldots, A_m) \in \mathcal{B}(\mathcal{H})^m$ *where* $\dim \mathcal{H} = \infty$, *then for any positive integer* p,

$$\Lambda_{p,q}^{ess}(\boldsymbol{A}) = W_{ess}(q : \boldsymbol{A})$$

is non-empty, compact and convex.

It turns out the the essential numerical range is connected to the algebra q-matricial range of $\mathbf{A} \in \mathcal{B}(\mathcal{H})^m$ defined as follows.

DEFINITION 7.4. Define the *algebra q-matricial range* of $\mathbf{A} \in \mathcal{B}(\mathcal{H})^m$ by

$$V_q(\mathbf{A}) = \{(\Phi(\pi(A_1)), \ldots, \Phi(\pi(A_m))) : \Phi \text{ is a unital completely positive linear map from } \mathcal{B}(\mathcal{H})/\mathcal{K}(\mathcal{H}) \text{ to } M_q\},$$

where π is the canonical surjection from $\mathcal{B}(\mathcal{H})$ to the Calkin algebra $\mathcal{B}(\mathcal{H})/\mathcal{K}(\mathcal{H})$.

The following results were obtained in [**24**].

THEOREM 7.5. *Let* $\boldsymbol{A} \in \mathcal{S}(\mathcal{H})^m$, *where* $\dim \mathcal{H} = \infty$, *and* p, q *be positive integers. Then there is* $\mathbf{K} \in \mathcal{K}(\mathcal{H})^m \cap \mathcal{S}(\mathcal{H})^m$ *such that*

$$\Lambda_{p,r}^{ess}(\boldsymbol{A}) = \text{cl}(\Lambda_{p,r}(\boldsymbol{A} + \mathbf{K})) = V_r(\boldsymbol{A}) \quad \textit{for all } r = 1, \ldots, p.$$

THEOREM 7.6. *Let* $\boldsymbol{A} \in \mathcal{S}(\mathcal{H})^m$ *be such that* $W_{ess}(1 : \boldsymbol{A})$ *is a simplex in* $\mathbb{R}^m$, *where* $\dim \mathcal{H} = \infty$. *Then there is* $\mathbf{K} \in \mathcal{K}(\mathcal{H})^m \cap \mathcal{S}(\mathcal{H})^m$ *such that*

$$\Lambda_{p,q}^{ess}(\boldsymbol{A}) = \text{cl}(\Lambda_{p,q}(\boldsymbol{A} + \mathbf{K})) = V_q(\boldsymbol{A}) \quad \textit{for all } p, q \in \mathbb{N}.$$

It is known that Theorem 7.6 does not hold for general $\mathbf{A} \in \mathcal{B}(\mathcal{H})^m$ if $m \geq 4$. The case $m = 1$ is covered by the theorem. The cases for $m = 2, 3$ are open. One may see the references in [**24**] for more background of this problem.

References

[1] T. Ando, *Structure of operators with numerical radius one*, Acta Sci. Math. (Szeged) **34** (1973), 11–15. MR0318920

[2] T. Ando and C.-K. Li, *Operator radii and unitary operators*, Oper. Matrices **4** (2010), no. 2, 273–281, DOI 10.7153/oam-04-14. MR2667338

[3] W. Arveson, *Subalgebras of C^*-algebras. II*, Acta Math. **128** (1972), no. 3-4, 271–308, DOI 10.1007/BF02392166. MR0394232

[4] Y. H. Au-Yeung and Y. T. Poon, *A remark on the convexity and positive definiteness concerning Hermitian matrices*, Southeast Asian Bull. Math. **3** (1979), no. 2, 85–92. MR564798

[5] P. Binding, D. R. Farenick, and C.-K. Li, *A dilation and norm in several variable operator theory*, Canad. J. Math. **47** (1995), no. 3, 449–461, DOI 10.4153/CJM-1995-025-5. MR1346148

[6] M. D. Choi, *Completely positive linear maps on complex matrices*, Linear Algebra and Appl. **10** (1975), 285–290, DOI 10.1016/0024-3795(75)90075-0. MR0376726

[7] M.-D. Choi, M. Giesinger, J. A. Holbrook, and D. W. Kribs, *Geometry of higher-rank numerical ranges*, Linear Multilinear Algebra **56** (2008), no. 1-2, 53–64, DOI 10.1080/03081080701336545. MR2378301

[8] M.-D. Choi, J. A. Holbrook, D. W. Kribs, and K. Życzkowski, *Higher-rank numerical ranges of unitary and normal matrices*, Oper. Matrices **1** (2007), no. 3, 409–426, DOI 10.7153/oam-01-24. MR2344684

[9] M.-D. Choi, D. W. Kribs, and K. Życzkowski, *Higher-rank numerical ranges and compression problems*, Linear Algebra Appl. **418** (2006), no. 2-3, 828–839, DOI 10.1016/j.laa.2006.03.019. MR2260232

[10] M.-D. Choi, D. W. Kribs, and K. Życzkowski, *Quantum error correcting codes from the compression formalism*, Rep. Math. Phys. **58** (2006), no. 1, 77–91, DOI 10.1016/S0034-4877(06)80041-8. MR2273568

[11] M.-D. Choi and C.-K. Li, *Numerical ranges and dilations*, Linear and Multilinear Algebra **47** (2000), no. 1, 35–48, DOI 10.1080/03081080008818630. MR1752164

[12] M.-D. Choi and C.-K. Li, *Constrained unitary dilations and numerical ranges*, J. Operator Theory **46** (2001), no. 2, 435–447. MR1870416

[13] W. F. Chuan, *The unitary equivalence of compact operators*, Glasgow Math. J. **26** (1985), no. 2, 145–149, DOI 10.1017/S0017089500005917. MR798741

[14] C. Davis, *The Toeplitz-Hausdorff theorem explained*, Canad. Math. Bull. **14** (1971), 245–246, DOI 10.4153/CMB-1971-042-7. MR0312288

[15] K. Fan and G. Pall, *Imbedding conditions for Hermitian and normal matrices*, Canad. J. Math. **9** (1957), 298–304, DOI 10.4153/CJM-1957-036-1. MR0085216

[16] D. R. Farenick, *Matricial extensions of the numerical range: a brief survey*, Linear and Multilinear Algebra **34** (1993), no. 3-4, 197–211, DOI 10.1080/03081089308818222. MR1304608

[17] P. R. Halmos, *Numerical ranges and normal dilations*, Acta Sci. Math. (Szeged) **25** (1964), 1–5. MR0171168

[18] F. Hausdorff, *Der Wertvorrat einer Bilinearform* (German), Math. Z. **3** (1919), no. 1, 314–316, DOI 10.1007/BF01292610. MR1544350

[19] E. Knill and R. Laflamme, *Theory of quantum error-correcting codes*, Phys. Rev. A (3) **55** (1997), no. 2, 900–911, DOI 10.1103/PhysRevA.55.900. MR1455854

[20] K. Kraus, *States, effects, and operations*, Lecture Notes in Physics, vol. 190, Springer-Verlag, Berlin, 1983. Fundamental notions of quantum theory; Lecture notes edited by A. Böhm, J. D. Dollard and W. H. Wootters. MR725167

[21] D. Kribs and R. Spekkens, *Quantum error-correcting subsystems are unitarily recoverable subsystems*, Physical Review A **74** (2006), no. 4, 042329.

[22] P.-S. Lau, C.-K. Li, Y.-T. Poon, and N.-S. Sze, *Convexity and star-shapedness of matricial range*, J. Funct. Anal. **275** (2018), no. 9, 2497–2515, DOI 10.1016/j.jfa.2018.03.018. MR3847477

[23] C.-K. Li, *A simple proof of the elliptical range theorem*, Proc. Amer. Math. Soc. **124** (1996), no. 7, 1985–1986, DOI 10.1090/S0002-9939-96-03307-2. MR1322932

[24] C.-K. Li, V. Paulsen, and Y.-T. Poon, *Preservation of the joint essential matricial range*, arXiv preprint arXiv:1805.10600 (2018).

[25] C.-K. Li and Y.-T. Poon, *Convexity of the joint numerical range*, SIAM J. Matrix Anal. Appl. **21** (1999), no. 2, 668–678, DOI 10.1137/S0895479898343516. MR1742818

[26] C.-K. Li and Y.-T. Poon, *Generalized numerical ranges and quantum error correction*, J. Operator Theory **66** (2011), no. 2, 335–351. MR2844468

[27] C.-K. Li, Y.-T. Poon, and N.-S. Sze, *Condition for the higher rank numerical range to be non-empty*, Linear Multilinear Algebra **57** (2009), no. 4, 365–368, DOI 10.1080/03081080701786384. MR2522848

[28] C.-K. Li, Y.-T. Poon, and N.-S. Sze, *Generalized interlacing inequalities*, Linear Multilinear Algebra **60** (2012), no. 11-12, 1245–1254, DOI 10.1080/03081087.2011.619534. MR2989760

[29] C.-K. Li and N.-S. Sze, *Canonical forms, higher rank numerical ranges, totally isotropic subspaces, and matrix equations*, Proc. Amer. Math. Soc. **136** (2008), no. 9, 3013–3023, DOI 10.1090/S0002-9939-08-09536-1. MR2407062

[30] C.-K. Li and N.-K. Tsing, *On the kth matrix numerical range*, Linear and Multilinear Algebra **28** (1991), no. 4, 229–239, DOI 10.1080/03081089108818047. MR1088420

[31] B. A. Mirman, *The numerical range of a linear operator, and its norm* (Russian), Voronež. Gos. Univ. Trudy Sem. Funkcional. Anal. **Vyp. 10** (1968), 51–55. MR0417814

[32] Y. Nakamura, *Numerical range and norm*, Math. Japon. **27** (1982), no. 1, 149–150. MR649031

[33] O. Toeplitz, *Das algebraische Analogon zu einem Satze von Fejér* (German), Math. Z. **2** (1918), no. 1-2, 187–197, DOI 10.1007/BF01212904. MR1544315

[34] H. J. Woerdeman, *The higher rank numerical range is convex*, Linear Multilinear Algebra **56** (2008), no. 1-2, 65–67, DOI 10.1080/03081080701352211. MR2378302

[35] F. Zhang, *Matrix theory*, 2nd ed., Universitext, Springer, New York, 2011. Basic results and techniques. MR2857760

Contemporary Mathematics
Volume **737**, 2019
https://doi.org/10.1090/conm/737/14857

Toeplitz kernels and finite rank truncated Toeplitz operators

M. Cristina Câmara and Carlos Carteiro

Abstract. We establish the existence of functions with specified zeroes in Toeplitz kernels and use it to study linearly independent sets of reproducing kernel functions and their analogues in model spaces. The results are applied to describe the ranges of a class of finite rank truncated Toeplitz operators.

1. Introduction

Toeplitz operators have been intensively studied in the past decades (see for example [**6**]). *Toeplitz operators on the Hardy space* $H^2(\mathbb{D})$, where $\mathbb{D}$ is the unit disk, are defined as compressions of a multiplication operator in $L^2(\mathbb{T})$, where $\mathbb{T}$ denotes the unit circle, to $H^2(\mathbb{D})$. *Truncated Toeplitz operators* are in their turn compressions of Toeplitz operators to closed subspaces of $H^2(\mathbb{D})$ called *model spaces.*

For an essentially bounded function $g \in L^\infty(\mathbb{T})$, the Toeplitz operator with *symbol* g is defined by

$$T_g = Pg_{|H^2(\mathbb{D})} \tag{1.1}$$

where P denotes the orthogonal projection from $L^2(\mathbb{T})$ onto $H^2(\mathbb{D})$.

The kernels of Toeplitz operators, also called *Toeplitz kernels*, possess many interesting properties and will play a key role in this work. The closed subspace $\ker T_g \subset H^2(\mathbb{D})$ can be defined as consisting of those functions $\phi_+ \in H^2(\mathbb{D})$ such that $g\phi_+ \in H^2_-$, where H^2_- denotes the orthogonal complement of $H^2(\mathbb{D}) =: H^2_+$ in $L^2(\mathbb{T})$, i.e., such that they provide a solution to the boundary value problem on $\mathbb{T}$

$$g\phi_+ = \phi_- \,, \qquad \phi_\pm \in H^2_\pm. \tag{1.2}$$

This is called a *Riemann-Hilbert problem*: to determine a sectionally analytic function in $\mathbb{C} \setminus \mathbb{T}$ given its jump across the curve $\mathbb{T}$, with a certain desired behaviour at ∞. We will say that we take a Riemann-Hilbert approach to Toeplitz kernels when their description is given in terms of the solutions to a Riemann-Hilbert problem of the form (1.2). By the *g-image* of a function in $\ker T_g$ we mean its image (in H^2_-) under multiplication by g, and the g-image of $\ker T_g$ is $g \ker T_g \subset H^2_-$.

This work was motivated by a paper by C. Gu and D.-O Kang [**5**] where explicit formulas were obtained for the rank of a class of finite rank truncated Toeplitz operators, in terms of certain poles appearing in their symbols, assuming the latter

2010 *Mathematics Subject Classification.* Primary 47B35; Secondary 30H10.

Key words and phrases. Toeplitz kernels, truncated Toeplitz operators, model spaces, reproducing kernel functions.

in the form $\bar{\theta}R_1 + \theta R_2$. Here θ is the inner function defining the model space and R_1, R_2 are rational functions without poles on $\mathbb{T}$. These results were obtained by using certain relations between truncated Toeplitz operators and products of Hankel operators.

We take here a different approach, viewing model spaces as kernels of Toeplitz operators of a particular kind and using several properties regarding the zeroes of functions in Toeplitz kernels and their g-images, obtained through a Riemann-Hilbert approach to Toeplitz kernels.

One of the main results of this paper is Theorem 3.4, presented in Section 3, stating that in any "big enough" Toeplitz kernel one can find a nonzero function with a zero of specified order $n_0 \in \mathbb{N} \cup \{0\} =: \mathbb{N}_0$ at any given point $z_0 \in \mathbb{D}$, zeroes of order greater or equal to $n_j \in \mathbb{N}_0\,(j = 1, 2, ..., k_1)$ at any given points $z_j \in \mathbb{D}$, respectively, and with its g-image possessing zeroes of order greater or equal to $n_s \in \mathbb{N}_0\,(s = 1, 2, ..., k_2)$ at any given points z_s in the exterior of the disk, respectively. Thus we can impose zeroes not only for the functions in a Toeplitz kernel, but also for their g-images.

These results are used in Section 4 to study linearly independent sets of reproducing kernel functions and their analogues, as well as their orthogonal vectors, in a model space. Functions of that type are shown in Section 5 to be generators of the ranges of a class of finite rank truncated Toeplitz operators which are called *truncated Toeplitz operators of type I* [**5**], in model spaces of high enough dimension. We thus obtain easy explicit formulas for their ranks, slightly generalizing the results of [**5**], and we describe their ranges in terms of a basis, as well as in terms of their orthogonal complements in K_θ. In particular we prove that the rank of these operators depends only on the number of poles in certain rational functions and that these poles determine the range.

Another goal of this paper is to present the results in such a way that they are accessible to the general public and in particular to newcomers to the topic. In order to achieve this and keep the paper as self contained as possible, we present in Section 2 a brief introduction to the subject, the main concepts, and the most important results underlying the study presented in the subsequent sections. Most of the results in this section can be found in [**4**].

2. Preliminaries

DEFINITION 2.1. We denote by L^2 the space $L^2(\mathbb{T}, m)$, where m is the normalized Lebesgue measure, i.e., the the space of all square-integrable functions on the unit circle $\mathbb{T}$ in the complex plane $\mathbb{C}$ such that

$$\int_0^{2\pi} |f(e^{i\theta})|^2 d\theta < \infty. \tag{2.1}$$

L^2 is a Hilbert space with the inner product

$$\langle f, g\rangle = \frac{1}{2\pi}\int_0^{2\pi} f(e^{i\theta})\overline{g(e^{i\theta})}d\theta. \tag{2.2}$$

Moreover, $\{z \mapsto z^n : n \in \mathbb{Z}\}$ is an orthonormal set in L^2, since

$$\langle \zeta^n, \zeta^m\rangle = \frac{1}{2\pi}\int_0^{2\pi} e^{i\theta n}e^{-i\theta m}d\theta = \delta_{n,m}. \tag{2.3}$$

Denoting the Fourier coefficients of $f \in L^2$ by

$$\widehat{f}(n) := \langle f, \zeta^n \rangle = \frac{1}{2\pi} \int_0^{2\pi} f(e^{i\theta}) e^{-i\theta n} d\theta, \tag{2.4}$$

we have the following theorem:

THEOREM 2.2 (Parseval's Theorem). *Let $f \in L^2$, then*

$$||f||^2 = \sum_{n \in \mathbb{Z}} |\widehat{f}(n)|^2. \tag{2.5}$$

Furthermore,

$$\lim_{N \to \infty} \left\| f - \sum_{n=-N}^{N} \widehat{f}(n) z^n \right\| = 0 \tag{2.6}$$

From Parseval's theorem, one concludes that the set $\{\zeta \mapsto \zeta^n : n \in \mathbb{Z}\}$ is in fact an orthonormal basis for L^2, and that the Fourier series $\sum_{n \in \mathbb{Z}} \widehat{f}(n) \zeta^n$ converges to f in the L^2 norm.

DEFINITION 2.3. Let f be an analytic function on the unit disc $\mathbb{D}$, with Taylor series $\sum_{n=0}^{\infty} a_n z^n$, then f belongs to the Hardy space of $\mathbb{D}$, $H^2(\mathbb{D})$, if and only if the sequence (a_n) is absolutely square summable:

$$\sum_{n=0}^{\infty} |a_n|^2 < \infty \tag{2.7}$$

We denote by H^2_+ the space $H^2(\mathbb{D})$, that is a Hilbert space with the inner product

$$\langle f, g \rangle = \sum_{n=0}^{\infty} a_n \overline{b_n}, \tag{2.8}$$

where $f(z) = \sum_{n=0}^{\infty} a_n z^n$ and $g(z) = \sum_{n=0}^{\infty} b_n z^n$.

The space H^2_+ can be understood as a subspace of L^2 as follows.

PROPOSITION 2.4. *Let f be an analytic function on the unit disc $\mathbb{D}$ with Taylor series $\sum_{n=0}^{\infty} a_n z^n$. The following statements are equivalent:*

(1) $\sum_{n=0}^{\infty} |a_n|^2 < \infty$.

(2) $\sup_{0 \le r < 1} \int_0^{2\pi} |f(re^{i\theta})|^2 d\theta < \infty$

(3) *The radial limit $\widetilde{f}(e^{i\theta}) = \lim_{r \to 1^-} f(re^{i\theta})$ exists a.e. on $\mathbb{T}$, $\widetilde{f} \in L^2$, the Fourier coefficients $\widehat{f}(n) = \frac{1}{2\pi} \int_0^{2\pi} \widetilde{f}(e^{i\theta}) e^{-in\theta} d\theta$ vanish for $n < 0$ and $\widehat{f}(n) = a_n$ for $n \ge 0$.*

Moreover, if $f \in H^2_+$, then

$$\|f\|^2 = \sum_{n=0}^{\infty} |a_n|^2 = \sup_{0 \leq r < 1} \frac{1}{2\pi} \int_0^{2\pi} |f(re^{i\theta})|^2 d\theta = \sum_{n=-\infty}^{\infty} |\widehat{f}(n)|^2 \tag{2.9}$$

Therefore, we can identify the space H^2_+ with the subspace of L^2 whose functions have null Fourier coefficients for negative indexes. And its orthogonal complement is the subspace of L^2 whose functions have null Fourier coefficients for non-negative indexes.

We denote by $\mathbb{D}^e := \mathbb{C} \backslash \overline{\mathbb{D}}$ the exterior of the unit disc and by $\overline{f}$, the function defined on $\mathbb{D}^e$ by

$$\overline{f}(z) = \overline{f\left(\frac{1}{\overline{z}}\right)}, z \in \mathbb{D}^e, \tag{2.10}$$

where f is a function defined on $\mathbb{D}$.

By Schwarz reflection principle, if $f \in H^2_+$, then $\overline{f}$ is analytic on $\mathbb{D}^e$, and we have that

$$\overline{f}(\zeta) = \overline{f(\zeta)}, \text{ a.e. on } \mathbb{T}. \tag{2.11}$$

DEFINITION 2.5. Let f be an analytic function on the exterior of the unit disc $\mathbb{D}^e$. Then f belongs to the Hardy space of $\mathbb{D}^e$, $H^2(\mathbb{D}^e)$, if and only if $\lim_{z \to \infty} f(z) = 0$, f has Taylor series $\sum_{n=-\infty}^{-1} a_n z^n$ around ∞, and the sequence (a_n) is absolutely square summable.

We denote by H^2_- the space $H^2(\mathbb{D}^e)$, and it is an Hilbert space with inner product

$$\langle f, g \rangle = \sum_{n=-\infty}^{-1} a_n \overline{b_n}, \tag{2.12}$$

where $f(z) = \sum_{n=-\infty}^{-1} a_n z^n$ and $g(z) = \sum_{n=-\infty}^{-1} b_n z^n$.

Analogously to H^2_+, H^2_- can also be identified with a subspace of L^2 using the Fourier coefficients, and we have

PROPOSITION 2.6. *$H^2_- = \overline{zH^2_+}$ and $L^2 = H^2_+ \oplus H^2_-$.*

DEFINITION 2.7. Let $L^\infty := L^\infty(\mathbb{T})$ be the set of all essentially bounded analytic functions on $\mathbb{T}$, equipped with the essential supremum norm

$$\|f\|_\infty = \operatorname{ess-sup}_{\zeta \in \mathbb{T}} |f(\zeta)| \tag{2.13}$$

DEFINITION 2.8. Let $H^\infty_+ := H^\infty(\mathbb{D})$ be the set of all analytic and bounded functions on $\mathbb{D}$, with the norm

$$\|f\|_\infty = \sup_{z \in \mathbb{D}} |f(z)| \tag{2.14}$$

Both L^∞ and H^∞_+ are Banach spaces. And similarly to L^2 and H^2_+, we can also identify H^∞_+ with a subspace of L^∞.

PROPOSITION 2.9. *Let $f \in H^\infty_+$. The radial limit*

$$\lim_{r\to 1^-} f(r\zeta) \tag{2.15}$$

exists for almost every $\zeta \in \mathbb{T}$.

Furthermore, let $\widetilde{f}(\zeta) = \lim_{r\to 1^-} f(r\zeta)$. Then $\widetilde{f} \in L^\infty$ and

$$\|f\|_\infty = \sup_{z\in\mathbb{D}} |f(z)| = \operatorname{ess-sup}_{\zeta\in\mathbb{T}} |\widetilde{f}(\zeta)| = \|\widetilde{f}\|_\infty \tag{2.16}$$

It is natural to define the analogous space of bounded functions on $\mathbb{D}^e$.

DEFINITION 2.10. Let $H^\infty_- := H^\infty(\mathbb{D}^e)$ be the set of all analytic and bounded functions on $\mathbb{D}^e$, with the norm

$$\|f\|_\infty = \sup_{z\in\mathbb{D}^e} |f(z)| \tag{2.17}$$

One can also identify H^∞_- as a subspace of L^∞ through radial limits, and we have $H^\infty_- = \overline{H^\infty_+}$. However, since all constant functions are elements of both H^∞_+ and H^∞_-, we have that $L^\infty \neq H^\infty_+ \oplus H^\infty_-$.

An important class of bounded functions is that of inner functions.

DEFINITION 2.11. We say that θ is an *inner function* if and only if $\theta \in H^\infty_+$ and $|\theta(\zeta)| = 1$ a.e. on $\mathbb{T}$. A rational inner function is a *finite Blaschke product.*

We denote by

$$B^n_{z_+}(z) := \left(\frac{z - z_+}{1 - \overline{z_+}\, z}\right)^n = \bar{z}^n \left(\frac{z - z_+}{\bar{z} - \overline{z_+}}\right)^n, \quad \text{for } z \in \mathbb{T}, \tag{2.18}$$

the finite Blaschke product with zero of order n at $z_+ \in \mathbb{D}$.

Moreover,

$$\overline{B^n_{z_+}}(z) = \overline{B^n_{z_+}\left(\frac{1}{\bar{z}}\right)} = \left(\frac{\frac{1}{z} - \overline{z_+}}{1 - z_+\frac{1}{z}}\right)^n = \left(\frac{1 - \overline{z_+}z}{z - z_+}\right)^n \in H^\infty_-. \tag{2.19}$$

Denoting by

$$B^n_{z_-}(z) := \left(\frac{z - z_-}{1 - \overline{z_-}z}\right)^n, \quad \text{for } z \in \mathbb{D}^e, \tag{2.20}$$

where $z_- \in \mathbb{D}^e$, we have

(2.21)

$$B^n_{z_-}(z) = \frac{1}{\overline{z_-}^n}\left(\frac{z - z_-}{\frac{1}{\overline{z_-}} - z}\right)^n = \left(\frac{z_-}{\overline{z_-}}\right)^n \left(\frac{1 - \frac{z}{z_-}}{z - \frac{1}{\overline{z_-}}}\right)^n = \left(\frac{z_+}{\overline{z_+}}\right)^n \overline{B^n_{z_+}}(z), \quad \text{for } z \in \mathbb{T},$$

where $z_+ = \frac{1}{\overline{z_-}} \in \mathbb{D}$.

To the orthogonal decomposition of the L^2 space in Proposition 2.6, there corresponds an orthogonal projection from L^2 onto H^2_+.

DEFINITION 2.12. Let P denote the so called *Riesz projection*, which is the orthogonal projection from L^2 onto H^2_+.

$$P(f)(\zeta) = \sum_{n=0}^{\infty} \widehat{f}(n)\zeta^n \tag{2.22}$$

The Riesz projection can also be given by the integral formula

$$P(f)(\zeta) = \frac{1}{2\pi i} \int_{\mathbb{T}} \frac{f(t)}{t-\zeta} dt, \tag{2.23}$$

for all $f \in L^2$ and $\zeta \in \mathbb{T}$.

Now we have what is needed to define Toeplitz operators on H^2_+.

DEFINITION 2.13. Let $g \in L^\infty$. Then, the Toeplitz operator with symbol g, $T_g : H^2_+ \to H^2_+$, is defined by

$$T_g f = P(gf) \tag{2.24}$$

In particular, for $g(z) = z$ we have the shift operator $S = T_z = M_z$, where M_z denotes the operator of multiplication by z.

We enumerate some basic properties of Toeplitz operators.

PROPOSITION 2.14. *Let $f, g \in L^\infty$ and $\alpha, \beta \in \mathbb{C}$. Then:*

1. $T_{\alpha f + \beta g} = \alpha T_f + \beta T_g$.
2. $T_g^* = T_{\overline{g}}$.
3. $\|T_g\| = \|g\|_\infty$.
4. $T_f = T_g$ *if and only if* $f = g$.
5. T_g *is compact if and only if* $g = 0$.

The kernels of Toeplitz operators, also called Toeplitz kernels, will play a very significant role on the next sections. The following characterization of these spaces will be used.

PROPOSITION 2.15. *Let $g \in L^\infty$ and $\phi_+ \in H^2_+$. Then, $\phi_+ \in \ker T_g$ if and only if*

$$g\,\phi_+ = \phi_-, \qquad \text{a.e. on } \mathbb{T}, \tag{2.25}$$

for some $\phi_- \in H^2_-$, where ϕ_+ and ϕ_- are identified with their boundary functions in L^2.

The image under multiplication by $g \in L^\infty$ of $\ker T_g$, which is a closed subspace of H^2_-, will be called its g-image.

Toeplitz kernels have been the object of much interest (see for example [**6, 7**]). They possess some very special and interesting properties, amongst which that of near invariance with respect to multiplication by functions in a wide class ([**3**]).

DEFINITION 2.16. Let E be a proper subspace of H^2_+ and η a complex valued function defined a.e. on $\mathbb{T}$. We say that the space E is nearly η-invariant if and only if

$$\phi \in E,\ \eta\phi \in H^2_+ \Rightarrow \eta\phi \in E. \tag{2.26}$$

In particular all Toeplitz kernels are nearly η-invariant if $\eta \in H^\infty_-$ and, in particular, nearly $\bar{\theta}$-invariant, where θ is an inner function ([**3**]). As a consequence, following the same reasoning as in the proof of Proposition 2.4 in [**3**], we have:

PROPOSITION 2.17. *Let K be a non zero Toeplitz kernel. Then, given any $z_0 \in \mathbb{D}$, there exists $\phi \in K$ such that $\phi(z_0) \neq 0$.*

More generally, K is not contained in θH^2_+ for any inner function.

Another important property is the following.

THEOREM 2.18 ([**2**]). *Let $g \in L^\infty$ and let θ be a finite Blaschke product. Then*

$$\dim \ker T_g < \infty \Leftrightarrow \dim \ker T_{\theta g} < \infty. \tag{2.27}$$

Moreover, if $\dim \ker T_g < \infty$*, then*

$$\dim \ker T_{\theta g} = \max\{0\,,\, \dim \ker T_g \;-\; k\} \tag{2.28}$$

where k is the number of zeroes of θ, counting their multiplicities.

Also important are model spaces, a special class of Toeplitz kernels, which provide the domain on which truncated Toeplitz operators, that will be relevant in the work that follows, and are defined below, act.

DEFINITION 2.19. Let θ be an inner function. The model space K_θ is defined as

$$K_\theta = \ker T_{\overline{\theta}} = \{\phi_+ \in H^2_+ : \overline{\theta}\phi_+ = \phi_-, \text{ for some } \phi_- \in H^2_-\} \tag{2.29}$$

An equivalent and more frequently used definition of the model space corresponding to an inner function θ is that K_θ is the orthogonal complement of the shift invariant subspace θH^2_+ in H^2_+:

$$K_\theta = H^2_+ \ominus \theta H^2_+ = H^2_+ \cap \theta H^2_-. \tag{2.30}$$

A model space K_θ is finite dimensional if and only if θ is a rational function, i.e., a finite Blaschke product. In that case, $\dim K_\theta = \deg \theta$, where by $\deg \theta$ we denote the number of zeroes of θ (in $\mathbb{D}$), counting their multiplicities.

Now, in order to define truncated Toeplitz operators on model spaces, which are subspaces of H^2_+, analogously to Toeplitz operators, we need orthogonal projections onto model spaces.

PROPOSITION 2.20. *Let θ be an inner function. The operator $P_\theta : L^2 \to K_\theta$ defined by*

$$P_\theta(f) = \theta P^-(\overline{\theta}P(f)), \quad f \in L^2, \tag{2.31}$$

is the orthogonal projection from L^2 onto K_θ, where $P^- = Id - P$ is the orthogonal projection from L^2 onto H^2_-.

Moreover, $P_\theta = \theta P^- \overline{\theta} P = P \theta P^- \overline{\theta}$.

Another important operator on K_θ is the conjugation operator C_θ.

DEFINITION 2.21. Let $\mathcal{H}$ be an Hilbert space, and $C : \mathcal{H} \to \mathcal{H}$. The function C is a conjugation if it satisfies the following conditions:

(1) C is anti-linear: $C(ax + by) = \overline{a}Cx + \overline{b}Cy$, for all $a, b \in \mathbb{C}$ and $x, y \in \mathcal{H}$.
(2) C is an involution: $C^2 = Id$.
(3) C is an isometry: $\|Cx\| = \|x\|$, for all $x \in \mathcal{H}$.

Since model spaces are closed subspaces of H^2_+, they are also Hilbert spaces, so it makes sense to talk about conjugations on model spaces.

PROPOSITION 2.22. *Let θ be an inner function, the operator $C_\theta : K_\theta \to K_\theta$ defined by*

$$C_\theta f = \theta \overline{zf}, \quad f \in K_\theta, \tag{2.32}$$

is a conjugation.

We now define truncated Toeplitz operators on model spaces.

DEFINITION 2.23. Let θ be an inner function and $g \in L^\infty$. The truncated Toeplitz operator on K_θ with symbol g, $A_g^\theta : K_\theta \to K_\theta$, is defined by

$$A_g^\theta f = P_\theta(gf), \quad f \in K_\theta. \tag{2.33}$$

More generally, if $g \in L^2$, we define

$$A_g^\theta f = P_\theta(gf), \quad f \in H_+^\infty \cap K_\theta. \tag{2.34}$$

A_g^θ is a closed, densely defined operator and, if it is bounded, it admits a unique bounded extension to K_θ that we also denote by A_g^θ.

In spite of their similarities as regards the definition, Toeplitz operators on H_+^2 and truncated Toeplitz operators on model spaces possess quite different properties. Namely, the symbol of a truncated Toeplitz operator is not unique. We have the following

PROPOSITION 2.24. *Let θ be an inner function and $g \in L^\infty$. Then, A_g^θ is the zero operator if and only if $g \in \theta H_+^2 + \overline{\theta H_+^2}$.*

3. Zeroes of functions in Toeplitz kernels and their g-images

We start by proving some auxiliary results:

PROPOSITION 3.1. *If ϕ is analytic in $\mathbb{D}$ and $z_- \in \mathbb{D}^e$, then*

$$\overline{\phi}'(z_-) = -\frac{1}{z_-^2}\,\overline{\phi'\left(\frac{1}{\overline{z_-}}\right)}. \tag{3.1}$$

PROOF. By definition,

$$\begin{aligned}
\overline{\phi}'(z_-) &= \lim_{w \to z_-} \frac{\overline{\phi}(w) - \overline{\phi}(z_-)}{w - z_-} = \\
&= \lim_{w \to z_-} \frac{\overline{\phi\left(\frac{1}{\overline{w}}\right)} - \overline{\phi\left(\frac{1}{\overline{z_-}}\right)}}{w - z_-} = \\
&= \lim_{w \to z_-} \overline{\left(\frac{\phi\left(\frac{1}{\overline{w}}\right) - \phi\left(\frac{1}{\overline{z_-}}\right)}{\frac{1}{\overline{w}} - \frac{1}{\overline{z_-}}}\right)} \frac{\frac{1}{w} - \frac{1}{z_-}}{w - z_-} = \\
&= \overline{\left(\lim_{u \to \frac{1}{\overline{z_-}}} \frac{\phi(u) - \phi\left(\frac{1}{\overline{z_-}}\right)}{u - \frac{1}{\overline{z_-}}}\right)} \lim_{w \to z_-} \frac{\frac{1}{w} - \frac{1}{z_-}}{w - z_-} = \\
&= \overline{\phi'\left(\frac{1}{\overline{z_-}}\right)}\left(-\frac{1}{z_-^2}\right)
\end{aligned}$$

□

By induction, we get:

COROLLARY 3.2. *If ϕ is analytic in $\mathbb{D}$, $z_- \in \mathbb{D}^e$ and $n \in \mathbb{N}$, then:*

$$\overline{\phi}^{(n)}(z_-) = \overline{\phi_I^{(n)}(\overline{z_-})}, \tag{3.2}$$

where $\phi_I(w) = \phi\left(\frac{1}{w}\right)$, for $w \in \mathbb{D}^e$.

PROOF. The result holds for $n = 0$, by definition of $\overline{\phi}$. For $n > 0$, using the induction hypothesis,

$$\overline{\phi}^{(n)}(z_-) = \lim_{w\to z_-} \frac{\overline{\phi}^{(n-1)}(w) - \overline{\phi}^{(n-1)}(z_-)}{w - z_-} =$$
$$= \lim_{w\to z_-} \overline{\left(\frac{\phi_I^{(n-1)}(\overline{w}) - \phi_I^{(n-1)}(\overline{z_-})}{\overline{w} - \overline{z_-}}\right)} =$$
$$= \overline{\phi_I^{(n)}(\overline{z_-})}$$

□

Note that, for $n = 1$, we get $\overline{\phi_I'(\overline{z_-})} = \overline{\phi'(\frac{1}{\overline{z_-}})(\frac{-1}{\overline{z_-}^2})} = \frac{-1}{z_-^2}\overline{\phi'(\frac{1}{\overline{z_-}})}$ as in Proposition 3.1.

COROLLARY 3.3. *If ϕ is analytic in $\mathbb{D}$, $z_- \in \mathbb{D}^e$ and $n \in \mathbb{N}$, then $\overline{\phi}$ has a zero of order n at z_- if and only if ϕ has a zero of order n at $\frac{1}{\overline{z_-}}$.*

PROOF. Suppose that $\frac{1}{\overline{z_-}}$ is a zero of order n of ϕ. Taking into account that

$$\phi_I^{(k)}(w) = \sum_{j=0}^{k-1} \left(\phi^{(n-j)}\left(\frac{1}{w}\right)\frac{a_{k,j}}{w^{2k-j}}\right),$$

for $w \in \mathbb{D}^e$, $k \in \mathbb{N}$ and $a_{k,j} \in \mathbb{N}$, we have that

$$\overline{\phi}(z_-) = \overline{\phi}'(z_-) = \cdots = \overline{\phi}^{(n-1)}(z_-) = 0 \text{ and } \overline{\phi}^{(n)}(z_-) = \overline{\phi^{(n)}\left(\frac{1}{\overline{z_-}}\right)\frac{a_{n,0}}{\overline{z_-}^{2n}}} \neq 0.$$

The converse can be proved analogously. □

For $g \in L^\infty$, the kernel of the Toeplitz operator T_g can be described by Proposition 2.15, and the question of finding solutions to Equation (2.25) is called a *Riemann Hilbert problem.* Since, given any $g \in L^\infty$, one can find $\widetilde{g} \in L^\infty$ with $|\widetilde{g}(\zeta)| = 1$ a.e. on $\mathbb{T}$, such that $\ker T_g = \ker T_{\widetilde{g}}$ ([**3, 7**]), we can assume in this case that the coefficient g in Equation (2.25) satisfies the condition $|g(\zeta)| = 1$ a.e. on $\mathbb{T}$.

We now prove that, in any Toeplitz kernel of a high enough dimension, we can choose a non-zero element ϕ_+ with specified zeroes in $\mathbb{D}$, and such that $\phi_- = g\phi_+$ has any specified zeros in $\mathbb{D}^e$.

THEOREM 3.4. *Let $g \in L^\infty$, n_0, $n_j\,(j = 1, \cdots, N_+)$, $m_l\,(l = 1, \cdots, N_-)$ be given non-negative integers and assume that*

$$\dim\ker T_g > n_0 + \sum_{j=1}^{N_+} n_j + \sum_{l=1}^{N_-} m_l. \tag{3.3}$$

Then there exists $\phi_+ \in \ker T_g, \phi_+ \neq 0$, such that:

(1) *ϕ_+ has a zero of order n_0 at $z_0^+ \in \mathbb{D}$*
(2) $\dfrac{\phi_+}{(z - z_j^+)^{n_j}} \in H_+^2$, *for all $j = 1, \cdots, N_+$*
(3) $\dfrac{\phi_-}{(z - z_l^-)^{m_l}} \in H_-^2$, *for all $l = 1, \cdots, N_-$,*

where $\phi_- = g\phi_+$, *the points* $z_0^+, z_j^+ (j = 1, \cdots, N_+)$ *are any given points in* $\mathbb{D}$ *and* $z_l^- (l = 1, \cdots, N_-)$ *are any given points in* $\mathbb{D}^e$.

PROOF. Let

$$B_+ = B_{z_0^+}^{n_0} \prod_{j=1}^{N_+} B_{z_j^+}^{n_j}, \quad B_- = \prod_{l=1}^{N_-} B_{\frac{1}{z_l^-}}^{m_l}, \quad \text{and} \quad \widetilde{g} = B_- B_+ g.$$

Since Equation (3.3) holds, by Theorem 2.18 we have that $\ker T_{\widetilde{g}} \neq \{0\}$. Therefore, by Proposition 2.17 there exists a nonzero element $\psi_+ \in \ker T_{\widetilde{g}}$ with $\psi_+(z_0) \neq 0$ and we have

$$\widetilde{g}\psi_+ = \psi_- \iff gB_+\psi_+ = \overline{B_-}\psi_-,$$

with $\psi_- \in H^2_-$. Defining $\phi_+ = B_+\psi_+$, it is clear that ϕ_+ has the desired properties. □

COROLLARY 3.5. *With the same assumptions as in Theorem 3.4, there exists* $\psi_+ \in \ker T_g, \psi_+ \neq 0$, *such that:*

(1) $\psi_- = g\psi_+$ *has a zero of order* n_0 *at* $z_0^- \in \mathbb{D}^e$,
(2) $\dfrac{\psi_-}{(z - z_l^-)^{m_l}} \in H^2_-$, *for all* $l = 1, \cdots, N_-$,
(3) $\dfrac{\psi_+}{(z - z_j^+)^{n_j}} \in H^2_+$, *for all* $j = 1, \cdots, N_+$.

PROOF. Assuming $|g| = 1$, we have from Proposition 2.15, that $\psi_+ \in \ker T_g$ if and only if $g\psi_+ = \psi_-$ with $\psi_- \in H^2_-$. Thus,

$$\psi_+ \in \ker T_g \iff \psi_+ = \overline{g}\psi_- \iff g(\overline{z\psi_-}) = \overline{z\psi_+} \iff g\phi_+ = \phi_-,$$

where $\phi_+ = \overline{z\psi_-} \in H^2_+, \phi_- = \overline{z\psi_+} \in H^2_-$.

By Theorem 3.4, there exists $\phi_+ \in \ker T_g$ such that ϕ_+ has a zero of order n_0 at $\frac{1}{\overline{z_0^-}} \in \mathbb{D}$, $\dfrac{\phi_+}{(z - \frac{1}{\overline{z_l^-}})^{m_l}} \in H^2_+$ for all $l = 1, \cdots, N_-$, and $\dfrac{\phi_-}{(z - \frac{1}{\overline{z_j^+}})^{n_j}} \in H^2_-$ for all $j = 1, \cdots, N_+$. By Corollary 3.3 it follows that $\psi_+ = \overline{z\phi_-} \in \ker T_g$ satisfies (1)-(3). □

We now apply the previous results to model spaces, since for any inner function θ we have $K_\theta = \ker T_{\overline{\theta}}$.

COROLLARY 3.6. *With the same notation as in Theorem 3.4 with* $g = \overline{\theta}$ *where* θ *is an inner function, if* $\dim K_\theta > n_0 + \sum_{j=1}^{N_+} n_j + \sum_{l=1}^{N_-} m_l$ *there exists* $\phi_\theta \in K_\theta$ *such that:*

(1) ϕ_θ *has a zero of order* n_0 *at* $z_0^+ \in \mathbb{D}$,
(2) ϕ_θ *has a zero of order greater or equal to* n_j *at* $z_j^+ \in \mathbb{D}$, *for all* $j = 1, \cdots, N_+$,
(3) $C_\theta\phi_\theta$ *has a zero of order greater or equal to* m_l *at* $w_l^+ = \frac{1}{\overline{z_l^-}} \in \mathbb{D}$, *for all* $l = 1, \cdots, N_-$.

PROOF. From Theorem 3.4, there exists $\phi_\theta \in K_\theta$ such that (1) and (2) hold and moreover $\overline{\theta}\phi_\theta$ has a zero of order greater or equal to m_l at $z_l^- = \frac{1}{\overline{w_l^+}} \in \mathbb{D}^e$, so the same must happen with $z\overline{\theta}\,\phi_\theta$. By Corollary 3.3, it follows that $\theta\,\overline{z\phi_\theta} = C_\theta\phi_\theta$ satisfies (3). □

COROLLARY 3.7. *With the same assumptions as in Corollary 3.6, there exists $\phi_\theta \in K_\theta$ such that:*

(1) *$C_\theta\phi_\theta$ has a zero of order n_0 at $z_0^+ \in \mathbb{D}$,*
(2) *ϕ_θ has a zero of order greater or equal to n_j at $z_j^+ \in \mathbb{D}$, for all $j = 1, \cdots, N_+$,*
(3) *$C_\theta\phi_\theta$ has a zero of order greater or equal to m_l at $w_l^+ = \frac{1}{\overline{z_l^-}} \in \mathbb{D}$, for all $l = 1, \cdots, N_-$.*

4. Reproducing kernel functions in model spaces and their conjugates

For any complex valued function f analytic in a neighbourhood of $\lambda \in \mathbb{C}$, and any $n \in \mathbb{N}_0 = \mathbb{N} \cup \{0\}$, we denote by $P^f_{n,\lambda}$ the Taylor polynomial of degree n

$$P^f_{n,\lambda}(z) = f(\lambda) + f'(\lambda)(z-\lambda) + \cdots + \frac{f^{(n)}(\lambda)}{n!}(z-\lambda)^n. \tag{4.1}$$

With this notation, if θ is an inner function and $\lambda \in \mathbb{D}$, let

$$\widetilde{k}^\theta_{n,\lambda}(z) := \frac{\theta(z) - P^\theta_{n,\lambda}(z)}{(z-\lambda)^{n+1}} \tag{4.2}$$

$$k^\theta_{n,\lambda}(z) := (C_\theta \widetilde{k}^\theta_{n,\lambda})(z) = \overline{z}\frac{1 - \overline{P^\theta_{n,\lambda}(z)}\,\theta(z)}{(\overline{z}-\overline{\lambda})^{n+1}} = z^n\frac{1 - \overline{P^\theta_{n,\lambda}(z)}\,\theta(z)}{(1-\overline{\lambda}z)^{n+1}} \tag{4.3}$$

For $n = 0$ we have the reproducing kernel functions of K_θ at the point λ,

$$k^\theta_{0,\lambda} = k^\theta_\lambda(z) = \frac{1 - \overline{\theta(\lambda)}\theta(z)}{1 - \overline{\lambda}z}, \tag{4.4}$$

such that for any $\phi_\theta \in K_\theta$ we have

$$\langle \phi_\theta, k^\theta_\lambda \rangle = \phi_\theta(\lambda). \tag{4.5}$$

We also have

$$\widetilde{k}^\theta_{0,\lambda}(z) = \widetilde{k}^\theta_\lambda(z) = \frac{\theta(z) - \theta(\lambda)}{z - \lambda} \tag{4.6}$$

and, for all $\phi_\theta \in K_\theta$,

$$\langle \phi_\theta, \widetilde{k}^\theta_\lambda \rangle = \overline{C_\theta\phi_\theta(\lambda)}. \tag{4.7}$$

The maps $\Phi_1 : \lambda \to \widetilde{k}^\theta_\lambda$ and $\Phi_2 : \overline{\lambda} \to k^\theta_\lambda$ are analytic on $\mathbb{D}$ with respect to λ and $\overline{\lambda}$, respectively. We have

$$\frac{\mathrm{d}^n \widetilde{k}^\theta_\lambda}{\mathrm{d}\lambda^n} = n!\widetilde{k}^\theta_{n,\lambda}, \quad \frac{\mathrm{d}^n k^\theta_\lambda}{\mathrm{d}\overline{\lambda}^n} = n! k^\theta_{n,\lambda} \tag{4.8}$$

and, for all $\phi_\theta \in K_\theta$,

$$\langle \phi_\theta, k^\theta_{n,\lambda} \rangle = \frac{\phi_\theta^{(n)}(\lambda)}{n!}, \quad \langle \phi_\theta, \widetilde{k}^\theta_{n,\lambda} \rangle = \frac{\overline{C_\theta\phi_\theta^{(n)}(\lambda)}}{n!}. \tag{4.9}$$

Every finite set of functions $k^\theta_{n,\lambda}$ with $n \in \mathbb{N}_0$ and $\lambda \in \mathbb{D}$ is linearly independent if $\dim K_\theta = \infty$, and analogously for any finite set of their conjugates $\widetilde{k}^\theta_{n,\lambda}$ ([**1**]).

Using the results of Section 3, we now prove that any finite subset of $\{k^\theta_{n,\lambda} : n \in \mathbb{N}_0, \lambda \in \mathbb{D}\} \cup \{\widetilde{k}^\theta_{n,\lambda} : n \in \mathbb{N}_0, \lambda \in \mathbb{D}\}$ is linearly independent if $\dim K_\theta$ is high enough.

We will use the following notation. Let θ be an inner function, $\lambda_j \in \mathbb{D}$ and $n_j \in \mathbb{N}_0$ for $j = 1, \cdots, N_+$, $\widetilde{\lambda}_l \in \mathbb{D}$ and $m_l \in \mathbb{N}_0$ for $l = 1, \cdots, N_-$, where $N_\pm$ are given natural numbers.

Now consider the pairs (λ_j, n_j) and $(\widetilde{\lambda}_l, m_l)$ corresponding to the functions

$$k^\theta_{n_j,\lambda_j}, j = 1, \cdots, N_+ \quad \text{and} \quad \widetilde{k}^\theta_{m_l,\widetilde{\lambda}_l}, l = 1, \cdots, N_-.$$

For any $\lambda \in \Lambda = \{\lambda_j : j = 1, \cdots, N_+\}$ let $M_\lambda = \max\{n_j : \lambda_j = \lambda\}$ and, analogously, for any $\widetilde{\lambda} \in \widetilde{\Lambda} = \{\widetilde{\lambda}_l : l = 1, \cdots, N_-\}$ let $M_{\widetilde{\lambda}} = \max\{m_l : \widetilde{\lambda}_l = \widetilde{\lambda}\}$. Then, we have the following.

THEOREM 4.1. *If* $\dim K_\theta \geq \sum_{\lambda \in \Lambda}(M_\lambda + 1) + \sum_{\widetilde{\lambda} \in \widetilde{\Lambda}}(M_{\widetilde{\lambda}} + 1)$, *then the set*

$$\{k^\theta_{n_j,\lambda_j} : j = 1, \cdots, N_+\} \cup \{\widetilde{k}^\theta_{m_l,\widetilde{\lambda}_l} : l = 1, \cdots, N_-\}$$

is linearly independent.

PROOF. Let $\alpha_j \in \mathbb{C}, j = 1, \cdots, N_+$ and $\beta_l \in \mathbb{C}, l = 1, \cdots, N_-$ be such that

$$\sum_{j=1}^{N_+} \alpha_j k^\theta_{n_j,\lambda_j} + \sum_{l=1}^{N_-} \beta_l \widetilde{k}^\theta_{m_l,\widetilde{\lambda}_l} = 0. \tag{4.10}$$

Then, by (4.9), for any $\phi_\theta \in K_\theta$ we have that

$$\sum_{j=1}^{N_+} \alpha_j \frac{\phi_\theta^{(n_j)}(\lambda_j)}{(n_j)!} + \sum_{l=1}^{N_-} \beta_l \overline{\frac{C_\theta \phi_\theta^{(m_l)}(\widetilde{\lambda}_l)}{(m_l)!}} = 0. \tag{4.11}$$

Given any particular $\lambda \in \Lambda$, we can assume without loss of generality that $\lambda = \lambda_1$ and $M_\lambda = n_1$, so that 4.11 can be written as

$$\alpha_1 \frac{\phi_\theta^{(M_\lambda)}(\lambda)}{(M_\lambda)!} + \sum_{j=2}^{N_+} \alpha_j \frac{\phi_\theta^{(n_j)}(\lambda_j)}{(n_j)!} + \sum_{l=1}^{N_-} \beta_l \overline{\frac{C_\theta \phi_\theta^{(m_l)}(\widetilde{\lambda}_l)}{(m_l)!}} = 0. \tag{4.12}$$

By Corollary 3.6, we can now choose $\phi_\theta \in K_\theta$ having a zero of order M_λ at λ, a zero of order at least $M_{\lambda_j} + 1$ at any $\lambda_j \neq \lambda$, and such that $C_\theta \phi_\theta$ has a zero of order at least $M_{\widetilde{\lambda}_l} + 1$ at any $\widetilde{\lambda}_l$. It follows that $\phi_\theta^{(M_\lambda)}(\lambda) \neq 0$, $\phi_\theta^{(n_j)}(\lambda_j) = 0$, $C_\theta \phi_\theta^{(m_l)}(\widetilde{\lambda}_l) = 0$ for all $j = 2, \cdots, N_+$, $l = 1, \cdots, N_-$, so from (4.12) we conclude that $\alpha_1 = 0$. Analogously, using Corollary 3.6, we conclude that all coefficients α_j and β_l must be zero. □

We will also need the following functions, which are related to the functions 4.3 by the proposition below.

For $\lambda_- \in \mathbb{D}^e$, $n \in \mathbb{N}_0$, we define

$$k^\theta_{n,\lambda_-}(z) = \frac{1 - P^{\overline{\theta}}_{n,\lambda_-}(z)\,\theta(z)}{(z - \lambda_-)^{n+1}}. \tag{4.13}$$

We have that $k^\theta_{n,\lambda_-} \in K_\theta$ because $\overline{\theta} k^\theta_{n,\lambda_-} = \dfrac{\overline{\theta} - P^{\overline{\theta}}_{n,\lambda_-}}{(z - \lambda_-)^{n+1}} \in H^2_-$. Using corollary 3.2, with $\phi = \theta$, we obtain:

PROPOSITION 4.2. *Let θ be an inner function, $n \in \mathbb{N}_0$, $z_- \in \mathbb{D}^e$ and $z_+ = \frac{1}{\overline{z_-}} \in \mathbb{D}$. Then*

$$k^\theta_{n,z_-} = \frac{(-1)^{n+1}}{z_-^{n+1}} \sum_{j=0}^{n} \binom{n}{j} \frac{1}{z_-^j} k^\theta_{j,\frac{1}{\overline{z_-}}} = (-\overline{z}_+)^{n+1} \sum_{j=0}^{n} \binom{n}{j} \overline{z}_+^j k^\theta_{j,z_+}, \tag{4.14}$$

$$k^\theta_{n,z_+} = \frac{(-1)^{n+1}}{\overline{z}_+^{n+1}} \sum_{j=0}^{n} \binom{n}{j} \frac{1}{\overline{z}_+^j} k^\theta_{j,\frac{1}{\overline{z}_+}} = (-z_-)^{n+1} \sum_{j=0}^{n} \binom{n}{j} z_-^j k^\theta_{j,z_-}, \tag{4.15}$$

As a consequence of Proposition 4.2 and Theorem 4.1 we have then:

COROLLARY 4.3. *With the same assumptions as in Theorem 4.1, and with $\lambda_j^- = \frac{1}{\overline{\lambda}_j} \in \mathbb{D}^e$, the set*

$$\{k^\theta_{n_j,\lambda_j^-} : j = 1, \cdots, N_+\} \cup \{\widetilde{k}^\theta_{m_l,\widetilde{\lambda}_l} : l = 1, \cdots, N_-\} \tag{4.16}$$

is linearly independent.

Another consequence of Proposition 4.2 and (4.9) is the following.

COROLLARY 4.4. *For $n \in \mathbb{N}_0$ and $z_- \in \mathbb{D}^e$, for all $\phi_\theta \in K_\theta$ we have*

$$\langle \phi_\theta, k^\theta_{n,z_-} \rangle = \left(\frac{-1}{\overline{z}_-}\right)^{n+1} \sum_{j=0}^{n} \binom{n}{j} \frac{1}{\overline{z}_-^j} \phi_\theta^{(j)} \left(\frac{1}{\overline{z}_-}\right). \tag{4.17}$$

We also have, from(4.9) and Corollary 4.4:

PROPOSITION 4.5. *For $z_+ \in \mathbb{D}$, $z_- \in \mathbb{D}^e$, $\phi_\theta \in K_\theta$, where θ is an inner function,*

(1) $\langle \phi_\theta, k^\theta_{j,z_+} \rangle = 0$ *for all* $j = 0, 1, \cdots, n-1 \iff \phi_\theta$ *has a zero of order greater or equal to n at z_+,*
(2) $\langle \phi_\theta, k^\theta_{j,z_-} \rangle = 0$ *for all* $j = 0, 1, \cdots, n-1 \iff \phi_\theta$ *has a zero of order greater or equal to n at $\frac{1}{\overline{z}_-}$,*
(3) $\langle \phi_\theta, \widetilde{k}^\theta_{j,z_+} \rangle = 0$ *for all* $j = 0, 1, \cdots, n-1 \iff C_\theta\phi_\theta$ *has a zero of order greater or equal to n at z_+.*

Consequently, we can write:

$$K_\theta \ominus \text{span}\{k^\theta_{j,z_+} : j = 0, 1, \cdots, n-1\} = K_\theta \cap B^n_{z_+} H^2_+, \tag{4.18}$$

$$K_\theta \ominus \text{span}\{k^\theta_{j,z_-} : j = 0, 1, \cdots, n-1\} = K_\theta \cap B^n_{\frac{1}{\overline{z}_-}} H^2_+, \tag{4.19}$$

$$K_\theta \ominus \text{span}\{\widetilde{k}^\theta_{j,z_+} : j = 0, 1, \cdots, n-1\} = K_\theta \cap B^n_{\frac{1}{\overline{z}_+}} \theta H^2_- = K_\theta \cap \overline{B^n_{z_+}} \theta H^2_-, \tag{4.20}$$

taking into account that $H^2_- = \overline{zH^2_+}$ and

$$C_\theta\phi_\theta = B^n_{z_+}\psi_+ \text{ with } \psi_+ \in H^2_+ \iff \phi_\theta = \overline{B^n_{z_+}}\theta\overline{z}\overline{\psi}_+ \text{ with } \psi_+ \in H^2_+$$

Using the same reasoning as in the proof of Theorem 3.4, we can write the relations (4.18) - (4.20) in yet another form, expressing the orthogonal complements on the left hand side of those expressions in terms of kernels of Toeplitz operators. In fact we have

$$\begin{aligned}
&\phi_+ \in K_\theta \cap B^n_{z_+} H^2_+ &&\iff\\
&\phi_+ \in H^2_+\,,\ \overline{\theta}\phi_+ = \phi_- \in H^2_- \quad \text{and} \quad \phi_+ = B^n_{z_+}\psi_+ \text{ with } \psi_+ \in H^2_+ &&\iff\\
&\phi_+ = B^n_{z_+}\psi_+ \text{ with } \psi_+ \in H^2_+ \quad \text{and} \quad \overline{\theta}B^n_{z_+}\psi_+ = \phi_- \in H^2_- &&\iff\\
&\phi_+ = B^n_{z_+}\psi_+ \text{ with } \psi_+ \in \ker T_{\overline{\theta}B^n_{z_+}}\,.
\end{aligned}$$

Analogously,

$$\begin{aligned}
&\phi_+ \in K_\theta \cap \overline{B^n_{z_+}}\theta H^2_- &&\iff\\
&\phi_+ \in H^2_+\,,\ \overline{\theta}\phi_+ = \phi_- \in H^2_- \quad \text{and} \quad \phi_+ = \overline{B^n_{z_+}}\theta\psi_- \text{ with } \psi_- \in H^2_- &&\iff\\
&\phi_+ \in H^2_+\,,\ \overline{\theta}B^n_{z_+}\phi_+ = \psi_- \in H^2_- &&\iff\\
&\phi_+ \in \ker T_{\overline{\theta}B^n_{z_+}} \subset K_\theta.
\end{aligned}$$

Therefore,

$$K_\theta \ominus \operatorname{span}\{k^\theta_{j,z_+} : j = 0, 1, \cdots, n-1\} = B^n_{z_+} \ker T_{\overline{\theta}B^n_{z_+}}\,, \tag{4.21}$$

$$K_\theta \ominus \operatorname{span}\{k^\theta_{j,z_-} : j = 0, 1, \cdots, n-1\} = B^n_{\frac{1}{\overline{z}_-}} \ker T_{\overline{\theta}B^n_{\frac{1}{\overline{z}_-}}}\,, \tag{4.22}$$

$$K_\theta \ominus \operatorname{span}\{\widetilde{k}^\theta_{j,z_+} : j = 0, 1, \cdots, n-1\} = \ker T_{\overline{\theta}B^n_{z_+}}\,. \tag{4.23}$$

5. Ranges of finite rank truncated Toeplitz operators

In spite of both being defined as compressions of a multiplication operator to a closed subspace of L^2, Toeplitz operators and truncated Toeplitz operators present quite different properties. For instance, the symbol of a Toeplitz operator is unique and the only compact Toeplitz operator is the zero operator. By contrast, truncated Toeplitz operators have non unique symbols (Proposition 2.24) and there are many non zero truncated Toeplitz operators that are compact. In particular, if we allow h_- and h_+ in the symbol

$$g = \overline{\theta}h_- + \theta h_+ \tag{5.1}$$

to be meromorphic, with a finite number of poles in $\mathbb{D}^e$ and $\mathbb{D}$, respectively, then A^θ_g is a non-zero finite rank (and therefore compact) operator ([**8**]). In fact, if the symbol takes the form (5.1) with $h_- \in \overline{H^2_+} + \mathcal{R}, h_+ \in H^2_+ + \mathcal{R}$, where $\mathcal{R}$ denotes the set of all rational functions without poles on $\mathbb{T}$, then we can write

$$A^\theta_g = A^\theta_{\overline{\theta}r_1 + \theta r_2} \quad r_1, r_2 \in \mathcal{R}; \tag{5.2}$$

the class of truncated Toeplitz operators of this form coincides with the class of finite rank truncated Toeplitz operators of type I ([**5**]), i.e., with the class of all finite linear combinations of operators of the form

$$A^\theta_{\frac{\overline{\theta}}{(\overline{z}-\overline{\lambda})^n}} \quad \text{or} \quad A^\theta_{\frac{\theta}{(z-\lambda)^n}} \quad (\lambda \in \mathbb{D}, n \in \mathbb{N})$$

([**1,5,8**]).

For any operator of the form (5.2) we can write

$$A^\theta_g = A^\theta_{\widetilde{g}} \quad \text{with } \widetilde{g} = \overline{\theta}(R_+ + P) + \theta R_-, \tag{5.3}$$

where $R_+ \in H^\infty_+ \cap \mathcal{R}$ and $R_- \in H^\infty_- \cap \mathcal{R}$ vanish at ∞ and P is a polynomial with $P(0) = 0$.

The ranges of truncated Toeplitz operators of type I were studied, in particular, in [**5**], where explicit formulas for their ranks, in terms of their symbols, were obtained, as well as explicit descriptions of the range spaces in certain cases. These results were based on relations between truncated Toeplitz operators and products of Hankel operators.

We use here a different approach, using the results of the previous sections, and viewing the model spaces on which the truncated Toeplitz operators act as Toeplitz kernels of a special type.

We start by decribing the ranges of some particular truncated Toeplitz operators. The following result can be obtained by using [**8**, Theorem 6.1] and taking (4.9) and Proposition 4.2 into account. We present a slightly different proof below.

PROPOSITION 5.1. *Let θ be an inner function, $z_+ \in \mathbb{D}, z_- \in \mathbb{D}^e, n \in \mathbb{N}$ and $\phi_\theta \in K_\theta$. Then*

(1) $P_\theta\left(\dfrac{\theta}{(z-z_+)^n}\phi_\theta\right) = \displaystyle\sum_{j=0}^{n-1} \frac{\phi_\theta^{(j)}(z_+)}{j!}\,\widetilde{k}^\theta_{n-j-1,z_+}$,

(2) $P_\theta\left(\dfrac{\overline{\theta}}{(z-z_-)^n}\phi_\theta\right) = \displaystyle\sum_{j=0}^{n-1} \frac{(\overline{\theta}\phi_\theta)^{(j)}(z_-)}{j!}\,k^\theta_{n-j-1,z_-}$,

(3) $P_\theta(\overline{\theta}z^n\phi_\theta) = \displaystyle\sum_{j=0}^{n-1} \frac{(C_\theta\phi_\theta)^{(j)}(0)}{j!}\,k^\theta_{n-j-1,0}$.

PROOF. (1)

$$
\begin{aligned}
P_\theta \frac{\theta\phi_\theta}{(z-z^+)^n} &= P\theta P^- \overline{\theta}\frac{\theta\phi_\theta}{(z-z^+)^n}\\
&= P\theta P^- \frac{\phi_\theta}{(z-z^+)^n}\\
&= P\theta P^- \left(\frac{\phi_\theta - P^{\phi_\theta}_{n-1}(z^+)}{(z-z^+)^n} + \frac{P^{\phi_\theta}_{n-1}(z^+)}{(z-z^+)^n}\right)\\
&= P\theta \frac{P^{\phi_\theta}_{n-1}(z^+)}{(z-z^+)^n}\\
&= \sum_{j=0}^{n-1} \frac{\phi_\theta^{(j)}(z^+)}{j!} P\frac{\theta}{(z-z^+)^{n-j}}\\
&= \sum_{j=0}^{n-1} \frac{\phi_\theta^{(j)}(z^+)}{j!} P\left(\frac{\theta - P^\theta_{n-j-1}(z^+)}{(z-z^+)^{n-j}} + \frac{P^\theta_{n-j-1}(z^+)}{(z-z^+)^{n-j}}\right)\\
&= \sum_{j=0}^{n-1} \frac{\phi_\theta^{(j)}(z^+)}{j!}\widetilde{k}^\theta_{n-j-1,z^+}
\end{aligned}
$$

(2)

$$\begin{aligned}
P_\theta\ \frac{\overline{\theta}\phi_\theta}{(z-z^-)^n} &= \theta P^- \overline{\theta} P \frac{\overline{\theta}\phi_\theta}{(z-z^-)^n} \\
&= \theta P^- \overline{\theta} \frac{P_{n-1}^{\overline{\theta}\phi_\theta}(z^-)}{(z-z^-)^n} \\
&= \sum_{j=0}^{n-1} \frac{(\overline{\theta}\phi_\theta)^{(j)}(z^-)}{j!} \theta P^- \frac{\overline{\theta}}{(z-z^-)^{n-j}} \\
&= \sum_{j=0}^{n-1} \frac{(\overline{\theta}\phi_\theta)^{(j)}(z^-)}{j!} \theta \frac{\overline{\theta} - P_{n-j-1}^{\overline{\theta}}(z^-)}{(z-z^-)^{n-j}} \\
&= \sum_{j=0}^{n-1} \frac{(\overline{\theta}\phi_\theta)^{(j)}(z^-)}{j!} \frac{1 - P_{n-j-1}^{\overline{\theta}}(z^-)\ \theta}{(z-z^-)^{n-j}} \\
&= \sum_{j=0}^{n-1} \frac{(\overline{\theta}\phi_\theta)^{(j)}(z^-)}{j!} k_{n-j-1,z^-}^{\theta}
\end{aligned}$$

(3) Using the identities $P\phi = \overline{zP^-\overline{z}\overline{\phi}}$, $P^-\phi = \overline{zP\overline{z}\overline{\phi}}$ and $P_\theta\phi = \overline{\theta z P \theta P^- \overline{z}\overline{\phi}}$:

$$\begin{aligned}
P_\theta\ z^n\overline{\theta}\phi_\theta &= \theta\overline{z}\overline{P\theta P^- \overline{z}^n z\theta\overline{\phi_\theta}} \\
&= \theta\overline{z}\overline{P\theta P^- \frac{C_\theta\phi_\theta}{z^n}} \\
&= \theta\overline{z}\overline{P\theta \frac{P_{n-1}^{C_\theta\phi_\theta}(0)}{z^n}} \\
&= \sum_{j=0}^{n-1} \overline{\frac{(C_\theta\phi_\theta)^{(j)}(0)}{j!}} \theta\overline{z}\overline{P\frac{\theta}{z^{n-j}}} \\
&= \sum_{j=0}^{n-1} \overline{\frac{(C_\theta\phi_\theta)^{(j)}(0)}{j!}} \theta\overline{z}\overline{\left(\frac{\theta - P_{n-j-1}^{\theta}(0)}{z^{n-j}}\right)} \\
&= \sum_{j=0}^{n-1} \overline{\frac{(C_\theta\phi_\theta)^{(j)}(0)}{j!}} \overline{z}\frac{1 - \overline{P_{n-j-1}^{\theta}(0)}\theta}{\overline{z}^{n-j}} \\
&= \sum_{j=0}^{n-1} \overline{\frac{(C_\theta\phi_\theta)^{(j)}(0)}{j!}} k_{n-j-1,0}^{\theta}
\end{aligned}$$

□

Let now

$$g = \overline{\theta}(P + R_+) + \theta R_- \tag{5.4}$$

where P is a polynomial of degree $p \in \mathbb{N}$, vanishing at 0, $R_+ \in H_+^\infty \cap \mathcal{R}$ and $R_- \in H_-^\infty \cap \mathcal{R}$, both vanishing at ∞. Let $\mathcal{Z}_- \subset \mathbb{D}^e$ and $\mathcal{Z}_+ \subset \mathbb{D}$ denote the set of poles of R_+ and R_-, respectively, and for each $z_\mp \in \mathcal{Z}_\mp$ let $n_{z_\mp}$ denote its order as a pole of $R_\pm$. With this notation, we have:

THEOREM 5.2. *If* $\dim K_\theta \geq n = p + \sum_{z_+ \in \mathcal{Z}_+} n_{z_+} + \sum_{z_- \in \mathcal{Z}_-} n_{z_-}$ *then* A_g^θ *has rank* n *and we have*

$$\text{range}\, A_g^\theta = \text{span}\, K_0 \oplus \text{span}\, K_- \oplus \text{span}\, K_+ \tag{5.5}$$

where

$$K_0 = \{k_{j,0}^\theta : j = 0, \cdots, p-1\}, \tag{5.6}$$

$$K_- = \{k_{j,z_-}^\theta : z_- \in \mathcal{Z}_-, j = 0, \cdots, n_{z_-} - 1\}, \tag{5.7}$$

$$K_+ = \{\widetilde{k}_{j,z_+}^\theta : z_+ \in \mathcal{Z}_+, j = 0, \cdots, n_{z_+} - 1\}. \tag{5.8}$$

PROOF. Firstly,

$$A_g^\theta = A_{\overline{\theta}P}^\theta + A_{\overline{\theta}R_+}^\theta + A_{\theta R_-}^\theta,$$

and P, R_+ and R_- are linear combinations of

$$z^s, \quad \frac{1}{(z-z_-)^m} \quad \text{and} \quad \frac{1}{(z-z_+)^l},$$

respectively, where $s = 1, \cdots, p$, $z_- \in \mathcal{Z}_-$, $m = 1, \cdots, n_{z_-}$, and $z_+ \in \mathcal{Z}_+$, $l = 1, \cdots, n_{z_+}$.

Therefore, $A_{\overline{\theta}P}^\theta \phi_\theta$, $A_{\overline{\theta}R_+}^\theta \phi_\theta$ and $A_{\theta R_-}^\theta \phi_\theta$ are linear combinations of $P_\theta\, z^s \overline{\theta} \phi_\theta$, $P_\theta\, \frac{\overline{\theta}\phi_\theta}{(z-z_-)^m}$ and $P_\theta\, \theta \frac{1}{(z-z_+)^l}$, respectively, where $\phi_\theta \in K_\theta$. And by Proposition 5.1, we have that,

$$\text{range}\, A_{\overline{\theta}P}^\theta \subset \text{span}\, K_0,$$
$$\text{range}\, A_{\overline{\theta}R_+}^\theta \subset \text{span}\, K_-,$$
$$\text{range}\, A_{\theta R_-}^\theta \subset \text{span}\, K_+$$

Thus,

$$\text{range}\, A_g^\theta \subset \text{span}\, K_0 \oplus \text{span}\, K_- \oplus \text{span}\, K_+$$

On the other hand, by Corollary 3.7, there exists $\phi_\theta \in K_\theta$ such that $C_\theta \phi_\theta$ has a zero of order $p-1$ at 0 and a zero of order greater or equal to n_{z_-} at each point $\frac{1}{\overline{z}_-}$ with $z_- \in \mathcal{Z}_-$ and ϕ_θ has a zero of order greater or equal to n_{z_+} at each $z_- \in \mathcal{Z}_-$. Therefore, by (5.4) and Proposition 5.1,

$$A_g^\theta \phi_\theta = a_p \frac{\overline{(C_\theta\phi_\theta)^{(p-1)}(0)}}{(p-1)!} k_{0,0}^\theta$$

with $(C_\theta\phi_\theta)^{(p-1)}(0) \neq 0$, where $P(z) = a_0 + a_1 z + \cdots + a_p z^p$, $a_p \neq 0$, and we conclude that $k_{0,0}^\theta \in \text{range}\, A_g^\theta$. If $p > 1$, we now choose $\phi_\theta \in K_\theta$ such that ϕ_θ has a zero of order greater or equal to n_{z_+} at each $z_+ \in \mathcal{Z}_+$, $C_\theta\phi_\theta$ has a zero of order $p-2$ at 0 and a zero of order greater or equal to n_{z_-} at each point $\frac{1}{\overline{z}_-}$ with $z_- \in \mathcal{Z}_-$. Then

$$A_g^\theta \phi_\theta = \left(a_p \frac{\overline{(C_\theta\phi_\theta)^{(p-1)}(0)}}{(p-1)!} + a_{p-1} \frac{\overline{(C_\theta\phi_\theta)^{(p-2)}(0)}}{(p-2)!} \right) k_{0,0}^\theta + a_p \frac{\overline{(C_\theta\phi_\theta)^{(p-2)}(0)}}{(p-2)!} k_{1,0}^\theta,$$

where $(C_\theta\phi_\theta)^{(p-2)}(0) \neq 0$ and $a_p \neq 0$. Since $k_{0,0}^\theta \in \text{range}\, A_g^\theta$, we conclude that $k_{1,0}^\theta \in \text{range}\, A_g^\theta$. Analogously, using Proposition 5.1, Corollary 3.5, Corollary 3.6 and Corollary 3.7, we conclude that all elements of $K_0 \cup K_+ \cup K_-$ belong to the range of A_g^θ. □

It follows from Theorem 5.2 that, writing the symbol in the form (5.4), the rank of A_g^θ depends only on the degree of P and the numbers of poles of R_+ and R_-, counting their multiplicities, while the range depends only on the set of poles and the respective orders, including ∞ if $P \neq 0$.

An alternative description of the range of A_g^θ is the following.

COROLLARY 5.3. *With the same assumptions as in Theorem 5.2,*

$$\operatorname{range} A_g^\theta = K_\theta \ominus \left(K_\theta \cap B_+ H_+^2 \cap B_- \theta H_-^2\right)$$

where $B_+ = z^p \prod_{z_- \in \mathcal{Z}_-} B_{\frac{1}{\overline{z_-}}}^{n_{z_-}}$, $B_- = \prod_{z_+ \in \mathcal{Z}_+} \overline{B_{z_+}^{n_{z_+}}}$.

In particular, if $g = \overline{\theta}(P + R_+)$ or $g = \theta R_-$ we recover the result from [**5**, Theorem 11].

Using (4.18) - (4.20) we can also write these relations in terms of Toeplitz kernels, as follows.

COROLLARY 5.4. *With the same assumptions as in Theorem 5.2,*

$$\operatorname{range} A_g^\theta = K_\theta \ominus B_+ \ker T_{\overline{\theta} B_+ \overline{B_-}}.$$

EXAMPLE 5.5. Let $w_+ \in \mathbb{D}$, $w_- \in \mathbb{D}^e$ and $g = \bar{z}^n \frac{A}{z - w_-} + C\bar{z}^{n-1} + z^n \frac{B}{z - w_+}$ where $n \geq 3$ and $A, B, C \in \mathbb{C} \setminus \{0\}$. By Theorem (4.2), $A_g^{z^n}$ has rank 3 and range equal to *span*$\{k_0^{z^n}, \tilde{k}_{w_+}^{z^n}, k_{w_-}^{z^n}\}$. An independent verification of this result, which also has an interest of its own, is provided by describing the kernel of the adjoint $A_{\bar{g}}^{z^n} = A_{\tilde{g}}^{z^n}$ with

$$\tilde{g} = \bar{z}^n \frac{B_1}{z - z_-} + C_1 z^{n-1} + z^n \frac{A_1}{z - z_+}$$

where $z_\pm = (\overline{w_\mp})^{-1}$, $z_+ \in \mathbb{D}$, $z_- \in \mathbb{D}^e$ and A_1, B_1, C_1 are nonzero complex numbers. We have that $\ker A_{\tilde{g}}^{z^n}$ consists of all $\phi_1^+ \in K_{z^n}$ such that $P_{z^n}(\tilde{g}\phi_1^+) = 0$, i.e., such that ([**3**])

$$\bar{z}^n \phi_1^+ = \phi_1^- \tag{5.9}$$

$$\tilde{g}\phi_1^+ + z^n \phi_2^+ = \phi_2^- \tag{5.10}$$

where $\phi_2^+ \in H_+^2$, $\phi_1^-, \phi_2^- \in H_-^2$. From (5.10) we have that

$$z^{n-1}[A_1 z(z - z_-)\phi_1^+ + C_1(z - z_-)(z - z_+)\phi_1^+ + z(z - z_+)(z - z_-)\phi_2^+]$$
$$= -B_1(z - z_+)\phi_1^- + (z - z_-)(z - z_+)\phi_2^-$$

where we took (5.9) into account. By an easy generalization of Liouville's Theorem, and taking into account that ϕ_1^- and ϕ_2^- tend to zero at ∞, we conclude that both sides of the previous equality are equal to a polynomial of degree at most 1, so

$$z^{n-1}[A_1 z(z - z_-)\phi_1^+ + C_1(z - z_-)(z - z_+)\phi_1^+ + z(z - z_+)(z - z_-)\phi_2^+] = a + bz \tag{5.11}$$

$$-B_1(z - z_+)\phi_1^- + (z - z_-)(z - z_+)\phi_2^- = a + bz. \tag{5.12}$$

Since the left hand side of (5.11) has a zero of order greater or equal to 2 at the point 0, we must have $a = b = 0$. Therefore

$$[A_1 z(z - z_-)\phi_1^+ + C_1(z - z_-)(z - z_+)\phi_1^+ + z(z - z_+)(z - z_-)\phi_2^+] = 0$$

and it follows that we must have

$$\phi_1^+(0) = 0 \tag{5.13}$$

$$\phi_1^+(z_+) = 0 \tag{5.14}$$

while from (5.12) it follows that

$$\phi_1^-(z_-) = 0. \tag{5.15}$$

Since we can write $\phi_1^+(z) = K_0 + K_1 z + ... + K_{n-1} z^{n-1}$ and $\phi_1^-(z) = \bar{z}^n \phi_1^+ = \frac{K_0}{z^n} + \frac{K_1}{z^{n-1}} + ... + \frac{K_{n-1}}{z}$ with $K_0, K_1, ..., K_{n-1} \in \mathbb{C}$ (and with $n \geq 3$), the conditions (5.13)-(5.15) are equivalent to

$$\begin{aligned} K_0 &= 0 \\ K_1 + ... + K_{n-1} z_+^{n-2} &= 0 \\ \frac{K_1}{z_-^{n-2}} + ... + K_{n-1} &= 0 \end{aligned}$$

or, equivalently,

$$\begin{aligned} K_0 &= 0 \\ K_1 + ... + K_{n-1} z_+^{n-2} &= 0 \\ K_1 + ... + K_{n-1} z_-^{n-2} &= 0 \end{aligned}$$

where $z_+ \neq z_-$. We conclude that $\dim \ker A_{\bar{g}}^{z^n} = n - 3$, $\dim \operatorname{range} A_g^{z^n} = 3$ and the range of $A_g^{z^n}$ is the orthogonal complement of the subspace of K_θ whose elements are defined by the conditions (5.13)-(5.15), i.e.

$$\begin{aligned} &K_{z^n} \ominus \{\phi \in K_{z^n} : \phi(0) = 0, \phi(z_+) = 0, (\overline{z^n}\phi)(z_-) = 0\} = \\ &K_{z^n} \ominus (K_{z^n} \cap z B_{z_+}^{z^n} B_{\frac{1}{\overline{z_-}}}^{z^n} K_{z^n}) = \operatorname{span}\{k_0^{z^n}, \tilde{k}_{z_+}^{z^n}, k_{z_-}^{z^n}\}. \end{aligned}$$

Acknowledgments

The authors would like to thank Jonathan R. Partington for helpful discussions on this topic. The work of the first author was partially supported by FCT/Portugal through UID/MAT/04459/2013.

References

[1] R. V. Bessonov, *Truncated Toeplitz operators of finite rank*, Proc. Amer. Math. Soc. **142** (2014), no. 4, 1301–1313, DOI 10.1090/S0002-9939-2014-11861-2. MR3162251

[2] M. C. Câmara, M. T. Malheiro, and J. R. Partington, *Model spaces and Toeplitz kernels in reflexive Hardy space*, Oper. Matrices **10** (2016), no. 1, 127–148, DOI 10.7153/oam-10-09. MR3460059

[3] M. C. Câmara and J. R. Partington, *Near invariance and kernels of Toeplitz operators*, J. Anal. Math. **124** (2014), 235–260, DOI 10.1007/s11854-014-0031-8. MR3286053

[4] S. R. Garcia, J. Mashreghi, and W. T. Ross, *Introduction to model spaces and their operators*, Cambridge Studies in Advanced Mathematics, vol. 148, Cambridge University Press, Cambridge, 2016. MR3526203

[5] C. Gu and D.-O Kang, *Rank of truncated Toeplitz operators*, Complex Anal. Oper. Theory **11** (2017), no. 4, 825–842, DOI 10.1007/s11785-016-0571-2. MR3626676

[6] A. Hartmann and M. Mitkovski, *Kernels of Toeplitz operators*, Recent progress on operator theory and approximation in spaces of analytic functions, Contemp. Math., vol. 679, Amer. Math. Soc., Providence, RI, 2016, pp. 147–177. MR3589674

[7] D. Sarason, *Kernels of Toeplitz operators*, Toeplitz operators and related topics (Santa Cruz, CA, 1992), Oper. Theory Adv. Appl., vol. 71, Birkhäuser, Basel, 1994, pp. 153–164. MR1300218

[8] D. Sarason, *Algebraic properties of truncated Toeplitz operators*, Oper. Matrices **1** (2007), no. 4, 491–526, DOI 10.7153/oam-01-29. MR2363975

Center for Mathematical Analysis, Geometry, and Dynamical Systems, Departamento de Matemática, Instituto Superior Técnico, 1049-001 Lisboa, Portugal
Email address: `ccamara@math.ist.utl.pt`

Instituto Superior Técnico, 1049-001 Lisboa, Portugal
Email address: `carlos.carteiro@tecnico.ulisboa.pt`

Contemporary Mathematics
Volume **737**, 2019
https://doi.org/10.1090/conm/737/14858

Commuting maps with the Mean Transform

F. Chabbabi and M. Mbekhta

Abstract. Given a bounded operator $T \in \mathcal{B}(H)$, H a (complex) Hilbert space, let $T = V|T|$ be the polar decomposition of T. The Mean transform of the operator T is defined by $\mathcal{M}(T) := \frac{1}{2}(V|T| + |T|V)$. In the present paper, we give a complete characterization of the bijective maps $\Phi : B(H) \to \mathcal{B}(K)$, where H, K are Hilbert spaces, that commutes with the mean transform under product. More precisely we show that :

$$\mathcal{M}(\Phi(A)\Phi(B)) = \Phi(\mathcal{M}(AB)) \text{ for all } A, B \in \mathcal{B}(H)$$

if and only if

$$\Phi(T) = UTU^* \text{ for every } T \in \mathcal{B}(H),$$

where $U : H \to K$ is a unitary or anti-unitary operator.

1. Introduction

Let H and K be two complex Hilbert spaces and let $\mathcal{B}(H, K)$ be the Banach space of all bounded linear operators from H into K. In the case $K = H$, $\mathcal{B}(H, H)$ is simply denoted by $\mathcal{B}(H)$ and it is a Banach algebra.

For an arbitrary operator $T \in \mathcal{B}(H, K)$, we denote by $\mathcal{R}(T)$, $\mathcal{N}(T)$ and T^* the range, the null subspace and the adjoint operator of T respectively. For $T \in \mathcal{B}(H)$, the spectrum of T is denoted by $\sigma(T)$.

An operator $T \in \mathcal{B}(H, K)$ is a *partial isometry* when T^*T is an orthogonal projection (or, equivalently $TT^*T = T$). In particular T is an *isometry* if $T^*T = I$, and T is *unitary* if it is a surjective isometry. As usual, we denote the module of $T \in \mathcal{B}(H)$ by $|T| = (T^*T)^{1/2}$, and $T = V|T|$ is the unique polar decomposition of T, where V is a partial isometry satisfying $\mathcal{N}(V) = \mathcal{N}(T)$. From the polar decomposition, the λ-Aluthge transform is defined in ([**1, 13**]), by

$$\Delta(T) = |T|^{\lambda} V |T|^{1-\lambda}, \quad \lambda \in [0, 1].$$

The Aluthge transform has been well studied, it is a good tool for studying other operators (see [**3, 4, 7–9, 17**]). In the same way, the mean transform of the operator T was recently introduced in [**12**], by

$$\mathcal{M}(T) := \frac{1}{2}(V|T| + |T|V).$$

2010 *Mathematics Subject Classification.* 47A05, 47A10, 47B49, 46L40.

Key words and phrases. normal, quasi-normal operators, polar decomposition, mean transform.

The mean transform has been studied by many authors, it has nice properties, for example see [**5**, **10**, **12**].

The paper is organized as follows. In Section 2, we provide necessary definitions, we fix main notation, and we recall the important proprieties of polar decomposition, which plays an important role in this paper. Also we give known or new results on the mean transform of operators.

In Section 3, we give a complete characterization of bijective linear maps that commute with the mean transformation. Precisely, we prove the following theorem :

THEOREM 1.1. Let H and K be two Hilbert spaces of dimensional greater than 2 , and let $\Phi : \mathcal{B}(H) \to B(K)$ be a bijective linear map. The following assertions are equivalent:

(1) $\mathcal{M}(\Phi(T)) = \Phi(\mathcal{M}(T))$ for all $T \in \mathcal{B}(H)$;
(2) There exist a unitary operator U and $0 \neq c \in \mathbb{C}$ such that

$$\Phi(T) = cUTU^*, \quad \forall T \in \mathcal{B}(H).$$

In tsection 4, we give the form of bijective maps $\Phi : \mathcal{B}(H) \to \mathcal{B}(K)$, (not assumed linear) which satisfies the following equality :

$$\mathcal{M}(\Phi(A)\Phi(B)) = \Phi(\mathcal{M}(AB)) \text{ for all } A, B \in \mathcal{B}(H).$$

Precisely, we prove the following theorem :

THEOREM 1.2. Let H and K be two complex Hilbert space, with $\dim H \geq 3$. Let $\Phi : \mathcal{B}(H) \to \mathcal{B}(K)$ be a bijective map. Then

$$\mathcal{M}(\Phi(A)\Phi(B)) = \Phi(\mathcal{M}(AB)) \text{ for all } A, B \in \mathcal{B}(H)$$

if and only if there exists a unitary or anti-unitary operator $U : H \to K$, such that

$$\Phi(T) = UTU^* \qquad \text{for every } T \in \mathcal{B}(H).$$

Observe that, even if the hypothesis on the map Φ is purely algebraic, the conclusion gives automatically the continuity of the map. Also, the linearity of Φ is not assumed, we get it automatically.

2. The Mean Transform

In this section we collect several results on the mean transform of an operator. These results will be used in the proof of the main theorems.

The polar decomposition of bounded operators on a Hilbert space, plays a very important role in the definition of the Aluthge and mean transformations of operators. So we start, this section, by recalling some known results on the polar decomposition (see [**6**, **11**]). If $T = T = V|T|$ is the polar decomposition of T, then:

$$V : \overline{\mathcal{R}(|T|)} = \mathcal{N}(T)^{\perp} :\longrightarrow \overline{\mathcal{R}(T)} \text{ isometry and } V_{|\mathcal{N}(T)} = 0.$$

It follows that

$$V^*V = P_{\overline{\mathcal{R}(|T|)}} = P_{\overline{\mathcal{R}(T^*)}} \quad VV^* = P_{\overline{\mathcal{R}(|T^*|)}} = P_{\overline{\mathcal{R}(T)}}$$

Therefore

$$V \text{ is unitary} \iff T \text{ and } T^* \text{ are injective.}$$

And

$$V^*VT^* = T^*, \;\; V^*V|T| = |T|, \;\; VV^*T = T \text{ and } VV^*|T^*| = |T^*|.$$

Other important properties

(1) $T = V|T| = |T^*|V = VT^*V$;

(2) $T^* = V^*|T^*| = |T|V^* = V^*TV^*$;

(3) $|T| = V^*T = T^*V = V^*|T^*|V$;

(4) $|T^*| = VT^* = TV^* = V|T|V^*$.

Moreover, according to (1), we have:

$|T^*|V = V|T|$. Hence for any polynomial $P \in \mathbb{C}[X]$, we have $P(|T^*|)V = VP(|T|)$. Using the Stone-Weierstrass' theorem, we deduce that for all $\lambda \in [0,1]$

$$T = |T^*|^\lambda V|T|^{1-\lambda} = |T^*|^{1-\lambda}V^*|T|^\lambda \text{ and } T^* = |T|^\lambda V|T^*|^{1-\lambda} = |T|^{1-\lambda}V^*|T^*|^\lambda.$$

Now we recall some basic definitions to be used later. An operator $T \in \mathcal{B}(H)$ is normal if $T^*T = TT^*$, and it is quasi-normal if it commutes with T^*T (i.e. $TT^*T = T^*T^2$), or equivalently if $|T|$ and V commute where $T = V|T|$ is a polar decomposition of T. In finite dimensional spaces every quasi-normal operator is normal. It is easy to see that if T is quasi-normal, then T^2 is also quasi-normal, but the converse is false. This can be shown by considering nonzero nilpotent operators. Also, quasi-normal operators are exactly the fixed points of the mean transform (see [**5**]).

$$T \text{ quasi-normal} \iff T|T| = |T|T \iff \mathcal{M}(T) = T. \tag{2.1}$$

Contrary to what happens with the Aluthge transform, the mean transform does depend on the polar decomposition of the given operator.

For example, consider $T = \begin{pmatrix} 0 & 1 \\ 0 & 0 \end{pmatrix}$ acting on $\mathbb{C}^2$. The canonical polar decomposition of T is $T = V|T|$, where $|T| = \sqrt{T^*T} = \begin{pmatrix} 0 & 0 \\ 0 & 1 \end{pmatrix}$ and $V = \begin{pmatrix} 0 & 1 \\ 0 & 0 \end{pmatrix}$.

On the other hand, we can also write $T = U_{max}|T|$, where $U_{max} = \begin{pmatrix} 0 & 1 \\ 1 & 0 \end{pmatrix}$ is unitary. This is the so-called *maximal* polar decomposition of T, since the partial isometry is unitary. In this case,

$$U_{max}|T| + |T|U_{max} = \begin{pmatrix} 0 & 1 \\ 1 & 0 \end{pmatrix} \neq V|T| + |T|V = \begin{pmatrix} 0 & 1 \\ 0 & 0 \end{pmatrix},$$

which shows that the mean transform depends on the polar decomposition. In what follows, we will always use the canonical polar decomposition when dealing with the mean transform.

PROPOSITION 2.1. ([**5**]) Let $T \in \mathcal{B}(H)$ be an arbitrary operator. Then we have

$$\mathcal{N}(\mathcal{M}(T)) = \mathcal{N}(T).$$

In particular $\mathcal{M}(T) = 0$ if and only if $T = 0$.

PROPOSITION 2.2. ([**5**]) Let $T \in \mathcal{B}(H)$. Then the following properties hold.

(*i*) For all $\alpha \in \mathbb{C}$, $\mathcal{M}(\alpha T) = \alpha\mathcal{M}(T)$.

(*ii*) For every unitary or anti-unitary operator $U : H \to H$, we have

$$\mathcal{M}(UTU^*) = U\mathcal{M}(T)U^*.$$

(*iii*) $\mathcal{M}(T) = I \iff T = I$.

THEOREM 2.1. ([**5**]) Let $T \in \mathcal{B}(H)$. Then the following statements are equivalent.

(i) T is invertible.
(ii) $\mathcal{M}(T)$ is invertible and $\mathcal{R}(T)$ is closed.

REMARK 2.1. In Theorem 2.1 (ii), the condition "$\mathcal{R}(T)$ is closed" is required; without it, the reverse implication is false, as shown by the following example.

EXAMPLE 2.1. Let us denote by $(e_n)_{n\in\mathbb{Z}}$ the canonical basis of $\ell^2(\mathbb{Z})$, and by $T : \ell^2(\mathbb{Z}) \to \ell^2(\mathbb{Z})$ the weighted bilateral shift defined by $Te_n = \alpha_n e_{n+1}$ for all $n \in \mathbb{Z}$, where

$$\alpha_n = \begin{cases} 1 & \text{if } n \text{ even} \\ \frac{1}{n^2} & \text{if } n \text{ odd} . \end{cases}$$

The mean transform $\mathcal{M}(T)$ is also a weighted shift, and we have $\mathcal{M}(T)e_n = \widehat{\alpha}_n e_{n+1}$ for $n \in \mathbb{Z}$, where

$$\widehat{\alpha}_n = \frac{\alpha_n + \alpha_{n+1}}{2} = \begin{cases} \dfrac{1 + \frac{1}{(n+1)^2}}{2} & \text{if } n \text{ even} \\ \dfrac{1 + \frac{1}{n^2}}{2} & \text{if } n \text{ odd} . \end{cases}$$

Clearly, $Te_{2n+1} \underset{n\to\infty}{\longrightarrow} 0$, and therefore the operator T is not invertible.

On the other hand, we have $1 \geq \widehat{\alpha}_n \geq \frac{1}{2}$ for all $n \in \mathbb{Z}$, and from this it follows that $\mathcal{M}(T)$ is invertible. □

REMARK 2.2. In general we have:
(1) $\sigma(T) \neq \sigma(\mathcal{M}(T))$ (see [**12**]);
(2) $(\mathcal{M}(T))^{-1} \neq \mathcal{M}(T^{-1})$ (see [**10**]).

An idempotent self adjoint operator $P \in \mathcal{B}(H)$ is called an orthogonal projection. Clearly quasi-normal idempotents are orthogonal projections.
Two projections $P, Q \in \mathcal{B}(H)$ are said to be orthogonal if $PQ = QP = 0$, and in this case we write $P \perp Q$. Note that

$$P \perp Q \quad \Longleftrightarrow \quad P + Q \text{ orthogonal projection.}$$

A partial ordering between orthogonal projections is defined as follows:

$$Q \leq P \quad \text{if} \quad PQ = QP = Q.$$

We also have $Q \leq P \iff PQP = Q$.

For $x, y \in H$ we denote by $x \otimes y$ the rank one operator (or 0) defined by

$$(x \otimes y)u = < u, y > x \quad \text{for} \quad u \in H.$$

Every rank one operator has the previous form, and $x \otimes y$ is an orthogonal projection if and only if $x = y$ and $\|x\| = 1$.

The following proposition can be found in [**5**].

PROPOSITION 2.3. ([**5**]) *Let $x, y \in H$ be two nonzero vectors. Let $T = x \otimes y$, then*

$$\mathcal{M}(T) = \mathcal{M}(x \otimes y) = \frac{1}{2}(x + \frac{< x, y >}{\|y\|^2} y) \otimes y.$$

Next lemma gives a characterization of the nilpotent operator of order two.

LEMMA 2.1. Let $T \in \mathcal{B}(H)$. Then

$$\mathcal{M}(T) = \frac{T}{2} \text{ if and only if } T^2 = 0.$$

PROOF. Let $T = V|T|$ be the polar decomposition of T. First suppose that $\mathcal{M}(T) = \frac{T}{2}$. Then $\frac{T}{2} = \frac{1}{2}(T + |T|V)$ and hence $|T|V = 0$. It then fowolls that $T^2 = V|T|V|T| = 0$.

Conversely, suppose that $T^2 = 0$. Then $V|T|V|T| = T^2 = 0$. Hence $V^*V|T|V|T| = 0$, which implies $|T|V|T| = 0$ (since V^*V is the projection onto $\overline{\mathcal{R}(|T|)}$). Whence $|T|V$ vanishes on $\overline{\mathcal{R}(|T|)}$. Furthermore $\mathcal{N}(V) = \mathcal{N}(|T|)$, from which one concludes $|T|V$ vanishes on $\mathcal{N}(|T|)$. Consequently, $|T|V = 0$ and thus $\mathcal{M}(T) = \frac{1}{2}(T + |T|V) = \frac{T}{2}$. □

PROPOSITION 2.4. with Let P, Q be two orthogonal projections, we have the following assertion are equivalent :

(1) $PQ = QP = P$ (we denote it by $P \leq Q$),
(2) $\mathcal{M}(PQ) = P$,
(3) $\mathcal{M}(QP) = P$.

PROOF. Clearly, (1) implies both (2) and (3). First, we show that (2) $\Rightarrow$ (1). Indeed, let $PQ = V_{PQ}|PQ|$ be the polar decomposition of PQ, where V_{PQ} the associate partial isometry. We have $V_{PQ}^* = V_{QP}$.

From the assumption we have $\frac{1}{2}(PQ + |PQ|V_{PQ}) = P$. It follows that

$$\frac{1}{2}(|PQ| + V_{PQ}^*|PQ|V_{PQ}) = \frac{1}{2}(V_{PQ}^*PQ + V_{PQ}^*|PQ|V_{PQ}) = V_{PQ}^*P.$$

Hence V_{PQ}^*P is a positive operator. In particular we have $V_{PQ}^*P = PV_{PQ}$. Therefore

$$V_{PQ}^*P = PV_{PQ}^*P = PV_{PQ}P = V_{PQ}P. \tag{2.2}$$

In the other hand, we have the following

$$\mathcal{R}(PQ) \subseteq \mathcal{R}(P) \quad \text{and} \quad \mathcal{R}(QP) = \mathcal{R}(V_{PQ}^*) = V_{PQ}^*(\mathcal{R}(PQ)) \subseteq V_{PQ}^*(\mathcal{R}(P)).$$

By (2.2), we get that

$$\mathcal{R}(QP) \subseteq V_{PQ}^*(\mathcal{R}(P)) \subseteq \mathcal{R}(P).$$

From the last inclusion and the assumption we get the equality

$$PQP = QP = PQ = P.$$

With the similar arguments, we show (3) $\Rightarrow$ (1). □

THEOREM 2.2. Let $T \in \mathcal{B}(H)$, the following assertions are equivalent:

(1) $\mathcal{M}(TR) = \mathcal{M}(RT)$ for all $R \in \mathcal{B}(H)$,
(2) $\mathcal{M}(TP) = \mathcal{M}(PT)$ for all rank one projections P,
(3) $T = \alpha I$, $\alpha \in \mathbb{C}$.

PROOF. Clearly, (3) $\Rightarrow$ (1) $\Rightarrow$ (2). We now show that (2) implies (3). Indeed, let $x \in H$ be a unit vector, and $P = x \otimes x$. To prove the statement, we need to show that T^*x and x are collinear. First, if $T^*x = 0$ then T^*x and x are obviously collinear. Hence, we assume that $T^*x \neq 0$, and we have

$$\mathcal{M}(Tx \otimes x) = \mathcal{M}(x \otimes T^*x).$$

Proposition 2.3 implies that

$$\frac{1}{2}(Tx + < Tx, x > x) \otimes x = \frac{1}{2}(x + \frac{< Tx, x >}{\|T^*x\|^2} T^*x) \otimes T^*x.$$

This shows that T^*x and x are linearly dependent for every $x \in H$. Consequently $T = \alpha I$, $\alpha \in \mathbb{C}$. This complete the proof. □

3. Commuting maps with the Mean Transform

The following theorem gives a complete description of bijective linear maps that commute with the Mean transform.

THEOREM 3.1. Let H and K be two Hilbert spaces such that $\dim H \geq 2$, and let $\Phi : \mathcal{B}(H) \to B(K)$ be a bijective linear map. Then the following assertions are equivalent:

(1) $\mathcal{M}(\Phi(T)) = \Phi(\mathcal{M}(T))$ for all $T \in \mathcal{B}(H)$;
(2) There exist a unitary operator U and $0 \neq c \in \mathbb{C}$ such that

$$\Phi(T) = cUTU^*, \quad \forall T \in \mathcal{B}(H). \tag{3.1}$$

As a direct consequence we have the following corollary :

COROLLARY 3.1. Let $\Phi : \mathcal{B}(H) \to B(K)$ be a bijective linear map. The following assertions are equivalent :

(1) Φ commutes with the Mean transform;
(2) Φ commutes with the λ-Aluthge transform;
(3) Φ preserves the set of quasi-normal operators;
(4) There exist a unitary operator U and $0 \neq c \in \mathbb{C}$ such that

$$\Phi(T) = cUTU^*, \quad \forall T \in \mathcal{B}(H).$$

To prove Theorem 3.1, we need the following lemma.

LEMMA 3.1. Let $T = U|T| \in \mathcal{B}(H)$ be an invertible operator, then : T is normal if and only if $|T|U$ is normal.
Moreover, in this case, we have $T = |T|U$.

PROOF. Since T is invertible, U is unitary. Now If T is normal, then T is quasi-normal. Hence $|T|U = U|T| = T$ is normal.
Conversely, If $|T|U$ is normal, then

$$T^*T = |T|^2 = |T|UU^*|T| = U^*|T|^2U = U|T|^2U^* = TT^*.$$

□

PROOF. The implication (2) $\Rightarrow$ (1) is a direct consequence of Proposition 2.2.We need to prove the other implication. Let us consider the following cases:

Case 1: Assume H is infinite dimensional. First we show that Φ preserves the quasi-normal operator. Let $T \in \mathcal{B}(H)$ be a quasi-normal operator. Then by (2.1) we have

$$\mathcal{M}(\Phi(T)) = \Phi(\mathcal{M}(T)) = \Phi(T).$$

Again by (2.1) the operator $\Phi(A)$ is quasi-normal. By [**2**, Proposition 7], Φ is necessarily of the form (3.1).

Case 2: Now, suppose that H is of finite dimensional. Therefore, K must be also of the same dimensional as H. We know that Φ preserves the quasi-normal elements. Since in finite dimensional, every quasi-normal operator is a normal operator, then Φ preserves the normal operators. On the other hand, by Lemma 2.1 Φ preserves also the square zero operator in both directions.

The form of all linear bijections on that space which preserve square-zero matrices is known (see [**2, 14**]). It implies that there are a nonzero scalar c and an bijective linear operator $A : H \to K$ such that Φ is either of the form

$$\Phi(T) = cATA^{-1} + dTr(T)I, \ \ \forall T \in \mathcal{B}(H),$$

or of the form

$$\Phi(T) = cAT^tA^{-1} + dTr(T)I, \ \ \forall T \in \mathcal{B}(H),$$

where T^t the transpose of the operator T in a fixed orthonormal basis $(e_n)_1^n$ of H.

Since Φ preserves the set of normal operators, then A must be a scalar multiple of a unitary operator. Without loss of generality, we can assume that $A = I$ and $c = 1$. Hence we have,

$$\Phi(T) = T + dTr(T)I, \ \ \forall T \in \mathcal{B}(H), \tag{3.2}$$

or

$$\Phi(T) = T^t + dTr(T)I, \ \ \forall T \in \mathcal{B}(H). \tag{3.3}$$

In order to complete the proof we have to show $d = 0$ and Φ has only form (3.2).

Let $x = \sum_{i=1}^n a_i e_i$ and $y = \sum_{i=1}^n b_i e_i$ be two linearly independent unit vectors of H such that $a_i > 0$ and $b_i > 0$ for all $i = 1, \cdots, n$ and $T = x \otimes y$.

Suppose that Φ has the form (3.2).

Since $\mathcal{M}(T) = \frac{1}{2}(x + <x, y> y) \otimes y$, we obtain that

$$\Phi(\mathcal{M}(T)) = \frac{1}{2}(x + <x, y> y) \otimes y + d <x, y> I. \tag{3.4}$$

On the other hand, let $\Phi(T) = x \otimes y + d <x, y> I = V_{x,y}|x \otimes y + d <x, y> I|$ be the polar decomposition of $\Phi(T)$. We have

$$\mathcal{M}(\Phi(T)) = \frac{1}{2}(x \otimes y + d <x, y> I + |x \otimes y + d <x, y> I| V_{x,y}). \tag{3.5}$$

From equations (3.4), (3.5) and $\Phi(\mathcal{M}(T)) = \mathcal{M}(\Phi(T))$, we deduce that

$$\begin{aligned} &\frac{1}{2}(x + <x, y> y) \otimes y + d <x, y> I \\ &\qquad = \frac{1}{2}(x \otimes y + d <x, y> I + |x \otimes y + d <x, y> I| V_{x,y}). \end{aligned} \tag{3.6}$$

It follows that

$$<x, y> (y \otimes y + dI) = |x \otimes y + d <x, y> I| V_{x,y} \tag{3.7}$$

In particular, the operator on the right is normal.

First, we prove $d \in \{0, -1\}$. Assume by way of contradiction that $d \notin \{0, -1\}$. Therefore, the operator $y \otimes y + dI$ is invertible. Hence $|x \otimes y + d < x, y > I|V_{x,y}$ is invertible and normal. By Lemma 3.1, we have

$$|x \otimes y + d < x, y > I|V_{x,y} = V_{x,y}|x \otimes y + d < x, y > I| = x \otimes y + d < x, y > I.$$

From (3.7), we have $x \otimes y =< x, y > y \otimes y$, which is in contradiction with the fact that x and y are linearly independent. So $d = 0$ or $d = -1$.

Now, assume that $d = -1$. Hence from (3.7), we obtain

$$< x, y > (y \otimes y - I) = |x \otimes y- < x, y > I|V_{x,y} \tag{3.8}$$

Observe that $x \in \mathcal{N}(x \otimes y- < x, y > I)$. It follows that $x \in \mathcal{N}(|x \otimes y- < x, y > I|V_{x,y})$ (since $\mathcal{N}(x \otimes y- < x, y > I) = \mathcal{N}(V_{x,y})$). From (3.8), $x \in \mathcal{N}(y \otimes y - I)$. Hence x and y are linearly dependent, contrary to our hypothesis. We conclude that $d = 0$.

With the same arguments, we get $d = 0$ in the case when Φ has the form (3.3). Consequently, Φ is the identity or the transposition map.

We seek a contradiction. Suppose that $\Phi(R) = R^t$ holds for all R. Using the rank one operator $T = x \otimes y$, then we have : $\Phi(T) = y \otimes x$ and thus

$$\mathcal{M}(\Phi(T)) = \frac{1}{2}(y+ < y, x > x) \otimes x, \tag{3.9}$$

and,

$$\Phi(\mathcal{M}(T)) = \frac{1}{2}y \otimes (x+ < x, y > y). \tag{3.10}$$

Since $\Phi(\mathcal{M}(T)) = \mathcal{M}(\Phi(T))$, we have $(y+ < y, x > x) \otimes x = y \otimes (x+ < x, y > y)$, which contradicts the fact that x and y are linearly independent. This complete the proof. □

4. Commuting maps with the Mean Transform under product

We start this section by establishing some results on the maps Φ satisfying (4.1). These results will be used in the proof of Theorem 1.2, which will be given at the end of this paragraph

Let $\Phi : \mathcal{B}(H) \to \mathcal{B}(K)$ be a bijective map. We consider the following equality

$$\mathcal{M}(\Phi(A)\Phi(B)) = \Phi(\mathcal{M}(AB)) \text{ for all } A, B \in \mathcal{B}(H). \tag{4.1}$$

As an immediate consequence of (4.1) and Lemma 2.1 and Theorem 2.2, we derive the following result :

LEMMA 4.1. If Φ is a bijective map that satisfies (4.1), then

(i) $\Phi(0) = 0$,
(ii) $\Phi(I) = I$.
(iii) $\Phi(\alpha I) = h(\alpha)I$, $\alpha \in \mathbb{C}$, where h is a bijection from $\mathbb{C}$ to it self.

PROOF. (i). Since Φ is onto, there is $A \in \mathcal{B}(H)$ such that $\Phi(A) = 0$. By (4.1), we have

$$0 = \mathcal{M}((\Phi(0)\Phi(A)) = \Phi(0).$$

Therefore, $\Phi(0) = 0$.
(ii). For simplicity, let us denote $T = \Phi(I)$. Pick $y \in K$ such that $Ty = 0$. Since Φ is onto, there exists $B \in \mathcal{B}(H)$ such that $\Phi(B) = y \otimes y$. By (4.1) we get

$$0 = \mathcal{M}(Ty \otimes y) = \mathcal{M}(\Phi(I)\Phi(B)) = \Phi(\mathcal{M}(B)).$$

Since Φ is bijective and $\Phi(0) = 0$, we have $\mathcal{M}(B) = 0$. From Proposition 2.1, we get that $B = 0$. Therefore $y \otimes y = \Phi(B) = 0$ and hence $y = 0$. This shows that T is one-to-one. Similarly we show that T^* is also one-to-one.

By taking $A = B = I$ in (4.1), we get

$$\mathcal{M}(T^2) = \Phi(I) = T. \tag{4.2}$$

Now, let $T^2 = V_2|T^2|$ be the polar decomposition of T^2. Since T^2 and T^{*2} are injective then V_2 is a unitary operator. From (4.2) we get that, $\frac{1}{2}(T^2 + |T^2|V_2) = T$. Multiplying this equation on the left by V_2^*, we obtain

$$\frac{1}{2}(|T^2| + V_2^*|T^2|V_2) = V_2^*T.$$

Consequently, V_2^*T is a positive operator. In particular we have $V_2^*T = T^*V_2$ and $V_2^*T^2 = |T^2| = T^*V_2T$ is positive. Since T has a dense range, V_2 is a unitary positive operator. Hence $V_2 = I$ and $T^2 = |T^2|$ is positive. From (4.2) , we get that $T^2 = T$ and thus $T = \Phi(I) = I$. This completes the proof for the statement (ii)
The statement (iii) follows from Theorem 2.2. □

As a consequence we get the following result.

LEMMA 4.2. Let $\Phi : \mathcal{B}(H) \to \mathcal{B}(K)$ be a bijective map satisfying (4.1). Then

(i) $\mathcal{M}(\Phi(B)) = \Phi(\mathcal{M}(B))$, for all $B \in \mathcal{B}(H)$.
In particular Φ preserves the set of quasi-normal operators in both directions.
(ii) $\Phi(A^2) = (\Phi(A))^2$ for all A quasi-normal.
(iii) Φ preserves the set of orthogonal projections.
(iv) Φ preserves the orthogonality between the projections:

$$P \perp Q \Leftrightarrow \Phi(P) \perp \Phi(Q).$$

(v) Φ preserves the order relation on the set of orthogonal projections in both directions:

$$Q \leq P \Leftrightarrow \Phi(Q) \leq \Phi(P).$$

(vi) $\Phi(P + Q) = \Phi(P) + \Phi(Q)$ for all orthogonal projections P, Q such that $P \perp Q$.
(vii) Φ preserves the set of rank one orthogonal projections in both directions.

PROOF. (i). Taking $A = I$ in (4.1), (i) is immediate.

(ii). Let A be a quasi-normal operator. Since Φ preserves the set of quasi-normal operators, then $\Phi(A), \Phi(A^2)$ and $(\Phi(A))^2$ are quasi-normal. By (4.1) with $B = A$, we get $\mathcal{M}((\Phi(A))^2) = \Phi(\mathcal{M}(A^2))$. Hence $(\Phi(A))^2 = \Phi(A^2)$ because the quasi-normal operators are exactly the fixed point of the Mean transform (see (2.1)).

(iii). It is an immediate consequence of (ii) and the fact that a idempotent quasi-normal is orthogonal projection.

Throughout the remaining of the proof P and Q are orthogonal projections.

(iv). Assume that P, Q are orthogonal (i.e. $PQ = 0$). Since Φ preserves the set of orthogonal projections, then $\Phi(P), \Phi(Q)$ are orthogonal projection. From (4.1) we get

$$\mathcal{M}(\Phi(P)\Phi(Q)) = \Phi(\mathcal{M}(PQ)) = \Phi(0) = 0.$$

Thus $\Phi(Q)\Phi(P) = 0$. The converse holds since the inverse Φ^{-1} satisfies the same condition as Φ.

(v). Suppose that $Q \leq P$ or equivalently $QP = PQ = Q$. Then by (4.1), we get

$$\mathcal{M}(\Phi(P)\Phi(Q)) = \mathcal{M}(\Phi(Q)\Phi(P)) = \Phi(Q). \tag{4.3}$$

Using Proposition 2.4 and the last relation, we obtain

$$\Phi(Q)\Phi(P) = \Phi(P)\Phi(Q) = \Phi(Q).$$

Since Φ^{-1} verified the same assumption as Φ, we deduce that Φ preserves the order relation between the orthogonal projections in both directions.

(vi). We have $P, Q \leq P+Q$. Then $\Phi(P), \Phi(Q) \leq \Phi(P+Q)$. So $\Phi(P)+\Phi(Q) \leq \Phi(P+Q)$. Since Φ and Φ^{-1} both satisfy the same conditions, it follows that $\Phi(P) + \Phi(Q) = \Phi(P+Q)$.

(vii). Let $P = x \otimes x$ be a rank one projection. Then $\Phi(P)$ is a non zero projection. Let $y \in K$ be a unit vector such that $y \otimes y \leq \Phi(P)$. Thus $\Phi^{-1}(y \otimes y) \leq P$. Since P is a minimal projection and $\Phi^{-1}(y \otimes y)$ is a non zero projection, then $\Phi^{-1}(y \otimes y) = P$. Therefore $\Phi(P) = y \otimes y$ is a rank one projection. This complete the proof. □

LEMMA 4.3. If $\Phi : \mathcal{B}(H) \to \mathcal{B}(K)$ be a bijective map satisfying (4.1), then Φ preserves the rank one operators in both directions.

PROOF. Let $A = x \otimes y \in \mathcal{B}(H)$ be a rank one operator defined by a nonzero vector x and a unit vector y in H. We can find $B \in \mathcal{B}(H)$ such that $By + < By, y > y = 2x$, for example we take $B = (2x - < x, y > y) \otimes y$. Since Φ preserves the rank one projections, then $\Phi(y \otimes y = z \otimes z$ for some unit vector $z \in K$. Now, from (4.1) we get the following,

$$\begin{aligned}
\Phi(x \otimes y) &= \Phi(\frac{1}{2}(By + < By, y > y) \otimes y) \\
&= \Phi(\mathcal{M}(By \otimes y)) = \mathcal{M}(\Phi(B)\Phi(y \otimes y)) \\
&= \mathcal{M}(\Phi(B)z \otimes z) \\
&= \frac{1}{2}(\Phi(B)z + < \Phi(B)z, z > z) \otimes z).
\end{aligned}$$

Consequently $\Phi(A)$ is of rank one. □

Let us denote by $\mathcal{P}_1(H)$ the set of all rank one projections on the Hilbert space H. For the proof of Theorem 1.2, we will need the following theorem, which a important generalization of Wigner's theorem [**16**].

THEOREM 4.1 (**Uhlhorn's theorem** [**15**]). Let H, K be two complex Hilbert spaces, with *dim* $H \geq 3$. Let $\Phi : \mathcal{P}_1(H) \to \mathcal{P}_1(K)$ be a bijective transformation which preserves the orthogonality between the elements of $\mathcal{P}_1(H)$ in both directions, i.e., assume that Φ has the property

$$PQ = 0 \iff \Phi(P)\Phi(Q) = 0, \quad P, Q \in \mathcal{P}_1(H).$$

Then Φ is of the form

$$\Phi(P) = UPU^*, \quad P \in \mathcal{P}_1(H),$$

for some unitary or antiunitary operator $U : H \to K$.

PROOF. **of Theorem 1.1** The "if" part of the theorem is an immediate consequence of Proposition 2.2. We show the "only if" part. By Lemma 4.2, Φ satisfied the hypothesis of Uhlhorn's theorem. Hence there exists a unitary or anti-unitary operator $U : H \to K$, such that $\Phi(P) = UPU^*, \quad P \in \mathcal{P}_1(H)$. Now, the map $A \to U^*\Phi(A)U, \ A \in \mathcal{B}(H)$, satisfies the hypothesis of Theorem 1.2. So we can assume that Φ is the identity on $\mathcal{P}_1(H)$. To complete the proof we need to show that Φ is the identity on $\mathcal{B}(H)$.

First we show that, for every unit vectors $x, y \in H$, there exists $\alpha_{x,y} \in \mathbb{C}$, such that

$$\Phi(x \otimes y) = \alpha_{x,y} x \otimes y \tag{4.4}$$

Suppose that $< x, y > \neq 0$ (the case $< x, y >= 0$ is obtained in the same way).

We start by showing that $\Phi(x \otimes y)x$ and x are linearly dependent. The equation displayed in (4.1), with $A = x \otimes y$ and $B = x \otimes x$, implies:

$$\begin{aligned} \mathcal{M}(\Phi(x \otimes y)x \otimes x) &= \mathcal{M}(\Phi(x \otimes y)\Phi(x \otimes x)) \\ &= \Phi(\mathcal{M}(< x, y > x \otimes x)) \\ &= \Phi(< x, y > x \otimes x) \\ &= h(< x, y > x \otimes x \qquad \text{(by Lemma 3.2 (iii))}. \end{aligned}$$

It follows that

$$\frac{1}{2}(\Phi(x \otimes y)x + < \Phi(x \otimes y)x, x > x) \otimes x = h(< x, y > x \otimes x, \tag{4.5}$$

which implies $\Phi(x \otimes y)x + < \Phi(x \otimes y)x, x > x$ and x are collinear. Thus $\Phi(x \otimes y)x$ and x are collinear.

Now we show $\Phi(x \otimes y)^*y$ and y are also linearly dependent. Again using (4.1), for $A = y \otimes y$ and $B = x \otimes y$ we get the following

$$\begin{aligned} \mathcal{M}((y \otimes y)\Phi(x \otimes y)) &= \mathcal{M}(\Phi(y \otimes y)\Phi(x \otimes y)) \\ &= \Phi(\mathcal{M}(y \otimes y)(x \otimes y))) \\ &= \Phi(\mathcal{M}(< x, y > y \otimes y)) \\ &= \Phi(< x, y > y \otimes y) \\ &= h(< x, y >)y \otimes y \qquad \text{(by Lemma 3.2 (iii))}. \end{aligned}$$

Therefore

$$\mathcal{M}((y \otimes \Phi(x \otimes y)^*y)) = h(< x, y >)y \otimes y. \tag{4.6}$$

Since $< x, y > \neq 0$, then $h(< x, y >) \neq 0$ and $\mathcal{M}((y \otimes \Phi(x \otimes y)^*y)) \neq 0$. By (3.5) we get

$$\frac{1}{2}((y + \frac{< \Phi(x \otimes y)y, y >}{\|\Phi(x \otimes y)^*y\|^2}\Phi(x \otimes y)^*y) \otimes \Phi(x \otimes y)^*y) = h(< x, y >)y \otimes y. \tag{4.7}$$

Consequently $\Phi(x \otimes y)^*y$ and y are linearly dependent.

On the other hand $\Phi(x \otimes y)$ is a rank one operator (by Lemma 4.3), then there exists $\alpha_{x,y} \in \mathbb{C}$ such that $\Phi(x \otimes y) = \alpha_{x,y} x \otimes y$. Hence (3.4) is proved.

In the next step we introduce a function $f : \mathcal{B}(H) \to \mathbb{C}$ satisfying the following

$$\Phi(A) = f(A)A \text{ for all } A \in \mathcal{B}(H).$$

Let $A \in \mathcal{B}(H)$, by (4.1) for every unit vector $x \in H$ we have the following,

$$\mathcal{M}(\Phi(A)x \otimes x) = \mathcal{M}(\Phi(A)\Phi(x \otimes x)) = \Phi(\mathcal{M}(Ax \otimes x)). \tag{4.8}$$

By Proposition 2.3

$$\mathcal{M}(\Phi(A)x \otimes x) = \frac{1}{2}(\Phi(A)x + < \Phi(A)x, x > x) \otimes x) \tag{4.9}$$

On the other hand, by (4.4), there exists $\alpha_{A,x}$ such that

$$\begin{aligned} \Phi(\mathcal{M}(Ax \otimes x)) &= \Phi(\frac{1}{2}(Ax + < Ax, x > x) \otimes x) \\ &= \frac{1}{2}\alpha_{A,x}(Ax + < Ax, x > x) \otimes x. \end{aligned}$$

From (4.8) and (4.9) we have

$$\frac{1}{2}(\Phi(A)x + < \Phi(A)x, x > x) \otimes x) = \frac{1}{2}\alpha_{A,x}(Ax + < Ax, x > x) \otimes x)). \tag{4.10}$$

Therefore

$$\Phi(A)x + < \Phi(A)x, x > x = \alpha_{A,x}(Ax + < Ax, x > x), \tag{4.11}$$

this follows

$$< (\Phi(A)x + < \Phi(A)x, x > x), x >= \alpha_{A,x} < (Ax + < Ax, x > x), x > .$$

Thus

$$< \Phi(A)x, x >= \alpha_{A,x} < Ax, x > . \tag{4.12}$$

By equations (4.11) and (4.12), we get that $\Phi(A)x = \alpha_{A,x}Ax$ for every $x \in H$. In particular $\Phi(A)x$ and Ax are linearly dependent for every $x \in H$. Hence there exists a scalar number α_A such that $\Phi(A) = \alpha_A A$ for all $A \in \mathcal{B}(H)$. We defined the function $f : \mathcal{B}(H) \ni A \mapsto \alpha_A \in \mathbb{C}$. Therefore, $\Phi(A) = f(A)A$ for all $A \in \mathcal{B}(H)$.

To finish the proof, we have to show $f(A) = 1$ for all $A \in \mathcal{B}(H)$. Indeed, by (4.1), we get that $f(AB) = f(A)f(B)$ for all $A, B \in \mathcal{B}(H)$. Since $\Phi(P) = P$ for all rank one projection P, then $f(P) = 1$. Now let P, Q two rank one projections such that $PQ = 0$, then $f(0) = f(PQ) = f(P)f(Q) = 1$. Now by $f(AB) = f(A)f(B)$, we get that $f(A) = 1$ for every $A \in \mathcal{B}(H)$. Therefore, $\Phi(T) = UTU^*$ for all $T \in \mathcal{B}(H)$. This complete the proof of our main theorem. □

Acknowledgment. We are grateful to Fernanda Botelho for useful discussion and comments. The second author is also very grateful for the kind hospitality during his stay in Memphis.

This work was supported in part by the Labex CEMPI (ANR-11-LABX-0007-01).

References

[1] A. Aluthge, *On p-hyponormal operators for* $0 < p < 1$, Integral Equations Operator Theory **13** (1990), no. 3, 307–315, DOI 10.1007/BF01199886. MR1047771

[2] F. Botelho, L. Molnár, and G. Nagy, *Linear bijections on von Neumann factors commuting with* λ*-Aluthge transform*, Bull. Lond. Math. Soc. **48** (2016), no. 1, 74–84, DOI 10.1112/blms/bdv092. MR3455750

[3] F. Chabbabi, *Product commuting maps with the* λ*-Aluthge transform*, J. Math. Anal. Appl. **449** (2017), no. 1, 589–600, DOI 10.1016/j.jmaa.2016.12.027. MR3595221

[4] F. Chabbabi and M. Mbekhta, *Jordan product maps commuting with the* λ*-Aluthge transform*, J. Math. Anal. Appl. **450** (2017), no. 1, 293–313, DOI 10.1016/j.jmaa.2017.01.036. MR3606169

[5] F. Chabbabi, R. E. Curto, and M. Mbekhta, *The mean transform and the mean limit of an operator*, Proc. Amer. Math. Soc. **147** (2019), no. 3, 1119–1133, DOI 10.1090/proc/14277. MR3896061

[6] T. Furuta, *Invitation to linear operators*, Taylor & Francis, Ltd., London, 2001. From matrices to bounded linear operators on a Hilbert space. MR1978629

[7] I. B. Jung, E. Ko, and C. Pearcy, *Aluthge transforms of operators*, Integral Equations Operator Theory **37** (2000), no. 4, 437–448, DOI 10.1007/BF01192831. MR1780122

[8] I. B. Jung, E. Ko, and C. Pearcy, *Spectral pictures of Aluthge transforms of operators*, Integral Equations Operator Theory **40** (2001), no. 1, 52–60, DOI 10.1007/BF01202954. MR1829514

[9] I. B. Jung, E. Ko, and C. Pearcy, *The iterated Aluthge transform of an operator*, Integral Equations Operator Theory **45** (2003), no. 4, 375–387, DOI 10.1007/s000200300012. MR1971744

[10] S. Jung, E. Ko, and S. Park, *Subscalarity of operator transforms*, Math. Nachr. **288** (2015), no. 17-18, 2042–2056, DOI 10.1002/mana.201500037. MR3434298

[11] T. Kato, *Perturbation theory for linear operators*, Classics in Mathematics, Springer-Verlag, Berlin, 1995. Reprint of the 1980 edition. MR1335452

[12] S. H. Lee, W. Y. Lee, and J. Yoon, *The mean transform of bounded linear operators*, J. Math. Anal. Appl. **410** (2014), no. 1, 70–81, DOI 10.1016/j.jmaa.2013.08.003. MR3109820

[13] K. Okubo, *On weakly unitarily invariant norm and the Aluthge transformation*, Linear Algebra Appl. **371** (2003), 369–375, DOI 10.1016/S0024-3795(03)00485-3. MR1997382

[14] P. Šemrl, *Linear mappings preserving square-zero matrices*, Bull. Austral. Math. Soc. **48** (1993), no. 3, 365–370, DOI 10.1017/S0004972700015811. MR1248039

[15] U. Uhlhorn, *Representation of symmetry transformations in quantum mechanics*, Ark. Fysik 23 (1962), 307-340.

[16] E.P. Wigner, *Gruppentheorie und ihre Anwendung auf die Quantenmechanik der Atomspektrum*, Fredrik Vieweg und Sohn, 1931.

[17] T. Yamazaki, *An expression of spectral radius via Aluthge transformation*, Proc. Amer. Math. Soc. **130** (2002), no. 4, 1131–1137, DOI 10.1090/S0002-9939-01-06283-9. MR1873788

Department of Mathematics, University of Lille, UMR-CNRS 8524 Laboratory P. Painlevé, 59655 Villeneuve d'Ascq Cedex, France

Email address: `fadil.chabbabi@univ-lille.fr`

Email address: `mostafa.mbekhta@univ-lille.fr`

Contemporary Mathematics
Volume **737**, 2019
https://doi.org/10.1090/conm/737/14859

Some geometric properties of relative Chebyshev centres in Banach spaces

Soumitra Daptari and Tanmoy Paul

ABSTRACT. In this paper we characterize Property-(R_1), a generalization of $1\frac{1}{2}$ ball property. As a necessary and sufficient condition of a subspace Y with Property-(R_1) we derive that $r(y,F) = rad_Y(F) + d(y, cent_Y(F))$ for any bounded subset F and $y \in Y$. We introduce the notion of modulus of relative chebyshev centre and characterize Property-(R_1) in terms of this modulus. It is observed that if Y is a finite co-dimensional strongly proximinal subspace of a L_1 predual space X and F is a finite subset of X then $rad_Y(F) = rad_X(F) + d(F,Y)$. We characterize continuity of $cent_V(.)$ in terms of the modulus of relative chebyshev centre.

1. Introduction

1.1. Notations and Definitions. By X we always mean a Banach space. For $x \in X$ and $r > 0$ $B(x,r)$ and $B[x,r]$ represent the open and closed ball centered at x and radius r respectively. By $CB(X), CL(X), CC(X), \mathcal{F}(X)$ we mean the set of all closed and bounded, closed, closed convex and finite subsets of X respectively. The underlying field for all the spaces is assumed to be Real.

For $x \in X, F \in CB(X)$ we define the following.

NOTATION. *(1)* $r(x,F) = \sup\{\|x-y\| : y \in F\}$
(2) $rad_V(F) = \inf\{r(x,F) : x \in V\}$
(3) $cent_V(F) = \{v \in V : r(v,F) = rad_V(F)\}$
(4) $\delta - cent_V(F) = \{v \in V : r(v,F) \leq rad_V(F) + \delta\}$
(5) *For* $B \subseteq X$ $B_\varepsilon = \{x \in X : d(x,B) \leq \varepsilon\}$.
(6) $S_\varepsilon(F) = \{x \in X : r(x,F) \leq \varepsilon\}$.

Note that,

(1) $cent_V(F) = \left\{\bigcap_{y\in F} B[y, rad_V(F)]\right\} \cap V$.

(2) $\delta - cent_V(F) = \left\{\bigcap_{y\in F} B[y, rad_V(F) + \delta]\right\} \cap V$.

(3) $B_\varepsilon = B + \varepsilon B_X$ and $S_\varepsilon(F) = \bigcap_{x\in F} B[x,\varepsilon]$.

Note that if $V \in CC(X)$ and $F \in \mathcal{F}$ the set $cent_V(F)$ may be empty, although the set $\delta - cent_V(F)$ is always nonempty for any $\delta > 0$. A pair $(V, \mathcal{F})$ where

2010 *Mathematics Subject Classification.* Primary 46B20, 46E15.
Key words and phrases. Property R_1, upper Hausdorff semi-continuity, $1\frac{1}{2}$ ball property.
The research was supported by Science and Engineering Research Board, India. Award No. MTR/2017/000061.

$V \in CL(X)$ and $\mathcal{F} \subseteq CB(X)$ is said to have *restricted center property* (in short r.c.p.) if for all $F \in \mathcal{F}, cent_V(F) \neq \emptyset$. $rad_V(F)$ represents the radius of the smallest ball (if it exists) in X centered in V which contains F, $cent_V(F)$ represents the possible points of centres of these balls. $\delta - cent_V(F)$ represents the points in V which are also in the δ perturbation of radius these balls.

We now introduce the central character of this paper to the reader.

DEFINITION 1.1. [**9**] Let $V \in CC(X)$ and $\mathcal{F} \subseteq CB(X)$, the triplet $(X, V, \mathcal{F})$ is said to have Property-(R_1) if for $v \in V, F \in \mathcal{F}, r_1, r_2 > 0$ the condition $B(v, r_1) \cap S_{r_2}(F) \neq \emptyset$ and $S_{r_2}(F) \cap V \neq \emptyset$ would imply that $V \cap B[v, r_1] \cap S_{r_2}(F) \neq \emptyset$.

The above definition is a set valued analogue of the following.

DEFINITION 1.2. [**12**] A subspace Y of a Banach space X is said to have $1\frac{1}{2}$ ball property if $y \in Y$ and $x \in X$ the open balls $B(x, r_1), B(y, r_2)$ intersects and the closed ball $B[x, r_1]$ has non empty intersection with the subspace Y then the intersection $Y \cap B[x, r_1] \cap B[y, r_2]$ is non empty.

It is clear that Property-(R_1) implies the property of $1\frac{1}{2}$ ball if $\mathcal{F}$ contains the set of all singletons. A list of examples are given in [**9**] having this property. Let us recall that a Banach space X is said to be L_1 predual if X^* is isometric with $L_1(\mu)$ for some measure space (Ω, Σ, μ).

Let us recall that a subspace Y of X is said to have *Best approximation property* (or *Proximinal*) if for any point $x \notin Y$ there exists $y \in Y$ such that $\|x - y\| = d(x, Y)$, where $d(x, Y) = \inf\{\|x - y\| : y \in Y\}$. For any such subspace Y one can define the set valued mapping $P_Y : X \to CB(Y)$ by $P_Y(x) = \{y \in Y : d(x, Y) = \|x - y\|\}$. P_Y is called the metric projection for the subspace Y.

It is well known that a subspace with $1\frac{1}{2}$ ball property also satisfies Best approximation property and also the metric projection $x \mapsto P_Y(x)$ satisfies a continuity criterian (it is *upper Hausdorff semi continuous*). One can define the following weaker notion than $1\frac{1}{2}$ ball property, subspace Y with which P_Y is upper Hausdorff semi continuous.

For a subspace Y and $\delta > 0$, let us define $P_Y(x, \delta) = \{y \in Y : \|x - y\| \leq d(x, Y) + \delta\}$.

DEFINITION 1.3. [**7**] A subspace Y of a Banach space X is said to be *Strongly proximinal* if given $\varepsilon > 0$ and $x \in X$ there exists a $\delta(\varepsilon, x) > 0$ such that $P_Y(x, \delta) \subseteq P_Y(x) + \varepsilon B_Y$.

In [**8**] the author observed that a subspace with $1\frac{1}{2}$ ball property also Strongly proximinal. A set valued analogue of Strong proximinality is introduced in [**9**] by the following.

DEFINITION 1.4. [**9**] Let $V \in CC(X)$ and $\mathcal{F} \subseteq CB(X)$, the triplet $(X, V, \mathcal{F})$ is said to have Property-(P_1) if for given $\varepsilon > 0$ and $F \in \mathcal{F}$ there exists a $\delta(\varepsilon, F) > 0$ such that $\delta - cent_V(F) \subseteq cent_V(F) + \varepsilon B_X$.

It is clear that if V is a subspace and if $\mathcal{F}$ contains all singletons then V is Strongly proximinal if the triplet $(X, V, \mathcal{F})$ has Property-(P_1).

We encounter various continuity of a set valued map in a normed linear space. Let us recall the following definition in this context.

DEFINITION 1.5. Let T be a topological space and $\Gamma : T \to CB(X)$ be a set valued map. Γ is said to be

(a) upper Hausdorff semi-continuous, abbreviated uHsc. (resp. lower Hausdorff semi-continuous, abbreviated lHsc) if for every $t_0 \in T$ and every $\varepsilon > 0$, there is a neighborhood N of t_0, such that $\Gamma(t) \subseteq \Gamma(t_0) + \varepsilon B_X$ (resp. $\Gamma(t_0) \subseteq \Gamma(t) + \varepsilon B_X$) for each $t \in N$.

(b) Γ is Hausdorff continuous, abbreviated H-continuous, if it is both uHsc and lHsc.

DEFINITION 1.6. A real valued map $\varphi : (X, \tau_X) \to \mathbb{R}$ is said to be upper semi continuous (usc) (lower semi continuous (lsc)) if for any real $\alpha \in \mathbb{R}$ the set $\{x \in X : \varphi(x) \geq \alpha\}$ ($\{x \in X : \varphi(x) \leq \alpha\}$) is closed.

1.2. A Brief review of the known results and the important outcomes of this investigation. History of (relative) Chebyshev Centre dates back to the papers [**1**–**4**, **9**] and also many more. A compactness criterion (like norm, weak or w^*-compactness) of the unit ball of the space makes sure of the non emptyness of the (relative) Chebyshev centre of any finite subset of the space. An ultimate result in this direction is available in [**1**] where the author proves that a uniformly convex Banach space always provides a non empty Chebyshev centre for any bounded subset and the set valued map which sends the bounded set to its centre is single valued and also uniformly continuous with respect to the Hausdorff metric. Authors in [**3**] and [**9**] prove this result into its full generalization. Now coming to the cases when the space is not reflexive (or the unit ball has no compactness criterion) we refer the reader to the articles [**6**] and [**11**]. In [**6**] the author observes that a space of type $C(K)$, with its standerd meaning, any of its finite co-dimensional subspace always provide a non empty Chebyshev centre for any compact subset of $C(K)$ if and only if the subspace is proximinal.

In [**11**] the authors establish some geometric properties of Chebyshev centre and relative Chebyshev centre in the space of type $C(K)$, in particular it is observed that $cent_G(F) = cent_{C(K)}(F) + R$, where $R = \inf\{\|x - g\| : x \in cent_{C(K)}(F), g \in G\}$ where G is a closed convex subset of $C(K)$. (see [**11**, Theorem 2.2]). We derive a similar result for a L_1 predual space when the above G is a finite co-dimensional subspace and F is a finite subset. Full generality of this result for the spaces of type L_1 is still unknown.

David Yost introduced the notion of $1\frac{1}{2}$ ball property (see [**12**]) of a closed subspace of a Banach space, a judicial modification of 2 ball property. In [**12**] Yost explored some geometric and analytic properties of subspaces of a Banach space having this property, viz. existence of nearest point property, continuity of metric projection, quasi additivity of metric projection, uniqueness of Hahn-Banach extension map from Y^* to X^* etc. Afterwards few articles appeared following the same line of investigation (see, [**8**, **10**, **12**–**14**]) which explore many interesting consequences of $1\frac{1}{2}$ ball property. Yost defined some other property viz. weak $1\frac{1}{2}$ ball property, which is in fact equivalent to $1\frac{1}{2}$ ball property, also proved in [**12**]. A hidden geometry of this property is that the closed unit ball of the whole space has a *flat face parallel* to the subspace. This refined geometry is established in [**10**, Proposition 1].

Yost's school of thought is adopted into the set-valued optimization technique by Pai and Nowroji in [**9**] in the name of Property-(R_1), see Definition 1.1. We continue this investigation in this paper and derive few characterizations of this Property.

A Section wise illustration of this work is given below. Most of our observations are related to the triplet $(X, V, \mathcal{F})$, where $V \in CC(X)$ and $\mathcal{F} \subseteq \mathcal{F}(X)$.

In Section 2 our prime objective is to establish some geometric properties of subspaces having Property-(R_1) and to find some new examples having this property. Few of these geometric properties are the set valued analogues of what was observed in **[8]** for $1\frac{1}{2}$ ball property.

In Section 3 we introduce the notion of *Modulus of Restricted Chebyshev centre* viz. $\varepsilon_V(F,t)$ corresponding to the triplet $(X, V, \mathcal{F})$ where $F \in \mathcal{F}$ and $0 < t < 1$. We characterize Property-(R_1) in terms of this modulus. We also derive that if Y is a finite codimensional subspace of a L_1 predual X which is strongly proximinal then for any finte subset F of X, $cent_Y(F) = cent_X(F) + d(F,Y)$. The result is known for spaces of type $C(K)$ endowed with the supremum norm, we derive it for the category of L_1 predual spaces.

The paper concludes with the fact that the set valued map $cent_V(.)$ is continuous at F if and only if $\varepsilon(.,t)$ are continuous at F for all $t > 0$, provided the subspace V has r.c.p. for all $F \in \mathcal{F}$.

2. Property-(R_1)

Let $V \in CL(X)$, $F \in CB(X)$ then for all $v \in V$ it is always true that $d(v, cent_V(F)) \geq r(v,F) - rad_V(F)$. In fact if $z \in cent_V(F)$, then $r(v,F) - rad_V(F) = r(v,F) - r(z,F) \leq \|v - z\|$, true for all $z \in cent_V(F)$, hence the result.

DEFINITION 2.1. We call the triplet $(X, V, \mathcal{F})$ has weak Property(R_1) if for $v \in V, F \in \mathcal{F}, r_1, r_2 > 0$ the condition $B[v, r_1] \cap S_{r_2}(F) \neq \emptyset$ and $S_{r_2}(F) \cap V \neq \emptyset$ would imply that $V \cap B[v, r_1 + \varepsilon] \cap S_{r_2+\varepsilon}(F) \neq \emptyset$, for all $\varepsilon > 0$.

We first show that the two different conditions viz. Property-(R_1) and weak Property-(R_1) are equivalent.

THEOREM 2.2. *For a triplet $(X, V, \mathcal{F})$ where $\mathcal{F} \subseteq CB(X)$, V is closed convex, $0 \in V$ and closed under translation. Then the following are equivalent.*

(*a*) *$(X, V, \mathcal{F})$ has Property-(R_1).*

(*b*) *$(X, V, \mathcal{F})$ has weak Property-(R_1).*

(*c*) *For $F \in \mathcal{F}$ $r > 0$ the condition $B(0, r) \cap S_1(F) \neq \emptyset$ and $S_1(F) \cap V \neq \emptyset$ would imply that $V \cap B[0, r] \cap S_1(F) \neq \emptyset$.*

(*d*) *For $F \in \mathcal{F}$ and $r > 0$ the condition $B[0, r] \cap S_1(F) \neq \emptyset$ and $S_1(F) \cap V \neq \emptyset$ would imply that $V \cap B[0, r + \varepsilon] \cap S_{1+\varepsilon}(F) \neq \emptyset$, for all $\varepsilon > 0$.*

PROOF. It is clear that $(a) \iff (c)$ and $(b) \iff (d)$ and also $(a) \implies (b)$. To complete the proof it suffices to show that $(b) \implies (a)$. Let $r(v,F) < r_1 + r_2$ and $rad_V(F) \leq r_2$.

CLAIM : $B[v, r_1] \cap S_{r_2}(F) \cap V \neq \emptyset$.

If $r(v,F) \leq r_2$ then $v \in B[v, r_1] \cap S_{r_2}(F) \cap V$ and we are done. Let $r(v,F) > r_2$, then we have $rad_V(F) \leq r_2 < r(v,F) < r_1 + r_2$.

Let $\varepsilon = \frac{1}{3}(r_1 + r_2 - r(v,F))$ then $\varepsilon > 0$ and $V \cap S_{r_2+\frac{\varepsilon}{2}}(F) \neq \emptyset$ (since $rad_V(F) \leq r_2$) also we have $r(v,F) \leq r_2 + (r_1 - 3\varepsilon)$. By weak Property-$(R_1)$ there exists $x_0 \in V \cap B[v, r_1 - 2\varepsilon] \cap S_{r_2+\varepsilon}(F)$. By induction we will construct a sequence $(x_n) \subseteq V$ such that $\|x_n - x_{n+1}\| \leq \frac{\varepsilon}{2^n}$ and $r(x_n, F) \leq r_2 + \frac{\varepsilon}{2^n}$.

Suppose $(x_i)_{i=1}^n$ are chosen. Hence we have $r(x_n, F) \leq r_2 + \frac{\varepsilon}{2^n} = r_2 + \frac{\varepsilon}{2^n}(\frac{3}{4} + \frac{1}{4}) = \frac{3\varepsilon}{4.2^n} + (r_2 + \frac{\varepsilon}{4.2^n})$. Again by weak Property-(R_1) there exists $x_{n+1} \in V \cap$

$B[x_n, \frac{3\varepsilon}{4.2^n} + \frac{\varepsilon}{4.2^n}] \cap S_{r_2 + \frac{\varepsilon}{4.2^n} + \frac{\varepsilon}{4.2^n}}$. Hence (x_n) is cauchy and hence $x_n \to x$ for some $x \in V$. We have $\|x - x_0\| \leq 2\varepsilon$ and $r(x, F) \leq r_2$. Finally, we have $\|x - v\| \leq \|x - x_0\| + \|v - x_0\| \leq 2\varepsilon + r_1 - 2\varepsilon = r_1$ which establishes the above CLAIM. □

PROPOSITION 2.3. *For $0 \leq \delta < \varepsilon$ and $F \in CB(X), V \in CC(X)$ we have,*

$$\delta - cent_V(F)_{\varepsilon - \delta} \subseteq \bigcap_{x \in F} B[x, rad_V(F) + \varepsilon] \cap V$$

In other words we have $\delta - cent_V(F)_{\varepsilon-\delta} \subseteq \varepsilon - cent_V(F)$.

PROOF. Let $y \in V$ such that $d(y, \delta - cent_V(F)) \leq \varepsilon - \delta$. Let $g_n \in \delta - cent_V(F)$ such that $\|y - g_n\| \to d(y, \delta - cent_V(F))$.

If $x \in F$ then

$$\begin{aligned}
\|x - y\| &\leq \|x - g_n\| + \|y - g_n\| \\
&\leq r(g_n, F) + \|y - g_n\| \\
&\leq rad_V(F) + \delta + \|y - g_n\| \\
&\to rad_V(F) + \delta + d(y, \delta - cent_V(F)) \\
&\leq rad_V(F) + \varepsilon
\end{aligned}$$

True for all $x \in F$, hence $r(y, F) \leq rad_V(F) + \varepsilon$. That is $y \in \varepsilon - cent_V(F)$. □

We now prove the main result of this Section.

THEOREM 2.4. *Let V be a subspace and $\mathcal{F} \subseteq CB(X)$ is closed under translation then for the triplet $(X, V, \mathcal{F})$ the following are equivalent*

(a) *$(X, V, \mathcal{F})$ has Property-(R_1).*
(b) *$d(v, cent_V(F)) = r(v, F) - rad_V(F)$, for all $v \in V, F \in \mathcal{F}$.*
(c) *$d(0, cent_V(F)) = r(0, F) - rad_V(F)$, for all $F \in \mathcal{F}$.*
(d) *$\delta - cent_V(F) = cent_V(F)_\delta \cap V$.*
(e) *For $F \in \mathcal{F}, r_1, r_2 \geq 0$ with $rad_V(F) \leq r_1 < r_2$ define $A_i = \{v \in V : r(v, F) = r_i\}, i = 1, 2$. Let $A_1 \neq \emptyset$ and $g_2 \in A_2$ then $d(g_2, A_1) = r_2 - r_1$.*

PROOF. $(b) \iff (d)$: Let $v \in \delta - cent_V(F)$. Then $r(v, F) = rad_V(F) + d(v, cent_V(F))$ but $r(v, F) \leq rad_V(F) + \delta$ also. Hence the result in (d) follows. Let $\delta = r(v, F) - rad_V(F), v \in V$. Then $v \in \delta - cent_V(F) = cent_V(F)_\delta \cap V$. That is $d(v, cent_V(F)) \leq \delta = r(v, F) - rad_V(F)$. Hence (b) follows.

$(b) \iff (c)$: Follows from obvious translation. Suppose (b) is true. Replace F by $F - v$. Clearly $r(0, F - v) = r(v, F)$.

$rad_V(F) = \inf\{r(v, F) : v \in V\} = \inf\{r(z, F - v) : z \in V\} = rad_V(F - v)$.

Now

$$\begin{aligned}
d(0, cent_V(F - v)) &= \inf\{\|x\| : x \in cent_V(F - v)\} \\
&= \inf\{\|x\| : r(x + v, F) = rad_V(F)\} \\
&= \inf\{\|x\| : x + v \in cent_V(F)\} \\
&= \inf\{\|y - v\| : y \in cent_V(F)\} \\
&= d(v, cent_V(F))
\end{aligned}$$

$(a) \implies (e)$: Suppose $d(g_2, A_1) \neq r_2 - r_1$, hence $d(g_2, A_1) > r_2 - r_1$. Choose $\varepsilon > 0$ such that $d(g_2, A_1) > r_2 - r_1 + \varepsilon$. Let $\overline{r_1} = r_2 - r_1 + \varepsilon$ and $\overline{r_2} = r_1$. Then $r(g_2, F) = r_2 < \overline{r_1} + \overline{r_2}$ and $\emptyset \neq A_1 \subseteq S_{\overline{r_2}}(F) \cap V$. By (a) there exists $y \in V \cap B[g_2, \overline{r_1}] \cap S_{\overline{r_2}(F)}$.

Since $r(y,F) \leq \overline{r_2} = r_1$ and $r(g_2,F) = r_2 > r_1$, there exists $g_1 \in [y, g_2]$, the line joining y and g_2, such that $r(g_1,F) = r_1$ then $g_1 \in A_1$ and $\|g_2 - g_1\| = \|g_2 - y\| - \|g_1 - y\| \leq \|g_2 - y\| \leq \overline{r_1} < d(g_2, A_1)$. This contradiction ensures the result.

$(e) \implies (d)$: Let $F \in \mathcal{F}, 0 \leq \varepsilon_1 < \varepsilon_2$ such that $\varepsilon_1 - cent_V(F) \neq \emptyset$. It remains to show that $\varepsilon_2 - cent_V(F) \subseteq \varepsilon_1 - cent_V(F)_{\varepsilon_2 - \varepsilon_1} \cap V$.

Let $v_0 \in \varepsilon_2 - cent_V(F)$, if $v_0 \in \varepsilon_1 - cent_V(F)$ then we are done. Hence assume that $v_0 \notin \varepsilon_1 - cent_V(F)$, then $r(v_0,F) > rad_V(F) + \varepsilon_1$. Let $r_1 = rad_V(F) + \varepsilon_1, r_2 = r(v_0,F)$. Define $A_i = \{v \in V : r(v,F) = r_i\}$, $i = 1,2$. We have $rad_V(F) \leq r_1 < r_2$ and $v_0 \in A_2$. Since $\varepsilon_1 - cent_V(F) \neq \emptyset$ we have $A_1 \neq \emptyset$. By $A_1 \subseteq \varepsilon_1 - cent_V(F)$, the hypothesis in (e) together with $v_0 \in \varepsilon_1 - cent_V(F)$ we have that,

$$\begin{aligned}
d(v_0, \varepsilon_1 - cent_V(F)) \leq d(v_0, A_1) &= r_2 - r_1 \\
&= r(v_0,F) - rad_V(F) - \varepsilon_1 \\
&\leq rad_V(F) + \varepsilon_2 - rad_V(F) - \varepsilon_1 \\
&= \varepsilon_2 - \varepsilon_1
\end{aligned}$$

$(d) \implies (a)$: Let $r_1, r_2 \geq 0$ and $x \in V$ satisfying $B(x, r_2) \cap S_{r_1}(F) \neq \emptyset$ and also $V \cap S_{r_1}(F) \neq \emptyset$. First, it is clear that $rad_V(F) \leq r_1$.

If $r(x,F) \leq r_1$ then $x \in V \cap B[x, r_2] \cap S_{r_1}(F)$. Let $r_1 < r(x,F)$ and $\varepsilon_1 = r_1 - rad_V(F)$ and $\varepsilon_2 = r(x,F) - rad_V(F)$. Then $0 \leq \varepsilon_1 < \varepsilon_2$ and $x \in \varepsilon_2 - cent_V(F)$. Since $\varepsilon_1 - cent_V(F) = V \cap (\bigcap_{z \in F} B[z, r_1]) \neq \emptyset$, hence by (d) and $r(x,F) < r_1 + r_2$ we have $d(x, \varepsilon_1 - cent_V(F)) \leq \varepsilon_2 - \varepsilon_1 = r(x,F) - r_1 < r_2$.

Let $y_1 \in \varepsilon_1 - cent_V(F)$ such that $\|x - y_1\| < r_2$, since $y_1 \in Y \cap (\cap_{z \in F} B[z, r_1])$ we have $r(y,F) \leq r_1$. And finally we have $y_1 \in B[x, r_2] \cap S_{r_1}(F) \cap V$, which ensures the non emptyness of the last set. □

As an easy consequence of Theorem 2.4 we get the following.

THEOREM 2.5. *Let $(X, V, \mathcal{F})$ has Property-(R_1) then the map $\delta - cent_V(.)$: $(\mathcal{F}, \tau_H) \to (CB(X), \tau_H)$ is Lipschitz continuous, for all $\delta \geq 0$.*

PROOF. We show that for $G, H \in \mathcal{F}, d_H(\delta - cent_V(G), \delta - cent_V(H)) \leq 2d_H(G,H)$, for $\delta \geq 0$.

CASE 1: When $\delta = 0$.

It is clear that $|rad_V(G) - rad_V(H)| \leq d_H(G,H)$ and $|r(v,G) - r(v,H)| \leq d_H(G,H)$, for all $v \in V$. In fact for any $g \in G$ and $\varepsilon > 0$ get a $h \in H$ such that $\|g - h\| < d_H(G,H) + \varepsilon$. Now $\|v - g\| \leq \|v - h\| + \|g - h\| < r(v,H) + d_H(G,H) + \varepsilon$. True for all $g \in G$ which leads to that $r(v,G) \leq r(v,H) + d_H(G,H) + \varepsilon$. Replacing G by H and varying $\varepsilon > 0$ we get $|r(v,F) - r(v,G)| \leq d_H(G,H)$.

The above inequality also gives, $rad_V(H) \leq r(v,H) \leq r(v,G) + d_H(H,G)$. And hence $|rad_V(H) - rad_V(G)| \leq d_H(H,G)$.

CASE 2: When $\delta > 0$.

Let $v \in \delta - cent_V(G)$ then

$$\begin{aligned}
r(v,G) &\leq rad_V(G) + \delta \\
\text{Hence } r(v,H) &\leq r(v,G) + d_H(G,H) \\
&\leq rad_V(G) + \delta + d_H(G,H) \\
&\leq rad_V(H) + \delta + 2d_H(G,H)
\end{aligned}$$

That is $v \in (2d_H(G,H) + \delta) - cent_V(H)$.

CLAIM: For $r_1, r_2 > 0$, $(r_1 + r_2) - cent_V(F) \subseteq (r_1 - cent_V(F))_{r_2}$.

Suppose for $v \in V, r(v, F) \leq rad_V(F) + r_1 + r_2$ and also we have $r_1 - cent_V(F) \neq \emptyset$ hence by Property $(R_1), V \cap B[v, r_2] \cap (r_1 - cent_V(F)) \neq \emptyset$. Which implies $d(v, r_1 - cent_V(F)) \leq r_2$.

Hence we get $d(v, \delta - cent_V(H)) \leq 2d_H(G, H)$. Interchanging G and H we get the result. □

Let us recall the Definition 1.4 of Property-(P_1). Here note that Property (P_1) is weaker than Property (R_1); Theorem 2.4 (d) says that in Property (P_1) if corresponding $\delta(\varepsilon, F) = \varepsilon$ always, then it is Property (R_1).

3. Modulus of restricted chebyshev centres

DEFINITION 3.1. Let $V \in CL(X)$ and $\mathcal{F} \subseteq CB(X)$ be such that $(V, \mathcal{F})$ has r.c.p., the modulus of restricted chebyshev centres $\varepsilon : \mathcal{F} \times \mathbb{R}_+ \to \mathbb{R}_+$ is defined by

$$\varepsilon_V(F, t) = \inf\{r > 0 : t - cent_V(F) \subseteq cent_V(F) + rB_X\}$$

It is clear from the definition that for $t_1 \geq t_2 > 0, \varepsilon(F, t_1) \geq \varepsilon(F, t_2)$ for all $F \in \mathcal{F}$.

LEMMA 3.2. *Let $V \in CL(X)$ and $\mathcal{F} \subseteq CB(X)$ be such that $(V, \mathcal{F})$ has r.c.p.. Then for all $F \in \mathcal{F}, \varepsilon_V(F, .)$ is continuous and increasing function of t whenever $t > 0$. Moreover if $rad_V(F) = d$ and $t > s$ then $d_H(t - cent_V(F), s - cent_V(F)) \leq (t - s)\frac{2d+t}{t}$.*

PROOF. Suppose $rad_V(F)$ and $t > s$, it is enough to prove that

$$t - cent_V(F) \subseteq s - cent_V(F) + (t - s)\frac{2d + t}{t}B_X.$$

Let $\eta = t - s$ and $y \in t - cent_V(F)$ then $r(y, F) \leq rad_V(F) + t = d + s + \eta$.

Let $y_0 \in cent_V(F)$ and $\overline{y} = (1 - \lambda)y + \lambda y_0$, where $\lambda = \frac{\eta}{s+\eta}$.

Now

$$\begin{aligned} r(\overline{y}, F) &= \sup_{z\in F} \|\overline{y} - z\| \\ &\leq \sup_{z\in F} \|(1 - \lambda)y + \lambda y_0 - z\| \\ &\leq \sup_{z\in F}(1 - \lambda)\|y - z\| + \sup_{z\in F} \|y_0 - z\| \\ &\leq (1 - \lambda)(d + s + \eta) + \lambda r(y_0, F) \\ &= (1 - \lambda)(d + s + \eta) + \lambda d = d + s. \end{aligned}$$

That is $\overline{y} \in s - cent_V(F)$.

Also $\|y - \overline{y}\| = \lambda\|y - y_0\| \leq \frac{\eta}{s+\eta}(\|y - x\| + \|x - y_0\|) \leq \eta\frac{2d+t}{t}$.

This completes the proof. □

Lemma 3.2 now enable us to characterize the geometric aspects defined in Section 1 viz. Property (P_1), Property (R_1).

THEOREM 3.3. *Let $V \in CL(X)$ and $\mathcal{F} \subseteq CB(X)$ be such that $(V, \mathcal{F})$ has r.c.p.. Then*

(a) $(V, \mathcal{F})$ has Property-(P_1) $\iff$ $\varepsilon(F, .)$ continuous at 0 for all $F \in \mathcal{F}$.
(b) $(V, \mathcal{F})$ has Property-(R_1) $\iff$ $\varepsilon(F, t) \leq t$ for all $F \in \mathcal{F}$.

PROOF. (b). Apply Theorem 2.4(d). □

We now derive few results related to the Chebyshev radius of a finite subset of X and X^{**} relative to a subspace Y of X and also relative to $Y^{\perp\perp}$ in X^{**} respectively. These are in fact some applications of Principle of Local Reflexivity. These results along with Theorem 3.3 will enable us to derive some geometric results of a subspace Y having $1\frac{1}{2}$-ball property. Let us recall the following result viz. *Principle of local reflexivity* (PLR in short) which is relevant in the proof of Proposition 3.5.

THEOREM 3.4. *Let X be a Banach space. For every finite dimensional subspace E of X^{**} and finite dimensional subspace F of X^* and $\varepsilon > 0$ there exists an isomorphism $T : E \to X$ such that $x^*(Tx^{**}) = x^{**}(x^*)$ for all $x^{**} \in E$ and for all $x^* \in F$. $T|_{E\cap X} = I$ and $\|T\|\|T^{-1}\| \leq 1 + \varepsilon$.*

For a bounded subset F of X and $t > 0$, $\varepsilon_X(F,t), \varepsilon_{X^{**}}(F,t)$ represent the modulus of restricted chebyshev centre of F in the corresponding space.

PROPOSITION 3.5. *Let Y be a finite codimensional subspace of a Banach space X and $\mathcal{F}$ represents the set of all finite subsets of X. Then,*
- (a) *$d(cent_X(F), Y) = d(cent_{X^{**}}(F), Y^{\perp\perp})$, for $F \in \mathcal{F}$, if $(X, Y, \mathcal{F})$ has Property-(R_1).*
- (b) *Let $(X, Y, \mathcal{F})$ has Property-(P_1) then $d(y, cent_Y(F)) = d(y, cent_{Y^{\perp\perp}}(F))$, for $y \in Y$.*
- (c) *Let $(X, Y, \mathcal{F})$ has Property-(P_1) then $\varepsilon_Y(F,t) \leq \varepsilon_{Y^{\perp\perp}}(F,t)$, for any $F \in \mathcal{F}$.*

PROOF. (a) We have $d(cent_X(F), Y) \geq d(cent_{X^{**}}(F), Y^{\perp\perp})$. Let us assume $d(cent_X(F), Y) > d(cent_{X^{**}}(F), Y^{\perp\perp})$ and hence there exist $\varepsilon, \delta > 0$ such that $d((cent_X(F))_\delta, Y) > d(cent_{X^{**}}(F), Y^{\perp\perp}) + \varepsilon$. Get $\Phi \in cent_{X^{**}}(F)$ and $y^{**} \in Y^{\perp\perp}$ such that $\|\Phi - y^{**}\| = d(cent_{X^{**}}(F), Y^{\perp\perp})$. Let $Z = span\{\{\Phi, y^{**}\} \cup F\}$ and $W = span\{y_i^* : 1 \leq i \leq n\}$. Then Z, W are finite dimensional subspace of X^{**} and X^* respectively. Choose $\eta > 0$ such that $\eta < \frac{\min\{\varepsilon, \delta/2\}}{\max\{\|\Phi - y^{**}\|, r(\Phi, F)\}}$ and by PLR get a $T : Z \to X$ an isomorphism into its range such that $\|T\|\|T^{-1}\| \leq 1 + \eta$ and also satisfying $T(z) = z$, $z \in F$ and also $f(T(x^{**})) = x^{**}(f)$ whenever $x^{**} \in Z$, $f \in W$. It is clear that $Ty^{**} \in Y$.

CLAIM: $T\Phi \in (cent_X(F))_\delta$.

$$
\begin{aligned}
\text{For } z \in F, \|T\Phi - z\| &= \|T(\Phi - z)\| \\
&\leq (1 + \eta)\|\Phi - z\| \\
&\leq r(\Phi, F) + \delta/2 \\
&= rad_{X^{**}}(F) + \delta/2 = rad_X(F) + \delta/2.
\end{aligned}
$$

Hence $r(T\Phi, F) < rad_X(F) + \delta/2$ and by Theorem 2.5 we have that $T\Phi \in cent_X(F) + \delta B_X = (cent_X(F))_\delta$.

Now $\|T\Phi - Ty^{**}\| \leq \|\Phi - y^{**}\|(1 + \eta) < d(cent_{X^{**}}(F), Y^{\perp\perp}) + \varepsilon$, this contradict the fact that $d(cent_{X^{**}}(F), Y^{\perp\perp}) + \varepsilon < d((cent_X(F))_\delta, Y^{\perp\perp})$. Hence the result follows.

(b) For a given $\varepsilon > 0$ and $F \in \mathcal{F}$ get $\delta > 0$ from the Property-(P_1) of $(X, Y, \mathcal{F})$. Choose $\eta > 0$ such that $\eta < \min\left\{\frac{(d(y, cent_Y(F)) - \varepsilon) - \|y - \Phi\|}{\|y - \Phi\|}, \frac{\delta}{rad_{Y^{\perp\perp}}(F)}\right\}$.

(c) If possible let for some $F \in \mathcal{F}$ and $t > 0$, $\varepsilon_{Y^{\perp\perp}}(F,t) < \varepsilon_Y(F,t)$. Get a $\varepsilon > 0$ and $y \in Y$ with $y \in t - cent_Y(F)$ such that $\varepsilon_{Y^{\perp\perp}}(F,t) < \varepsilon_Y(F,t) - \varepsilon$ and

$$d(y, cent_Y(F)) > \varepsilon_Y(F,t) - \varepsilon. \tag{1}$$

It is clear that $y \in t - cent_{Y^{\perp\perp}}(F)$ and

$$d(y, cent_{Y^{\perp\perp}}(F)) < \varepsilon_Y(F,t) - \varepsilon. \tag{2}$$

The last inequality follows from the fact that $\varepsilon_Y(F,t)-\varepsilon > \varepsilon_{Y^{\perp\perp}}(F,t)$, the modulus. The inequalities in equations 1, 2 contradicting each other. Hence the result follows. □

COROLLARY 3.6. *Let Y be a finite co-dimensional subspace of a L_1 predual space X which is strongly proximinal and F be a finite subset of X. Then $rad_Y(F) = rad_X(F) + d(cent_X(F), Y)$.*

PROOF. From Proposition 3.5(a) it follows that $d(cent_X(F), Y) = d(cent_{X^{**}}(F), Y^{\perp\perp})$. Since X^{**} is a space of type $C(\Omega)$ for some compact Hausdorff space Ω and $Y^{\perp\perp}$ is strongly proximinal we have from [**5**, Theorem 2.1] the linear functionals which determine the subspace $Y^{\perp\perp}$ are finitely supported. Hence from [**11**, Theorem 2.2, Proposition 3.3] we have $rad_{Y^{\perp\perp}}(F) = rad_{X^{**}}(F) + d(cent_{X^{**}}(F), Y^{\perp\perp})$. Since $rad_{Y^{\perp\perp}}(F) = rad_Y(F)$ and $rad_{X^{**}}(F) = rad_X(F)$ we have the result. □

THEOREM 3.7. *Let $V \in CL(X)$ and $\mathcal{F} \subseteq CB(X)$ be such that $(V, \mathcal{F})$ has r.c.p.. Then,*

- (a) *If $cent_V(.)$ is continuous at F then $\varepsilon_V(.,t)$ continuous at F for all $t > 0$.*
- (b) *If $(V, \mathcal{F})$ has Property-(P_1) and $\varepsilon_V(.,t)$ are continuous at F for all $t > 0$ then $cent_V(.)$ is continuous at F.*

PROOF. CASE 1: When $cent_V(F)$ is continuous.

CASE 1.1: $cent_V(.)$ is lHsc at F implies $\varepsilon_V(.,t)$ usc at F for all $t > 0$.

Fix $t > 0$ and let $F_n \xrightarrow{H} F$. We show that $\limsup_n \varepsilon_V(F_n,t) \leq \varepsilon(F,t)$.

Let $d_n = rad_V(F_n)$. Let $\alpha > \varepsilon_V(F,t)$, choose $r > 0$ such that $\alpha > \alpha - r > \varepsilon_V(F,t)$. Hence $t - cent_V(F) \subseteq cent_V(F) + (\alpha - r)B_X$.

For any β with $t > \beta > 0$ and n large we have

$$(t-\beta) - cent_V(F_n) \subseteq t - cent_V(F) \subseteq cent_V(F) + (\alpha - r)B_X.$$

Now for large $n, cent_V(F) \subseteq cent_V(F_n) + \frac{r}{2}B_X$ and hence from above $(t-\beta) - cent_V(F_n) \subseteq cent_V(F_n) + (\alpha - \frac{r}{2})B_X$.

Hence $\alpha - \frac{r}{2} \geq \varepsilon_V(F_n, t-\beta) \geq \varepsilon_V(F_n,t) - \beta\frac{2d_n+t}{t}$.

Taking $n \to \infty$ as $d_n \to d = rad_V(F)$ we have $\limsup_n \varepsilon_V(F_n,t) \leq \alpha$.

This establishes Case 1.1.

CASE 1.2: $cent_V(.)$ is uHsc at F implies $\varepsilon_V(.,t)$ is lsc at F for all $t > 0$.

Fix $t > 0$ and let $F_n \xrightarrow{H} F$. We show that $\liminf_n \varepsilon_V(F_n,t) \geq \varepsilon_V(F,t)$.

Let $\alpha < \varepsilon_V(F,t)$, choose $r > 0$ such that $\alpha < \alpha + r < \varepsilon_V(F,t)$.

If along some subsequence $\varepsilon_V(F_n,t) \leq \alpha$ then

$$t - cent_V(F_n) \subseteq cent_V(F_n) + (\alpha + \frac{r}{4})B_X.$$

Since $cent_V(.)$ is uHsc at F, for large n

$$cent_V(F_n) \subseteq cent_V(F) + \frac{r}{4}B_X.$$

Hence $t - cent_V(F_n) \subseteq cent_V(F) + (\alpha + \frac{r}{2})B_X$.

Now for any β with $t > \beta > 0$ and large n,

$$(t-\beta) - cent_V(F) \subseteq t - cent_V(F_n)$$

To prove the last statement let $p \notin t - cent_V(F_n)$ ie. $r(p, F_n) > rad_V(F_n) + t$ ie. $r(p, F) > r(p, F_n) - d_H(F_n, F) > rad_V(F_n) + t - d_H(F_n, F)$.

Since $rad_V(F_n) \to rad_V(F)$ we have for sufficiently large n

$$r(p, F) \geq rad_V(F) + (t-\beta) \text{ ie. } p \notin (t-\beta) - cent_V(F_n).$$

Hence $(t-\beta) - cent_V(F) \subseteq cent_V(F) + (\alpha + \frac{r}{2})B_X$. That is

$$(\alpha + \frac{r}{2}) \geq \varepsilon_V(F, t-\beta) \geq \varepsilon_V(F, t) - \beta\frac{2d+t}{t}$$

Since β is arbitrary we have $\varepsilon_V(F, t) \leq \alpha$, a contradiction. This establishes Claim 2.

Case 2: When $\varepsilon_V(., t)$ are all continuous at F for all $t > 0$.

By continuity of $\varepsilon_V(F, .)$ at 0 there exists $t_0 > 0$ such that $\varepsilon(F, t_0) < \frac{\varepsilon}{4}$. Now let $F_n \xrightarrow{H} F$, since $\varepsilon_V(., t_0)$ is continuous at F we have $\lim_n \varepsilon_V(F_n, t_0) = \varepsilon_V(F, t_0)$, hence for large n $\varepsilon_V(F_n, t_0) \leq \varepsilon_V(F, t_0) + \frac{\varepsilon}{4}$.

For large n,

$$cent_V(F_n) \subseteq t_0 - cent_V(F) \subseteq cent_V(F) + (\varepsilon_V(F, t_0) + \frac{\varepsilon}{4})B_X$$

and also

$$cent_V(F) \subseteq t_0 - cent_V(F_n) \subseteq cent_V(F_n) + \varepsilon_V(F_n, t_0)(1 + \frac{1}{n})B_X$$

Thus

$$d_H\left(cent_V(F), cent_V(F_n)\right) \leq \max\{\varepsilon_V(F, t_0) + \frac{\varepsilon}{4}, \varepsilon_V(F_n, t_0)(1 + \frac{1}{n})\} \leq \varepsilon.$$

Hence the result follows. □

References

[1] D. Amir, *Chebyshev centers and uniform convexity*, Pacific J. Math. **77** (1978), no. 1, 1–6. MR507615

[2] D. Amir and F. Deutsch, *Approximation by certain subspaces in the Banach space of continuous vector-valued functions*, J. Approx. Theory **27** (1979), no. 3, 254–270, DOI 10.1016/0021-9045(79)90108-4. MR555626

[3] D. Amir, J. Mach, and K. Saatkamp, *Existence of Chebyshev centers, best n-nets and best compact approximants*, Trans. Amer. Math. Soc. **271** (1982), no. 2, 513–524, DOI 10.2307/1998896. MR654848

[4] D. Amir and Z. Ziegler, *Relative Chebyshev centers in normed linear spaces. I*, J. Approx. Theory **29** (1980), no. 3, 235–252, DOI 10.1016/0021-9045(80)90129-X. MR597471

[5] S. Dutta and D. Narayana, *Strongly proximinal subspaces of finite codimension in C(K)*, Colloq. Math. **109** (2007), no. 1, 119–128, DOI 10.4064/cm109-1-10. MR2308830

[6] A. L. Garkavi, *The conditional Čebyšev center of a compact set of continuous functions* (Russian), Mat. Zametki **14** (1973), 469–478. MR0328443

[7] G. Godefroy and V. Indumathi, *Strong proximinality and polyhedral spaces*, Rev. Mat. Complut. **14** (2001), no. 1, 105–125, DOI 10.5209/rev_REMA.2001.v14.n1.17047. MR1851725

[8] G. Godini, *Best approximation and intersections of balls*, Banach space theory and its applications (Bucharest, 1981), Lecture Notes in Math., vol. 991, Springer, Berlin, 1983, pp. 44–54, DOI 10.1007/BFb0061557. MR714172

[9] D. V. Pai and P. T. Nowroji, *On restricted centers of sets*, J. Approx. Theory **66** (1991), no. 2, 170–189, DOI 10.1016/0021-9045(91)90119-U. MR1115277

[10] R. Payá and D. Yost, *The two-ball property: transitivity and examples*, Mathematika **35** (1988), no. 2, 190–197, DOI 10.1112/S0025579300015187. MR986628

[11] P. W. Smith and J. D. Ward, *Restricted centers in $C(\Omega)$*, Proc. Amer. Math. Soc. **48** (1975), 165–172, DOI 10.2307/2040710. MR0380227

[12] D. T. Yost, *Best approximation and intersections of balls in Banach spaces*, Bull. Austral. Math. Soc. **20** (1979), no. 2, 285–300, DOI 10.1017/S0004972700010972. MR557239

[13] D. Yost, *The n-ball properties in real and complex Banach spaces*, Math. Scand. **50** (1982), no. 1, 100–110, DOI 10.7146/math.scand.a-11947. MR664511

[14] D. Yost, *Intersecting balls in spaces of vector-valued functions*, Banach space theory and its applications (Bucharest, 1981), Lecture Notes in Math., vol. 991, Springer, Berlin, 1983, pp. 296–302, DOI 10.1007/BFb0061578. MR714193

DEPT. OF MATHEMATICS, INDIAN INSTITUTE OF TECHNOLOGY HYDERABAD, INDIA
Email address: ma17resch11003@iith.ac.in & tanmoy@iith.ac.in

Contemporary Mathematics
Volume **737**, 2019
https://doi.org/10.1090/conm/737/14860

2-local isometries on function spaces

Osamu Hatori and Shiho Oi

ABSTRACT. We study 2-local reflexivity of the set of all surjective isometries between certain function spaces. We do not assume linearity for isometries. We prove that a 2-local isometry in the group of all surjective isometries on the algebra of all continuously differentiable functions on the closed unit interval with respect to several norms is a surjective isometry. We also prove that a 2-local isometry in the group of all surjective isometries on the Banach algebra of all Lipschitz functions on the closed unit interval with the sum-norm is a surjective isometry.

1. Introduction

Motivated by the paper by Kowalski and Słodkowski [**16**], the concept of 2-locality was introduced by Šemrl,who obtained the first results on 2-local automorphisms and 2-local derivations [**25**]. Molnár [**23**] studied 2-local isometries on operator algebras. Given a metric space $\mathfrak{M}_j$ for $j = 1, 2$ an isometry from $\mathfrak{M}_1$ into $\mathfrak{M}_2$ is a distance preserving map. The set of all surjective isometries from $\mathfrak{M}_1$ onto $\mathfrak{M}_2$ is denoted by $\mathrm{Iso}(\mathfrak{M}_1, \mathfrak{M}_2)$, and $\mathrm{Iso}(\mathfrak{M})$ if $\mathfrak{M}_1 = \mathfrak{M}_2 = \mathfrak{M}$. We say a map $T : \mathfrak{M}_1 \to \mathfrak{M}_2$ is 2-local in $\mathrm{Iso}(\mathfrak{M}_1, \mathfrak{M}_2)$ if for every pair $x, y \in \mathfrak{M}_1$ there exists a surjective isometry $T_{x,y} \in \mathrm{Iso}(\mathfrak{M}_1, \mathfrak{M}_2)$ such that

$$T(x) = T_{x,y}(x) \text{ and } T(y) = T_{x,y}(y).$$

In this case we say that T is a 2-local isometry. It is obvious by the definition that a 2-local isometry is in fact an isometry, which needs not to be surjective. Hence a 2-local isometry T belongs to $\mathrm{Iso}(\mathfrak{M}_1, \mathfrak{M}_2)$ if T is surjective. We say that $\mathrm{Iso}(\mathfrak{M}_1, \mathfrak{M}_2)$ is 2-local reflexive if every 2-local isometry belongs to $\mathrm{Iso}(\mathfrak{M}_1, \mathfrak{M}_2)$.

If $\mathfrak{M}_j$ is a Banach space, linearity of the maps is a subject of consideration. Let $\mathrm{Iso}_{\mathbb{C}}(\mathfrak{M}_1, \mathfrak{M}_2)$ denote the set of all surjective *complex-linear* isometries. There exists an extensive literature on 2-local isometries in $\mathrm{Iso}_{\mathbb{C}}(\mathfrak{M}_1, \mathfrak{M}_2)$ and 2-iso-reflexivity of $\mathrm{Iso}_{\mathbb{C}}(\mathfrak{M}_1, \mathfrak{M}_2)$ (see, for example, [**1,2,4,7,11,12,17,22,23**]). Note that Hosseini showed that a 2-local *real-linear* isometry is in fact a surjective real-linear isometry on the algebra of n-times continuously differentiable functions on the interval $[0, 1]$ with a certain norm [**9**, Theorem 3.1]. She verified that a 2-local real-linear isometry defined on the Banach algebra $C(X)$ of all complex-valued continuous functions on a compact Hausdorff space X which is separable and first countable is in fact a surjective real-linear isometry on $C(X)$ [**9**, Proposition 3.2]. At this point we

2010 *Mathematics Subject Classification.* 46B04,46J15,46J10.
Key words and phrases. 2-local maps, surjective isometries, continuously differentiable maps.

emphasize that the situation is very different from that for the problem of 2-local isometries. We do not know if a 2-local isometry defined on $C[0,1]$ is a surjective isometry on $C[0,1]$ or not. The problem of whether the group of all surjective isometries (without assuming linearity) of $C(X)$ is 2-local reflexive or not has been raised by Molnár who has proved a related positive result concerning the group of all surjective isometries in the setting of operator algebras [**21**].

In this paper we study 2-local reflexivity for $\mathrm{Iso}(\mathfrak{M}_1, \mathfrak{M}_2)$ and we consider the question whether every 2-local isometry necessarily belongs to $\mathrm{Iso}(\mathfrak{M}_1, \mathfrak{M}_2)$, where $\mathfrak{M}_j$ is a certain space of continuous functions.

2. Preliminaries

Let X_j be a compact Hausdorff space for $j = 1, 2$. The algebra of all complex-valued continuous functions on X_j is denoted by $C(X_j)$. The supremum norm is denoted by $\|\cdot\|_\infty$. In the rest of the paper E_j is a subspace of $C(X_j)$ which contains the constant functions and separates the points of X_j. For $c \in \mathbb{C}$ we write the constant function which takes the value c by c. We assume that the norm $\|\cdot\|_j$ is defined on E_j (not necessary complete) and it satisfies that $\|c\|_j = |c|$ for every $c \in \mathbb{C}$. We assume that E_j is conjugate closed in the sense that $f \in E_j$ implies $\bar{f} \in E_j$, and that $\|f\|_j = \|\bar{f}\|_j$ for every $f \in E_j$. For an $\epsilon \in \{\pm 1\}$ and $f \in E_j$, $[f]^\epsilon = f$ if $\epsilon = 1$ and $[f]^\epsilon = \bar{f}$ if $\epsilon = -1$. Let $M(E_1, E_2)$ be the set of all maps from E_1 into E_2. Note that we say a map is a surjective isometry if it is just a distance preserving map, we do not assume complex nor real linearity on it. We abbreviate $\mathrm{Iso}(E_j, E_j)$ by $\mathrm{Iso}(E_j)$. Let Π denotes a non-empty set of (not always all) homeomorphisms from E_2 onto E_1. Let

$$\begin{aligned} G_\Pi(E_1, E_2) = \{T \in M(E_1, E_2) : &\text{ there exists a } \lambda \in E_2, \\ &\text{an } \alpha \in \mathbb{C} \text{ of unit modulus, a } \pi \in \Pi, \text{ and an } \epsilon \in \{\pm 1\} \\ &\text{such that } T(f) = \lambda + \alpha[f \circ \pi]^\epsilon \text{ for every } f \in E_1\}. \end{aligned}$$

We abbreviate $G_\Pi(E_j, E_j)$ by $G_\Pi(E_j)$. We usually abbreviate $G_\Pi(E_1, E_2)$ by G_Π if E_1 and E_2 are clear from the context. Let $\mathrm{Id}_{[0,1]} = \pi_0 : [0,1] \to [0,1]$ be the identity function and $\pi_1 = 1 - \mathrm{Id}_{[0,1]}$. Put $\Pi_0 = \{\pi_0, \pi_1\}$. Kawamura, Koshimizu and Miura [**14**] (cf. [**20**]) proved that $G_{\Pi_0}(C^1[0,1]) = \mathrm{Iso}(C^1[0,1], \|\cdot\|)$ with respect to several norms including $\|\cdot\|_\Sigma$. For a compact metric space K, let

$$\mathrm{Lip}(K) = \left\{ f \in C(K) : L_f = \sup_{x \neq y} \frac{|f(x) - f(y)|}{d(x,y)} < \infty \right\}$$

with the norm $\|f\|_\Sigma = \|f\|_\infty + L_f$ for $f \in \mathrm{Lip}(K)$. We say that L_f is the Lipschitz constant for f. With this norm $\mathrm{Lip}(K)$ is a unital semisimple commutative Banach algebra. We show in section 4 that $G_\Pi(\mathrm{Lip}(K_1), \mathrm{Lip}(K_2)) = \mathrm{Iso}(\mathrm{Lip}(K_1), \mathrm{Lip}(K_2))$, where Π is the set of all surjective isometries from K_2 onto K_1.

Let

$$\begin{aligned} W_j = \{f \in E_j : &\text{ if } S : \mathbb{C} \to \mathbb{C} \text{ is an isometry} \\ &\text{and } S(f(X_j)) = f(X_j), \text{ then } S \text{ is the identity function}\}. \end{aligned}$$

Suppose that $S : \mathbb{C} \to \mathbb{C}$ is an isometry. It is well known that there exist $a, b \in \mathbb{C}$ with $|a| = 1$ such that $S(z) = b + az$, $(z \in \mathbb{C})$ or $S(z) = b + a\bar{z}$, $(z \in \mathbb{C})$. The first case of S is a parallel translation by b if $a = 1$ and S is a rotation around

$b/(1-a)$. Hence there is no fixed point if $a = 1$ and $b \neq 0$, and $b/(1-a)$ is the unique fixed point if $a \neq 1$ for the first case. For the second case, denoting one of the square roots of a by $a^{\frac{1}{2}}$, S is a symmetric translation with respect to the line $t \mapsto a^{\frac{1}{2}}t + i(\mathrm{Im}(\overline{a^{\frac{1}{2}}}b)/2)$ $(t \in \mathbb{R})$ followed by the parallel translation by $\mathrm{Re}(\overline{a^{\frac{1}{2}}}b)/2)$ to the direction of $a^{\frac{1}{2}}$. Hence the fixed points exist and they are all the points on line $t \mapsto a^{\frac{1}{2}}t + i(\mathrm{Im}(\overline{a^{\frac{1}{2}}}b)/2)$ if and only if $\mathrm{Re}(\overline{a^{\frac{1}{2}}}b)/2) = 0$. We will prove that W_j for $E_j = C^1[0,1]$ is uniformly dense in $C[0,1]$ (see Proposition 3).

LEMMA 1. *If $T \in M(E_1, E_2)$ is 2-local in $G_\Pi \cap \mathrm{Iso}(E_1, E_2)$, then T is an isometry with respect to the metric induced by the norm $\|\cdot\|_j$; $\|T(f) - T(g)\|_2 = \|f - g\|_1$ for every pair $f, g \in E_1$. The map T is also an isometry with respect to the supremum norm $\|\cdot\|_\infty$.*

PROOF. Let $f, g \in E_1$. Then there exists $T_{f,g} \in G_\Pi \cap \mathrm{Iso}(E_1, E_2)$ such that

$$(1) \qquad T(f) = T_{f,g}(f), \qquad T(g) = T_{f,g}(g).$$

As $T_{f,g}$ is an isometry we have

$$\|T(f) - T(g)\|_2 = \|T_{f,g}(f) - T_{f,g}(g)\|_2 = \|f - g\|_1.$$

Thus T is an isometry. As $T_{f,g} \in G_\Pi \cap \mathrm{Iso}(E_1, E_2)$ there exists a $\lambda_{f,g} \in E_2$, an $\alpha_{f,g} \in \mathbb{C}$ of unit modulus, a $\pi \in \Pi$ and an $\epsilon_{f,g} \in \{\pm 1\}$ such that

$$(2) \qquad T_{f,g}(h) = \lambda_{f,g} + \alpha_{f,g}[h \circ \pi]^{\epsilon_{f,g}}, \quad h \in E_1.$$

Then by (1) and (2) we observe that

$$\begin{aligned}\|T(f) - T(g)\|_\infty &= \|T_{f,g}(f) - T_{f,g}(g)\|_\infty \\ &= \|\alpha_{f,g}[f \circ \pi]^\epsilon - \alpha_{f,g}[g \circ \pi]^\epsilon\|_\infty \\ &= \|[f \circ \pi]^\epsilon - [g \circ \pi]^\epsilon\|_\infty = \|f - g\|_\infty\end{aligned}$$

since π is a surjection. Thus T is an isometry with respect to the supremum norm. □

PROPOSITION 2. *Suppose that $T \in M(E_1, E_2)$ is 2-local in $G_\Pi \cap \mathrm{Iso}(E_1, E_2)$. Then there exists $\epsilon \in \{\pm 1\}$ and $\alpha \in \mathbb{C}$ of unit modulus such that for every $f \in W_1$ there exists a homeomorphism $\pi_f \in \Pi$ such that*

$$T(f) = T(0) + \alpha[f \circ \pi_f]^\epsilon.$$

Note that if we proved that π_f did not depend on f, then the map T were surjective, hence $T \in \mathrm{Iso}(E_1, E_2)$ by the Mazur-Ulam theorem. But it is not the case in general (cf. [**7**, Theorem 2.3]); $G_\Pi \cap \mathrm{Iso}(E_1, E_2)$ needs not be 2-local reflexive in $M(E_1, E_2)$.

PROOF OF PROPOSITION 2. Put $T_0 = T - T(0)$. We infer by a simple calculation that T_0 is 2-local in $G_\Pi \cap \mathrm{Iso}(E_1, E_2)$. Let $h \in E_1$. Since T_0 is 2-local in $G_\Pi \cap \mathrm{Iso}(E_1, E_2)$, there exist $T_{h,0} \in G_\Pi \cap \mathrm{Iso}(E_1, E_2), \lambda_{h,0} \in E_2, \alpha_{h,0} \in \mathbb{C}$ with $|\alpha_{h,0}| = 1$, a homeomorphism $\pi_{h,0} : X_2 \to X_2$ and $\epsilon_{h,0} \in \{\pm 1\}$ such that

$$(3) \qquad \begin{aligned} T_0(h) &= T_{h,0}(h) = \lambda_{h,0} + \alpha_{h,0}[h \circ \pi_{h,0}]^{\epsilon_{h,0}}, \\ 0 = T_0(0) &= T_{h,0}(0) = \lambda_{h,0}. \end{aligned}$$

Note that $T_{h,0}$ is represented by $T_{h,0}(\cdot) = \alpha_{h,0}[\cdot \circ \pi_{h,0}]^{\epsilon_{h,0}}$ on E_1. Hence $\lambda_{h,0} = 0$ and

$$T_0(h) = \alpha_{h,0}[h \circ \pi_{h,0}]^{\epsilon_{h,0}}. \tag{4}$$

In particular, if $h = c \in \mathbb{C}$, then

$$T_0(c) = \alpha_{c,0}[c]^{\epsilon_{h,0}}.$$

Thus we obtain $T_0(\mathbb{C}) \subset \mathbb{C}$. For every pair $c, d \in \mathbb{C}$, here exists $T_{c,d} \in G_\Pi \cap \mathrm{Iso}(E_1, E_2)$ such that

$$T_0(c) = T_{c,d}(c), \quad T_0(d) = T_{c,d}(d).$$

As $T_{c,d} \in \mathrm{Iso}(E_1, E_2)$ we infer that

$$\begin{aligned}|T_0(c) - T_0(d)| &= \|T_0(c) - T_0(d)\|_2 \\ &= \|T_{c,d}(c) - T_{c,d}(d)\|_2 = \|c - d\|_1 = |c - d|.\end{aligned}$$

As c, d are arbitrary, we have that $T_0|_{\mathbb{C}} : \mathbb{C} \to \mathbb{C}$ is an isometry. Applying a well known result about the form of an isometry on $\mathbb{C}$, there exists an $\alpha \in \mathbb{C}$ of modulus one such that

$$T_0(z) = \alpha z, \quad z \in \mathbb{C}, \tag{5}$$

or

$$T_0(z) = \alpha \bar{z}, \quad z \in \mathbb{C}. \tag{6}$$

Without loss of generality we may assume that (5) holds and $\alpha = 1$; $T_0(z) = z$ for every $z \in \mathbb{C}$. (Put $T_1 = [\bar{\alpha} T_0]^\epsilon$, where $\epsilon = 1$ if (5) holds and $\epsilon = -1$ if (6) holds. Since E_2 is conjugate closed, T_1 is well defined in the second case. Since $\|f\|_2 = \|\bar{f}\|_2$ for every $f \in E_2$, it is a routine work to see that T_1 is 2-local in $G_\Pi \cap \mathrm{Iso}(E_1, E_2)$. By the definition of T_1 we infer that $T_1(z) = z$ for every $z \in \mathbb{C}$. In this case, we can prove that for every $f \in W_1$ there exists a homeomorphism $\pi_f : X_2 \to X_1$ such that

$$T_1(f) = f \circ \pi_f.$$

If it is proved, then we have

$$T(f) = T(0) + \alpha[f \circ \pi_f]^\epsilon,$$

the desired form.)

We prove that for every $f \in W_1$ there exists a homeomorphism $\pi_f : X_2 \to X_1$ such that $T_0(f) = f \circ \pi_f$. Let $f \in W_1$ and $c \in \mathbb{C}$. As T_0 is 2-local in $G_\Pi \cap \mathrm{Iso}(E_1, E_2)$, there exists $\lambda_{f,c} \in E_2$ and $\alpha_{f,c} \in \mathbb{C}$ of modulus 1, a homeomorphism $\pi_{f,c} : X_2 \to X_1$, and $\epsilon_{f,c} \in \{\pm 1\}$ such that

$$T_0(f) = T_{f,c}(f) = \lambda_{f,c} + \alpha_{f,c}[f \circ \pi_{f,c}]^{\epsilon_{f,c}}, \tag{7}$$

$$c = T_0(c) = \lambda_{f,c} + \alpha_{f,c}[c]^{\epsilon_{f,c}}, \tag{8}$$

where $T_{f,c}(\cdot) = \lambda_{f,c} + \alpha_{f,c}[\cdot \circ \pi_{f,c}]^{\epsilon_{f,c}}$. By (8) we infer that $\lambda_{f,c}$ is a constant. By comparing (4) for f with (7) we get

$$\alpha_{f,0}[f \circ \pi_{f,0}]^{\epsilon_{f,0}} = \lambda_{f,c} + \alpha_{f,c}[f \circ \pi_{f,c}]^{\epsilon_{f,c}},$$

hence

$$f \circ \pi_{f,0} = [\overline{\alpha_{f,0}} \lambda_{f,c} + \overline{\alpha_{f,0}} \alpha_{f,c}[f \circ \pi_{f,c}]^{\epsilon_{f,c}}]^{\epsilon_{f,0}}. \tag{9}$$

Considering the range of the both side of (9) we get

(10) $$f(X_1) = [\overline{\alpha_{f,0}}\lambda_{f,c} + \overline{\alpha_{f,0}}\alpha_{f,c}[f(X_1)]^{\epsilon_{f,c}}]^{\epsilon_{f,0}}.$$

We have four possibility depending on (i)$\epsilon_{f,0} = 1$ and $\epsilon_{f,c} = 1$; (ii)$\epsilon_{f,0} = 1$ and $\epsilon_{f,c} = -1$; (iii)$\epsilon_{f,0} = -1$ and $\epsilon_{f,c} = -1$; (iv)$\epsilon_{f,0} = -1$ and $\epsilon_{f,c} = 1$. Then we have that at least one of the following (i) through (iv) holds.

$$\text{(i)}\quad f(X_1) = \overline{\alpha_{f,0}}\lambda_{f,c} + \overline{\alpha_{f,0}}\alpha_{f,c}f(X_1);$$

$$\text{(ii)}\quad f(X_1) = \overline{\alpha_{f,0}}\lambda_{f,c} + \overline{\alpha_{f,0}}\alpha_{f,c}\overline{f(X_1)};$$

$$\text{(iii)}\quad f(X_1) = \alpha_{f,0}\overline{\lambda_{f,c}} + \alpha_{f,0}\overline{\alpha_{f,c}}f(X_1);$$

$$\text{(iv)}\quad f(X_1) = \alpha_{f,0}\overline{\lambda_{f,c}} + \alpha_{f,0}\overline{\alpha_{f,c}}\overline{f(X_1)},$$

corresponding to the cases (i) through (iv) respectively. Let $S_j : \mathbb{C} \to \mathbb{C}$ be defined by $S_1(z) = \overline{\alpha_{f,0}}\lambda_{f,c} + \overline{\alpha_{f,0}}\alpha_{f,c}z$, $z \in \mathbb{C}$; $S_2(z) = \overline{\alpha_{f,0}}\lambda_{f,c} + \overline{\alpha_{f,0}}\alpha_{f,c}\overline{z}$, $z \in \mathbb{C}$; $S_3(z) = \alpha_{f,0}\overline{\lambda_{f,c}} + \alpha_{f,0}\overline{\alpha_{f,c}}z$, $z \in \mathbb{C}$; $S_4(z) = \alpha_{f,0}\overline{\lambda_{f,c}} + \alpha_{f,0}\overline{\alpha_{f,c}}\overline{z}$, $z \in \mathbb{C}$. Then S_j is an isometry on $\mathbb{C}$ for $j = 1, 2, 3, 4$. Using S_j we rewrite (i) by $f(X_1) = S_1(f(X_1))$, (ii) by $f(X_1) = S_2(f(X_1))$, (iii) by $f(X_1) = S_3(f(X_1))$ and (iv) by $f(X_1) = S_4(f(X_1))$. As $f \in W_1$ we see that (ii) and (iv) do not occur. We have $\overline{\alpha_{f,0}}\lambda_{f,c} = 0$ and $\overline{\alpha_{f,0}}\alpha_{f,c} = 1$ for the case (i). Hence $\lambda_{f,c} = 0$ and $\alpha_{f,0} = \alpha_{f,c}$ in this case. In the same way we have $\lambda_{f,c} = 0$ and $\alpha_{f,0} = \alpha_{f,c}$ for the case (iii). We conclude that only (i) or (iii) occur, and in any case

$$\lambda_{f,c} = 0 \text{ and } \alpha_{f,0} = \alpha_{f,c}.$$

To prove $\alpha_{f,0} = 1$, put $c = 1$. Then (i) or (iii) occur. If (i) occurs, then

$$1 = T_0(1) = T_{f,1}(1) = \alpha_{f,1}1 = \alpha_{f,0}1.$$

If (iii) occurs, then

$$1 = T_0(1) = T_{f,1}(1) = \alpha_{f,1}\bar{1} = \alpha_{f,0}1.$$

It follows that

(11) $$\alpha_{f,0} = 1.$$

Next we prove that (iii) does not occur for any $c \in \mathbb{C}$. Suppose that (iii) occurs for some $c_0 \in \mathbb{C}$. Then we have

$$c_0 = T_0(c_0) = T_{f,c_0}(c_0) = \alpha_{f,c_0}\overline{c_0} = \alpha_{f,0}\overline{c_0} = \overline{c_0}.$$

Hence c_0 is a real number. We also have

(12) $$T_0(f) = T_{f,c_0}(f) = \overline{f \circ \pi_{f,c_0}}$$

since $\alpha_{f,c_0} = \alpha_{f,0} = 1$. On the other hand, (iii) does not occur for $c_0 + i$ since $c_0 + i$ is not a real number. Thus (i) occurs and

(13) $$T_0(f) = T_{f,c_0+i}(f) = f \circ \pi_{f,c_0+i}.$$

It follows by (12) and (13) that

$$\overline{f(X_1)} = f(X_1),$$

which is a contradiction since $f \in W_1$. We conclude that (iii) does not occur for any $c \in \mathbb{C}$. It follows that only (i) occurs, hence we have

(14) $$\epsilon_{f,0} = 1.$$

Then by (4) for $h = f$ we have

$$T_0(f) = T_{f,0}(f) = f \circ \pi_{f,0}.$$

Letting $\pi_f = \pi_{f,0}$ we have the conclusion. □

3. Spaces of continuous functions on $[0,1]$

In this section A_j $(j = 1,2)$ is a subspace of $C[0,1]$ and a superspace of $C^1[0,1]$, the space of all complex-valued continuously differentiable functions on the interval $[0,1]$. We assume that A_j satisfies the following three conditions.

1) A_j is conjugate-closed in the sense that $f \in A_j$ implies $\bar{f} \in A_j$;
2) The norm $\|\cdot\|_j$ on A_j satisfies that $|c| = \|c\|_j$ for every $c \in \mathbb{C}$;
3) The norm $\|\cdot\|_j$ on A_j satisfies that $\|f\|_j = \|\bar{f}\|_j$ for every $f \in A_j$, where $\bar{\cdot}$ denotes the complex-conjugation.

We do not assume the completeness of $\|\cdot\|_j$. The space A_j satisfies the conditions for E_j in the previous section. The difference between A_j and E_j is that A_j is defined on $[0,1]$ and we assume that $C^1[0,1] \subset A_j$. The spaces $(C^1[0,1], \|\cdot\|_\Sigma)$, $(C^1[0,1], \|\cdot\|_M)$, $(\mathrm{Lip}[0,1], \|\cdot\|_\Sigma)$, $(\mathrm{Lip}[0,1], \|\cdot\|_M)$ and $(C[0,1], \|\cdot\|_\infty)$ are typical examples of A_j. Recall that π_0 is the identity function on $[0,1]$ and $\pi_1 = 1 - \pi_0$, $\Pi_0 = \{\pi_0, \pi_1\}$. Kawamura, Koshimizu and Miura [**14**] studied the space $C^1[0,1]$ with a variety of norms including $\|\cdot\|_\Sigma$ and $\|\cdot\|_M$. Recall that

$$W_1 = \{f \in A_1 : \text{if } S : \mathbb{C} \to \mathbb{C} \text{ is an isometry} \\ \text{and } S(f(X_1)) = f(X_1), \text{ then } S \text{ is the identity function}\}.$$

Put $P = \{p + iq : p \text{ and } q \text{ are polynomials of real-coefficients.}\}$. Many polynomials are in W_1, but some are not. For example $(t - \frac{1}{2})^4 + i(t - \frac{1}{2})^3 \notin W_1$. We do not know if $P \cap W_1$ is uniformly closed in P or not. We have that following. Let $cl(\cdot)$ denote the uniform closure on $[0,1]$.

PROPOSITION 3. *We have $P \subset cl(W_1)$. Hence $cl(W_1) = C[0,1]$.*

PROOF. Let $f = p + iq \in P$ and $\varepsilon > 0$. If p is not a constant, then put $p_\varepsilon = p$. If p is a constant, then put $p_\varepsilon = p + \varepsilon\pi_0$, where π_0 is the identity function on $[0,1]$. Let l be any positive integer greater than both of the degree of p and q. Put $q_\varepsilon = q + \varepsilon\pi_0^l$. Then p_ε is not a constant and there is no pair of complex numbers c and d such that $p_\varepsilon = cq_\varepsilon + d$ since the degree of the each side of the equation is different. We prove that $p_\varepsilon + iq_\varepsilon \in cl(W_1)$. Then $p + iq \in cl(W_1)$ follows since $p_\varepsilon + iq_\varepsilon$ uniformly converges on $[0,1]$ to $p + iq$ as $\varepsilon \to 0$.

Since p_ε is a non-constant polynomial, there exists a positive integer m_0 such that $p'_\varepsilon(\frac{1}{m}) \neq 0$ for every $m \geq m_0$. Let $m \geq m_0$. Put

$$f_m(t) = \begin{cases} iw\left(\frac{1}{m} - t\right) + \left(p'_\varepsilon\left(\frac{1}{m}\right) + iq'_\varepsilon\left(\frac{1}{m}\right)\right)\left(t - \frac{1}{m}\right) \\ \qquad\qquad + p_\varepsilon\left(\frac{1}{m}\right) + iq_\varepsilon\left(\frac{1}{m}\right), & 0 \leq t \leq \frac{1}{m}, \\ (p_\varepsilon + iq_\varepsilon)(t), & \frac{1}{m} \leq t \leq 1, \end{cases}$$

where

$$w(t) = \begin{cases} 0, & t = 0 \\ t^3 \sin\frac{1}{t}, & 0 < t \leq 1. \end{cases}$$

Then $f_m \in C^1[0,1]$ for every $m \geq m_0$. It is a routine work to prove that f_m converges uniformly to $p + iq$ on $[0,1]$ and a proof is omitted. We prove that $f_m \in W_1$ for every $m \geq m_0$.

Let K be a real number. We look at the number of the points t in $[0,1]$ such that the slope of the tangent on the curve $f_m([0,1])$ at the point $f_m(t)$ is equal to K. The curve $f_m([0,1])$ has a tangent with the slope K at the point $f_m(t)$ if and only if

$$K = \lim_{\delta \to 0} \frac{\operatorname{Im} f_m(t+\delta) - \operatorname{Im} f_m(t)}{\operatorname{Re} f_m(t+\delta) - \operatorname{Re} f_m(t)}.$$

Suppose that $0 \leq t \leq \frac{1}{m}$. Since $\operatorname{Re} f_m(t) = p'_\varepsilon\left(\frac{1}{m}\right)\left(t - \frac{1}{m}\right) + p_\varepsilon\left(\frac{1}{m}\right)$ and $\operatorname{Im} f_m(t) = w\left(\frac{1}{m} - t\right) + q'_\varepsilon\left(\frac{1}{m}\right)\left(t - \frac{1}{m}\right) + q_\varepsilon\left(\frac{1}{m}\right)$, we have

$$\lim_{\delta \to 0} \frac{\operatorname{Im} f_m(t+\delta) - f_m(t)}{\operatorname{Re} f_m(t+\delta) - \operatorname{Re} f_m(t)} = \frac{-w'\left(\frac{1}{m} - t\right) + q'_\varepsilon\left(\frac{1}{m}\right)}{p'_\varepsilon\left(\frac{1}{m}\right)}, \quad 0 \leq t \leq \frac{1}{m}.$$

Hence the curve $f_m([0,1])$ has a tangent line of the slope K at the point $f_m(t)$ for $0 \leq t \leq \frac{1}{m}$ if and only if

$$\frac{-w'\left(\frac{1}{m} - t\right) + q'_\varepsilon\left(\frac{1}{m}\right)}{p'_\varepsilon\left(\frac{1}{m}\right)} = K. \tag{15}$$

If $K \neq \frac{q'_\varepsilon\left(\frac{1}{m}\right)}{p'_\varepsilon\left(\frac{1}{m}\right)}$, the number of such points $0 \leq t \leq \frac{1}{m}$ is at most finite. (The reason is as follows. Suppose that $\frac{q'_\varepsilon\left(\frac{1}{m}\right)}{p'_\varepsilon\left(\frac{1}{m}\right)} \neq K = \frac{-w'\left(\frac{1}{m}-t\right)+q'_\varepsilon\left(\frac{1}{m}\right)}{p'_\varepsilon\left(\frac{1}{m}\right)}$. Then

$$w'\left(\frac{1}{m} - t\right) = p'_\varepsilon\left(\frac{1}{m}\right)\left(\frac{q'_\varepsilon\left(\frac{1}{m}\right)}{p'_\varepsilon\left(\frac{1}{m}\right)} - K\right). \tag{16}$$

On the other hand, a simple calculation shows that

$$\left|w'\left(\frac{1}{m} - t\right)\right| \leq 4\left(\frac{1}{m} - t\right). \tag{17}$$

We have $p'_\varepsilon\left(\frac{1}{m}\right)\left(\frac{q'_\varepsilon\left(\frac{1}{m}\right)}{p'_\varepsilon\left(\frac{1}{m}\right)} - K\right) \neq 0$ since $\frac{q'_\varepsilon\left(\frac{1}{m}\right)}{p'_\varepsilon\left(\frac{1}{m}\right)} \neq K$. By (17) there is no $t \leq \frac{1}{m}$ with

$$\frac{1}{m} - t < \frac{1}{4}\left|p'_\varepsilon\left(\frac{1}{m}\right)\left(\frac{q'_\varepsilon\left(\frac{1}{m}\right)}{p'_\varepsilon\left(\frac{1}{m}\right)} - K\right)\right|$$

such that (16) holds. It is easy to see that the number of $t \geq 0$ with

$$\frac{1}{4}\left|p'_\varepsilon\left(\frac{1}{m}\right)\left(\frac{q'_\varepsilon\left(\frac{1}{m}\right)}{p'_\varepsilon\left(\frac{1}{m}\right)} - K\right)\right| \leq \frac{1}{m} - t$$

such that (16) holds is at most finite.) On the other hand if $K = \frac{q'_\varepsilon\left(\frac{1}{m}\right)}{p'_\varepsilon\left(\frac{1}{m}\right)}$, then by (15) we infer that $w'\left(\frac{1}{m} - t\right) = 0$. By a calculation, for every positive integer k there exists a unique $k\pi < s_k < k\pi + \pi/2$ such that $w'\left(\frac{1}{s_k}\right) = 0$. Letting $t_k = \frac{1}{m} - \frac{1}{s_k}$ we have $w'\left(\frac{1}{m} - t_k\right) = 0$. Thus $K = \frac{q'_\varepsilon\left(\frac{1}{m}\right)}{p'_\varepsilon\left(\frac{1}{m}\right)}$ for $0 \leq t < \frac{1}{m}$ if and only if $t = t_k$ for some positive integer k. As $w'(0) = 0$, we see that $w'\left(\frac{1}{m} - t\right) = \frac{q'_\varepsilon\left(\frac{1}{m}\right)}{p'_\varepsilon\left(\frac{1}{m}\right)}$

if $t = \frac{1}{m}$. We conclude that the set of the all points in $f_m([0, \frac{1}{m}])$ at which $f_m([0,1])$ has a tangent line with the slope $\frac{q'_\varepsilon(\frac{1}{m})}{p'_\varepsilon(\frac{1}{m})}$ is $\{f_m(t_n)\}_{n\geq N} \cup \{f_m\left(\frac{1}{m}\right)\}$, where $N = \min\{k : \frac{1}{m} > t_k\}$.

Suppose that $\frac{1}{m} < t \leq 1$. We have $\operatorname{Re} f_m(t) = p_\varepsilon(t)$ and $\operatorname{Im} f_m(t) = q_\varepsilon(t)$. Therefore we have

$$\frac{\operatorname{Im} f_m(t+\delta) - \operatorname{Im} f_m(t)}{\operatorname{Re} f_m(t+\delta) - \operatorname{Re} f_m(t)} = \frac{q_\varepsilon(t+\delta) - q_\varepsilon(t)}{p_\varepsilon(t+\delta) - p_\varepsilon(t)}.$$

Hence the curve $f_m([0,1])$ has a tangent line of the slope K at $f_m(t)$ for $\frac{1}{m} < t \leq 1$ if and only if

$$\lim_{\delta\to 0} \frac{q_\varepsilon(t+\delta) - q_\varepsilon(t)}{p_\varepsilon(t+\delta) - p_\varepsilon(t)} = K.$$

If $p_\varepsilon(t) \neq 0$, then $\frac{q'_\varepsilon(t)}{p'_\varepsilon(t)} = K$. The number of such points $t \in \left(\frac{1}{m}, 1\right]$ is at most finite. Suppose not. Then $q'_\varepsilon(t) = Kp'_\varepsilon(t)$ for infinitely many t, hence $q'_\varepsilon = Kp'_\varepsilon$ on the interval $\left(\frac{1}{m}, 1\right]$ since p'_ε and q'_ε are polynomials. It follows that $q_\varepsilon = Kp_\varepsilon + c$ for some $c \in \mathbb{C}$, which contradicts to our hypothesis on p_ε and q_ε. We obtain that the number of $t \in \left(\frac{1}{m}, 1\right]$ such that $\frac{q'_\varepsilon(t)}{p'_\varepsilon(t)} = K$ is at most finite. The number of $t \in \left(\frac{1}{m}, 1\right]$ such that $p'_\varepsilon(t) = 0$ is at most finite since p_ε is a polynomial. We conclude that the number of points t such that $f_m([0,1])$ has a tangent line of the slope K at $f_m(t)$ is at most finite. In a similar way we see that the number of $t \in \left(\frac{1}{m}, 1\right]$ such that $f_m([0,1])$ has a tangent line at $f_m(t)$ which is parallel to the imaginary axis is at most finite. If $p'_\varepsilon(t) \neq 0$, then $f_m([0,1])$ has a tangent line which is not parallel to the imaginary axis. Hence if $f_m([0,1])$ has a tangent line with a point of tangency at $f_m(t)$ which is parallel to the imaginary axis, then $p'_\varepsilon(t) = 0$. Thus the number of such points is at most finite.

We conclude that for a real number $K \neq \frac{q'_\varepsilon(\frac{1}{m})}{p'_\varepsilon(\frac{1}{m})}$ the number of $t \in [0,1]$ such that $f_m([0,1])$ has a tangent line of the slope K at $f_m(t)$ is at most finite; the number of $t \in [0,1]$ such that $f_m([0,1])$ has a tangent line which is parallel to the imaginary axis at $f_m(t)$ is at most finite; the number of $t \in [0,1]$ such that $f_m([0,1])$ has a tangent line of the slope $\frac{q'_\varepsilon(\frac{1}{m})}{p'_\varepsilon(\frac{1}{m})}$ at $f_m(t)$ is countable, the number of such $t \in \left(\frac{1}{m}, 1\right]$ is at most finite, say $\{t_{-n}\}_{n=1}^{l}$, there is a sequence $\{t_k\}_{k\geq N}$ in $[0, \frac{1}{m})$ of such points that converges to $\frac{1}{m}$. Denote them as $\{t_k\} = \{t_k\}_{k\geq N} \cup \{t_\infty = \frac{1}{m}\} \cup \{t_{-n}\}_{n=1}^{l}$.

Suppose that $S : \mathbb{C} \to \mathbb{C}$ is an isometry such that $S(f_m([0,1])) = f_m([0,1])$. We prove that S is the identity so that $f_m \in W$. Since there are $a, b \in \mathbb{C}$ with $|a| = 1$ such that $S(z) = b + az$, $z \in \mathbb{C}$, or $S(z) = b + a\bar{z}$, $z \in \mathbb{C}$, $S(\ell_1)$ and $S(\ell_2)$ are parallel for every pair of parallel lines ℓ_1 and ℓ_2 in $\mathbb{C}$. Thus the parallel tangent line at $\{t_k\}$ are translated by S to a parallel tangent line. Hence we get

$$\{S(f_m(t_k))\} = \{f_m(t_k)\}.$$

As S is an isometry the unique cluster points t_∞ of $\{t_k\}$ translates to t_∞ by S; $S(t_\infty) = t_\infty$. As $\{f_m(t_{-k})\}_{k=1}^{l}$ is discrete, there is a positive integer M such that $\{S(f_m(t_k))\}_{k\geq M} \subset \{f_m(t_k)\}_{k\geq N}$. Hence if $n \geq M$, then there is an $n_1 \geq N$ such that

$$|f_m(t_\infty) - f_m(t_n)| = |S(f_m(t_\infty)) - S(f_m(t_n))| = |f_m(t_\infty) - f_m(t_{n_1})|.$$

If $n, n_1 \geq N$ and $n \neq n_1$, then by the definition of t_n and t_{n_1} we have

$$|f_m(t_\infty) - f_m(t_n)| \neq |f_m(t_\infty) - f_m(t_{n_1})|.$$

It follows that $t_n = t_{n_1}$. Since $S : \mathbb{C} \to \mathbb{C}$ is an isometry, the set of the fixed point of S is empty or a singleton or points on a straight line if S is not the identity. As $\{f_m(t_n)\}_{n \geq M}$ is a set of fixed points which are not on the line since $w\left(\frac{1}{m} - t_n\right) > 0$ when n is an even number and $w\left(\frac{1}{m} - t_n\right) < 0$ when n is an odd number, we conclude that S is the identity.

By the Weierstrass approximation theorem we see that $cl(W_1) = C[0,1]$. □

THEOREM 4. *$G_{\Pi_0} \cap \mathrm{Iso}(A_1, A_2)$ is 2-local reflexive in $M(A_1, A_2)$.*

PROOF. Suppose that $T \in M(A_1, A_2)$ is 2-local in $G_{\Pi_0} \cap \mathrm{Iso}(A_1, A_2)$. Then by Proposition 2 there exist an $\alpha \in \mathbb{C}$ of unit modulus and $\epsilon \in \{\pm 1\}$ which satisfies that for every f in W_1, there exists a $\pi_f \in \Pi_0$ such that

$$T(f) = T(0) + \alpha[f \circ \pi_f]^\epsilon.$$

We prove that π_f is independent of $f \in W_1$. Let $T_1 \in M(A_1, A_2)$ be defined by

$$T_1(h) = [\bar{\alpha}(T(h) - T(0))]^\epsilon, \quad h \in A_1.$$

Then T_1 is 2-local in $G_{\Pi_0} \cap \mathrm{Iso}(A_1, A_2)$. In the same way as in the proof of Proposition 2 there exists a $T_{f,0} \in G_{\Pi_0} \cap \mathrm{Iso}(A_1, A_2)$ such that

$$T_1(f) = T_{f,0}(f) = f \circ \pi_f \tag{18}$$

for every $f \in W_1$. Let $\varepsilon > 0$ be given. Then $g_\varepsilon = \pi_0 + i\varepsilon\pi_0^2 \in W_1$. Hence there exist $T_{g_\varepsilon,0} \in G_{\Pi_0} \cap \mathrm{Iso}(A_1, A_2)$ and $\pi_\varepsilon \in \Pi_0$ such that

$$T_1(g_\varepsilon) = T_{g_\varepsilon,0}(g_\varepsilon) = g_\varepsilon \circ \pi_\varepsilon. \tag{19}$$

Note that $T_{g_\varepsilon,0}(h) = h \circ \pi_\varepsilon$ for every $h \in A_1$ by the proof of Proposition 2. (In fact, due to the note just after (3) we have $T_{g_\varepsilon,0}(h) = \alpha_{g_\varepsilon,0}[h \circ \pi_{g_\varepsilon,0}]^{\epsilon_{g_\varepsilon,0}}$ for $h \in A_1$. Using (11), (14), the fact that $g_\varepsilon \in W_1$ and letting $\pi_{g_\varepsilon,0} = \pi_\varepsilon$, we see that $T_{g_\varepsilon,0}(h) = h \circ \pi_\varepsilon$ for every $h \in A_1$.) We prove that there exists an $\varepsilon_0 > 0$ such that $\pi_\varepsilon = \pi_{\varepsilon'}$ for every $0 < \varepsilon, \varepsilon' < \varepsilon_0$. Suppose not. Then there exist sequences $\{\varepsilon_n\}$ and $\{\varepsilon'_n\}$ of positive real numbers which converge to 0 respectively such that $\pi_{\varepsilon_n} \neq \pi_{\varepsilon'_n}$ for every n. By Lemma 1, T_1 is a isometry with respect to $\|\cdot\|_\infty$, hence we infer that

$$\|T_1(g_{\varepsilon_n}) - T_1(g_{\varepsilon'_n})\|_\infty = \|g_{\varepsilon_n} - g_{\varepsilon'_n}\|_\infty = \|\varepsilon_n - \varepsilon'_n\| = |\varepsilon_n - \varepsilon'_n| \to 0$$

as $n \to \infty$. On the other hand, as $\pi_{\varepsilon_n} \neq \pi_{\varepsilon'_n}$ for every n we have

$$\|T_1(g_{\varepsilon_n}) - T_1(g_{\varepsilon'_n})\|_\infty = \|g_{\varepsilon_n} \circ \pi_{\varepsilon_n} - g_{\varepsilon'_n} \circ \pi_{\varepsilon'_n}\|_\infty \geq \|\pi_{\varepsilon_n} - \pi_{\varepsilon'_n}\|_\infty - \varepsilon_n - \varepsilon'_n \to 1$$

as $n \to \infty$, which is a contradiction proving $\pi_\varepsilon = \pi_{\varepsilon'}$ for every $0 < \varepsilon, \varepsilon' < \varepsilon_0$ for some positive ε_0. Put the common π_ε as π. Letting $\varepsilon \to 0$ in (19) we get

$$T_1(\pi_0) = \pi_0 \circ \pi.$$

We prove that

$$T_1(f) = f \circ \pi \tag{20}$$

for every $f \in W$. We prove the case where $\pi = \pi_0$. A proof for the case where $\pi = \pi_1$ is similar, and is omitted.

Assume for a moment that we have already proved that $T_1(f) = f$ for every $f \in W_1$. Then T_1 is a surjective isometry. The reason is as follows. For a sufficiently small positive ε, we have proved $\pi_\varepsilon = \pi_0$ since we assume $\pi = \pi_0$ in (20). Then

$T_{g_\varepsilon,0} = T_1$ on W_1. Proposition 3 asserts that W_1 is uniformly dense in $C[0,1]$, hence in A_1. As T_1 is continuous with respect to $\|\cdot\|_\infty$ by Lemma 1, we conclude that $T_1 = T_{g_\varepsilon,0}$ on A_1. Since $T_{g_\varepsilon,0}$ is a surjective isometry we conclude that T_1 is a surjective isometry.

We prove that $T_1(f) = f$ for every $f \in W_1$. To prove it, suppose that there exists a $f_0 \in W_1$ such that $T_1(f_0) \neq f_0$. Then by (18) we have

$$T_1(f_0) = f_0 \circ \pi_1. \tag{21}$$

As T_1 is 2-local in $G_{\Pi_0} \cap \mathrm{Iso}(A_1, A_2)$, there exist a $\lambda_{f_0,\pi_0} \in A_2$, $\alpha_{f_0,\pi_0} \in \mathbb{C}$ of unit modulus and $\epsilon_{f_0,\pi_0} \in \{\pm 1\}$ such that one of

$$\begin{aligned} f_0 \circ \pi_1 = T_1(f_0) &= \lambda_{f_0,\pi_0} + \alpha_{f_0,\pi_0}[f_0]^{\epsilon_{f_0,\pi_0}}, \\ \pi_0 = T_1(\pi_0) &= \lambda_{f_0,\pi_0} + \alpha_{f_0,\pi_0}\pi_0 \end{aligned} \tag{22}$$

and

$$\begin{aligned} f_0 \circ \pi_1 = T_1(f_0) &= \lambda_{f_0,\pi_0} + \alpha_{f_0,\pi_0}[f_0 \circ \pi_1]^{\epsilon_{f_0,\pi_0}}, \\ \pi_0 = T_1(\pi_0) &= \lambda_{f_0,\pi_0} + \alpha_{f_0,\pi_0}\pi_1 \end{aligned} \tag{23}$$

holds. Thus

$$\lambda_{f_0,\pi_0}(t) = (1 - \alpha_{f_0,\pi_0})t, \quad t \in [0,1] \tag{24}$$

when (22) occurs and

$$\lambda_{f_0,\pi_0}(t) = (1 + \alpha_{f_0,\pi_0})t - \alpha_{f_0,\pi_0}, \quad t \in [0,1] \tag{25}$$

when (23) occurs.

We will prove that both of (22) and (23) are impossible. Suppose that (22) occurs. Rewriting the first equation of (22) using (24) we get

$$f_0(1-t) = (T_1(f_0))(t) = (1 - \alpha_{f_0,\pi_0})t + \alpha_{f_0,\pi_0}[f_0(t)]^{\epsilon_{f_0,\pi_0}}, \quad t \in [0,1]. \tag{26}$$

Suppose that $\alpha_{f_0,\pi_0} = 1$. Then

$$f_0(1-t) = f_0(t), \qquad t \in [0,1] \tag{27}$$

or

$$f_0(1-t) = \overline{f_0(t)}, \qquad t \in [0,1]. \tag{28}$$

If (27) holds, then $(T_1(f_0))(t) = f_0(1-t) = f_0(t)$, $t \in [0,1]$ by (21), which is against our choice of f_0. Thus (27) does not hold. Suppose that (28) holds. Then $f_0([0,1]) = \overline{f_0([0,1])}$ holds, which means that $f_0 \notin W_1$. Thus (28) does not hold. It follows that $\alpha_{f_0,\pi_0} \neq 1$. Suppose that $\varepsilon_{f_0,\pi_0} = 1$ for (26). Then we have

$$f_0(1-t) = (1 - \alpha_{f_0,\pi_0})t + \alpha_{f_0,\pi_0} f_0(t), \quad t \in [0,1]. \tag{29}$$

Changing $1-t$ by t we have

$$f_0(t) = (1 - \alpha_{f_0,\pi_0})(1-t) + \alpha_{f_0,\pi_0} f_0(1-t), \quad t \in [0,1]. \tag{30}$$

Applying (29) we have

$$f_0(t) = (1 - \alpha_{f_0,\pi_0})(1-t) + \alpha_{f_0,\pi_0}\left((1 - \alpha_{f_0,\pi_0})t + \alpha_{f_0,\pi_0} f_0(t)\right), \quad t \in [0,1]. \tag{31}$$

As $\alpha_{f_0,\pi_0} \neq 1$ we infer that

$$(1 + \alpha_{f_0,\pi_0}) f_0(t) = 1 - (1 - \alpha_{f_0,\pi_0})t, \qquad t \in [0,1]. \tag{32}$$

If $\alpha_{f_0,\pi_0} = -1$, then we have that $0 = 1 - 2t$ for every $t \in [0,1]$, which is a contradiction. Hence $\alpha_{f_0,\pi_0} \neq -1$. Then by (32) we have

$$f_0(t) = \frac{1}{(1+\alpha_{f_0,\pi_0})} - \frac{(1-\alpha_{f_0,\pi_0})}{(1+\alpha_{f_0,\pi_0})}t, \qquad t \in [0,1].$$

Hence $f_0 \notin W_1$, which is a contradiction. Suppose that $\epsilon_{f_0,\pi_0} = -1$ for (26). Then we have

$$f_0(1-t) = (1-\alpha_{f_0,\pi_0})t + \alpha_{f_0,\pi_0}\overline{f_0(t)}, \quad t \in [0,1]. \tag{33}$$

Substituting $1-t$ by t in (33), we have

$$f_0(t) = (1-\alpha_{f_0,\pi_0})(1-t) + \alpha_{f_0,\pi_0}\overline{f_0(1-t)}, \quad t \in [0,1]. \tag{34}$$

Substituting (33) in (34) we get

$$f_0(t) = (1-\alpha_{f_0,\pi_0})(1-t) + \alpha_{f_0,\pi_0}\overline{(1-\alpha_{f_0,\pi_0})t + \alpha_{f_0,\pi_0}\overline{f_0(t)}}, \quad t \in [0,1]. \tag{35}$$

Hence we get

$$0 = \left(\alpha_{f_0,\pi_0}\overline{(1-\alpha_{f_0,\pi_0})} - (1-\alpha_{f_0,\pi_0})\right)t + (1-\alpha_{f_0,\pi_0}), \qquad t \in [0,1].$$

We get $\alpha_{f_0,\pi_0} = 1$, which contradicts to $\alpha_{f_0,\pi_0} \neq 1$. We conclude that (22) does not occur.

Suppose that (23) holds. Rewriting (23) by applying (25) we get

$$f_0(1-t) = (T_1(f_0))(t) = (1+\alpha_{f_0,\pi_0})t - \alpha_{f_0,\pi_0} + \alpha_{f_0,\pi_0}[f_0(1-t)]^{\epsilon_{f_0,\pi_0}}. \tag{36}$$

Suppose that $\epsilon_{f_0,\pi_0} = 1$. By (36) we get

$$(1-\alpha_{f_0,\pi_0})f_0(1-t) = (1+\alpha_{f_0,\pi_0})t - \alpha_{f_0,\pi_0}, \qquad t \in [0,1]. \tag{37}$$

Then we have $0 = 1 - 2t$ for every $t \in [0,1]$ if $\alpha_{f_0,\pi_0} = 1$, which is impossible, so that $\alpha_{f_0,\pi_0} \neq 1$. Then by (37) we get

$$f_0(1-t) = \frac{1+\alpha_{f_0,\pi_0}}{1-\alpha_{f_0,\pi_0}}t - \frac{\alpha_{f_0,\pi_0}}{1-\alpha_{f_0,\pi_0}}, \qquad t \in [0,1],$$

so that

$$f_0(t) = \frac{1+\alpha_{f_0,\pi_0}}{1-\alpha_{f_0,\pi_0}}(1-t) - \frac{\alpha_{f_0,\pi_0}}{1-\alpha_{f_0,\pi_0}}, \qquad t \in [0,1],$$

which is a contradiction to $f_0 \in W_1$. We have that $\epsilon_{f_0,\pi_0} \neq 1$, hence $\epsilon_{f_0,\pi_0} = -1$. Then by (36) we get

$$f_0(1-t) = (1+\alpha_{f_0,\pi_0})t - \alpha_{f_0,\pi_0} + \alpha_{f_0,\pi_0}\overline{f_0(1-t)}, \qquad t \in [0,1]. \tag{38}$$

Thus

$$\begin{aligned} f_0(1-t) &= (1+\alpha_{f_0,\pi_0})t - \alpha_{f_0,\pi_0} \\ &\quad + \alpha_{f_0,\pi_0}\overline{((1+\alpha_{f_0,\pi_0})t - \alpha_{f_0,\pi_0} + \alpha_{f_0,\pi_0}\overline{f_0(1-t)})}, \quad t \in [0,1]. \end{aligned} \tag{39}$$

As $|\alpha_{f_0,\pi_0}| = 1$ we get

$$\begin{aligned} f_0(1-t) &= (1+\alpha_{f_0,\pi_0})t - \alpha_{f_0,\pi_0} \\ &\quad + \alpha_{f_0,\pi_0}\overline{(1+\alpha_{f_0,\pi_0})}t - 1 + f_0(1-t), \quad t \in [0,1]. \end{aligned} \tag{40}$$

Hence

$$0 = ((1+\alpha_{f_0,\pi_0}) + \alpha_{f_0,\pi_0}(1+\overline{\alpha_{f_0,\pi_0}}))\,t - (\alpha_{f_0,\pi_0} + 1), \qquad t \in [0,1]. \tag{41}$$

Thus $\alpha_{f_0,\pi_0} = -1$. Substituting $\alpha_{f_0,\pi_0} = -1$ into (36) we get

$$f_0(1-t) = 1 - \overline{f_0(1-t)}, \qquad t \in [0,1]$$

since $\epsilon_{f_0,\pi_0} = -1$. Then

$$f_0([0,1]) = 1 - \overline{f_0([0,1])},$$

which contradicts to $f_0 \in W_1$. It follows that (23) does not occur.

Assuming the existence of $f_0 \in W_1$ such that $T_1(f_0) \neq f_0$ we arrived at the contradiction. We conclude that $T_1(f) = f$ for every $f \in W_1$.

Let $g \in A_1$. Then by Proposition 3 there is a sequence $\{g_n\}$ in W_1 such that $\|g - g_n\|_\infty \to 0$ as $n \to \infty$. By the previous part of the proof we have

$$T_1(g_n) = g_n \tag{42}$$

for every n. By Lemma 1, T_1 is an isometry with respect to $\|\cdot\|_\infty$, we have $T_1(g) = g$ by letting $n \to \infty$ for (42). We conclude that $T_1(g) = g$ for every $g \in A_1$ if $T_1(\pi_0) = \pi_0$. It follows that

$$T(g) = T(0) + \alpha[g]^\epsilon, \qquad g \in A_1$$

if $T_1(\pi_0) = \pi_0$. Suppose that $T_1(\pi_0) = \pi_1$. As we have already described, we see that $T_1(g) = g \circ \pi_1$ for every $g \in A_1$. Hence we have

$$T(g) = T(0) + \alpha[g \circ \pi_1]^\epsilon, \qquad g \in A_1.$$

Thus we observed that $T \in G_{\Pi_0}$. As we have already proved that T_1 is a surjective isometry from A_1 onto A_2, we see that T is also a surjective isometry. Hence we conclude that $T \in G_{\Pi_0} \cap \mathrm{Iso}(A_1, A_2)$. □

4. Surjective real-linear isometries on Lip(K)

Jarosz and Pathak exhibited in [**10**, Example 8] the form of surjective *complex-linear* isometries on the Banach algebra $\mathrm{Lip}(K_j)$ with the norm $\|\cdot\|_\Sigma$ of the Lipschitz functions on a compact metric space K_j, thus answering the question posed by Rao and Roy [**24**]. After the publication of [**10**] some authors expressed their suspicion about the argument there and the validity of the statement there had not been confirmed until the correction [**8**, Corollary 15] was published by Hatori and Oi. In this section by applying [**8**, Lemmas 10,11] and [**5**, Proposition 7] we exhibit the form of surjective real-linear isometries between the Banach algebras of Lipschitz functions.

LEMMA 5. *Let $p, q \in \mathbb{C}$. Suppose that $|p + \lambda q| = 1$ for at least three different unimodular $\lambda \in \mathbb{C}$. Then $p = 0$ and $|q| = 1$, or $|p| = 1$ and $q = 0$.*

PROOF. Let $\lambda_1, \lambda_2, \lambda_3$ be three unimodular complex numbers such that $|p + \lambda_j q| = 1$ for $j = 1, 2, 3$. Suppose that $q \neq 0$. Then $|p/q + \lambda_j| = 1/|q|$, by which the circle of the center $-p/q$ and the radius $1/|q|$ is the circumscribed circle of the triangle defined by the three point $\lambda_1, \lambda_2, \lambda_3$. On the other hand, the unit circle is the circumscribed circle of these three points since $|\lambda_j| = 1$ for $j = 1, 2, 3$. By the uniqueness of the circumscribed circle, we see that $|q| = 1$ and $p/q = 0$, thus $p = 0$. On the other hand, if $q = 0$, then it is apparent that $|p| = 1$ □

THEOREM 6. *Let K_j be a compact metric space for $j = 1, 2$. Suppose that $U : \mathrm{Lip}(K_1) \to \mathrm{Lip}(K_2)$ is a surjective real-linear isometry with respect to the* *norm* $\|f\|_\Sigma = \|f\|_\infty + L_f$ *for* $f \in \mathrm{Lip}(K_1)$. *Then there exists a surjective isometry* $\pi : K_2 \to K_1$ *such that*

$$U(f) = U(1) f \circ \pi, \qquad f \in \mathrm{Lip}(K_1)$$

or

$$U(f) = U(1)\overline{f \circ \pi}, \qquad f \in \mathrm{Lip}(K_1).$$

PROOF. In the same way as in the proof of [**8**, Proposition 9] we apply Choquet's theory. Let $j = 1, 2$. Let $\mathfrak{M}_j$ be the Stone-Čech compactification of $\{(x, x') \in K_j^2 : x \neq x'\}$. For $f \in \mathrm{Lip}(K_j)$, let $D_j(f)$ denote the continuous extension to $\mathfrak{M}_j$ of the function $(f(x) - f(x'))/d(x, x')$ on $\{(x, x') \in K_j^2 : x \neq x'\}$. Then $D_j : \mathrm{Lip}(K_j) \to C(\mathfrak{M}_j)$ is well defined. We have $\|D_j(f)\|_\infty = L_f$ for every $f \in \mathrm{Lip}(K_j)$. Then $(K_j, \mathbb{C}, \mathrm{Lip}(K_j), \mathrm{Lip}(K_j))$ is an admissible quadruple of type L (see [**8**, Definition 4, Example 12]). Let $\mathbb{T} = \{z \in \mathbb{C} : |z| = 1\}$. For $j = 1, 2$, define a map

$$I_j : \mathrm{Lip}(K_j) \to C(K_j \times \mathfrak{M}_j \times \mathbb{T})$$

by $I_j(f)(x, m, \gamma) = f(x) + \gamma D_j(f)(m)$ for $f \in \mathrm{Lip}(K_j)$ and $(x, m, \gamma) \in K_j \times \mathfrak{M}_j \times \mathbb{T}$. As D_j is a complex-linear map, so is I_j. For simplicity we write $\tilde{f} = I_j(f)$ for $f \in \mathrm{Lip}(K_j)$. For every $f \in \mathrm{Lip}(K_j)$ we have

$$\|\tilde{f}\|_\infty = \|f\|_\infty + \|D_j(f)\|_\infty = \|f\|_\infty + L_f,$$

hence I_j is a surjective complex-linear isometry from $\mathrm{Lip}(K_j)$ onto $B_j = I_j(\mathrm{Lip}(K_j))$. We have $D_j(1) = 0$ and $\tilde{c} = c$ for every $c \in \mathbb{C}$, where the constant function taking the value c is denoted also by c. It follows that B_j is a complex-linear closed subspace of $C(K_j \times \mathfrak{M}_j \times \mathbb{T})$ which contains 1. A point $p = (x, m, \mathbb{T}) \in K_j \times \mathfrak{M}_j \times \mathbb{T}$ is in the Choquet boundary $\mathrm{Ch}\, B_j$ if the point evaluation δ_p is an extreme point of the closed unit ball $B_{j,1}^*$ of the (complex) dual space B_j^* of B_j. See the description just after [**8**, Proposition 9]. Define $\mathcal{S} : B_1 \to B_2$ by $\mathcal{S}(\tilde{f}) = \widetilde{U(f)}$ for $\tilde{f} \in B_1$. Then $\mathcal{S}$ is a surjective real-linear isometry from B_1 onto B_2.

Let $x_0 \in K_2$. Put $b_0(x) = 1 - \frac{d(x, x_0)}{d(K_2)}$ for $x \in K_2$, where $d(K_2)$ is the diameter of K_2. By a simple calculation we have $b_0 \in \mathrm{Lip}(K_2)$, $0 \le b_0 \le 1$ on K_2, and $b_0(x) = 1$ if and only if $x = x_0$. Then by [**8**, Lemma 10] there exists a pair $(m_0, \gamma_0) \in \mathfrak{M}_2 \times \mathbb{T}$ such that $(x_0, m_0, \gamma_0) \in \mathrm{Ch}\, B_2$. By [**8**, Lemma 11] we have that $p_\theta = (x_0, m_0, e^{i\theta}\gamma_0) \in \mathrm{Ch}\, B_2$ for every $0 < \theta < \pi/2$. For $\eta \in B_2^*$ we define $\mathcal{S}_*(\eta) \in B_1^*$ by

$$(\mathcal{S}_*(\eta))(\tilde{f}) = \mathrm{Re}\, \eta(\mathcal{S}(\tilde{f})) - i\, \mathrm{Re}\, \eta(\mathcal{S}(i\tilde{f})), \qquad \tilde{f} \in B_1.$$

Then $\mathcal{S}_* : B_2^* \to B_1^*$ is a surjective complex-linear isometry (cf. [**20**, (2.3)]). Denote the set of all extreme points in $\{\nu \in B_j^* : \|\nu\| \le 1\}$ by $\mathrm{ext}\, B_j^*$. As $\mathcal{S}_*$ is a surjective complex-linear isometry, we have that $\eta \in \mathrm{ext}\, B_2^*$ if and only if $\mathcal{S}_*(\eta) \in \mathrm{ext}\, B_1^*$. By the definition of the Choquet boundary, the point evaluation $\delta_{p_\theta} : B_2 \to \mathbb{C}$ defined by $\delta_{p_\theta}(\tilde{g}) = \tilde{g}(p_\theta)$, $\tilde{g} \in B_2$ is in $\mathrm{ext}\, B_2^*$. Then the Arens-Kelley theorem asserts that

$$\mathcal{S}_*(\delta_{p_\theta}) = \lambda_1 \delta_{p_1}$$

for a unimodular $\lambda_1 \in \mathbb{C}$ and a $p_1 \in \mathrm{Ch}\, B_1$. In the same way there exist a unimodular $\lambda_i \in \mathbb{C}$ and $p_i \in \mathrm{Ch}\, B_1$ such that

$$\mathcal{S}_*(i\delta_{p_\theta}) = \lambda_i \delta_{p_i}.$$

In the same way as in the proof of [**20**, Lemma 3.3] we have that $\lambda_i = i\lambda_1$ or $-i\lambda_1$. For the convenience of the readers we present the proof. As $p_\theta \in \operatorname{Ch} B_2$, $\frac{1+i}{\sqrt{2}}\delta_{p_\theta}$ is in $\operatorname{ext} B_2^*$. Then there exists a unimodular $\mu \in \mathbb{C}$ and a $q \in \operatorname{Ch} B_1$ such that

$$\frac{1}{\sqrt{2}}\left(\lambda_1\delta_{p_1} + \lambda_i\delta_{p_i}\right) = S_*\left(\frac{1+i}{\sqrt{2}}\delta_{p_\theta}\right) = \mu\delta_q.$$

Substituting $1 \in B_1$ in this equation we get

$$\frac{1}{\sqrt{2}}(\lambda_1 + \lambda_i) = \mu.$$

As $|\mu| = 1$ we have

$$|\lambda_1 + \lambda_i| = \sqrt{2}.$$

As $|\lambda_1| = |\lambda_i| = 1$ we conclude that $\lambda_i = i\lambda_1$ or $\lambda_i = -i\lambda_1$. Put $\varepsilon_0 = \lambda_i/\lambda_1$.

Let $c \in \mathbb{C}$ be arbitrary. For simplicity we also write c as the constant function which takes the value c. We have

$$\begin{aligned} (43) \qquad \left(\mathcal{S}_*(\delta_{p_\theta})\right)(\tilde{c}) &= \operatorname{Re}\delta_{p_\theta}(\mathcal{S}(\tilde{c})) - i\operatorname{Re}\delta_{p_\theta}(\mathcal{S}(i\tilde{c})) \\ &= \operatorname{Re}(\mathcal{S}(\tilde{c}))(p_\theta) - i\operatorname{Re}(\mathcal{S}(i\tilde{c}))(p_\theta), \end{aligned}$$

and

$$\begin{aligned} (44) \qquad \left(\mathcal{S}_*(i\delta_{p_\theta})\right)(\tilde{c}) &= \operatorname{Re} i\delta_{p_\theta}(\mathcal{S}(\tilde{c})) - i\operatorname{Re} i\delta_{p_\theta}(\mathcal{S}(i\tilde{c})) \\ &= -\operatorname{Im}(\mathcal{S}(\tilde{c}))(p_\theta) + i\operatorname{Im}(\mathcal{S}(i\tilde{c}))(p_\theta). \end{aligned}$$

As $\mathcal{S}_*(\delta_{p_\theta}) = \lambda_1\delta_{p_1}$, we have by (43) that

$$\operatorname{Re}(\mathcal{S}(\tilde{c}))(p_\theta) = \operatorname{Re}\left(\mathcal{S}_*(\delta_{p_\theta})\right)(\tilde{c}) = \operatorname{Re}\lambda_1(\tilde{c})(p_1) = \operatorname{Re}\lambda_1 c.$$

As $\mathcal{S}_*(i\delta_{p_\theta}) = i\varepsilon_0\lambda_1\delta_{p_i}$, we have by (44) that

$$\operatorname{Im}(\mathcal{S}(\tilde{c}))(p_\theta) = -\operatorname{Re}\left(\mathcal{S}_*(i\delta_{p_\theta})\right)(\tilde{c}) = -\operatorname{Re}\left(i\varepsilon_0\lambda_1(\tilde{c})(p_i)\right) = \operatorname{Im}\varepsilon_0\lambda_1 c.$$

Thus

$$(45) \qquad (\mathcal{S}(\tilde{c}))(p_\theta) = \operatorname{Re}\lambda_1 c + i\operatorname{Im}\varepsilon_0\lambda_1 c = \begin{cases} \lambda_1 c, & \text{if} \quad \varepsilon_0 = 1, \\ \overline{\lambda_1 c} & \text{if} \quad \varepsilon_0 = -1. \end{cases}$$

On the other hand we have by the definition of $\mathcal{S}$ that

$$(46) \qquad \mathcal{S}(\tilde{c})(p_\theta) = (U(c))(x_0) + e^{i\theta}\lambda_0(D_2(U(c)))(m_0).$$

Combining (45) and (46) we have

$$(47) \qquad \left|(U(c))(x_0) + e^{i\theta}\lambda_0(D_2(U(c)))(m_0)\right| = |c|$$

for every $c \in \mathbb{C}$. Substituting $c = 1$ in (47) we get

$$(48) \qquad \left|(U(1))(x_0) + e^{i\theta}\lambda_0(D_2(U(1)))(m_0)\right| = 1$$

for $0 < \theta < \pi/2$. By Lemma 5 we have $(U(1))(x_0) = 0$ or $(D_2(U(1)))(m_0) = 0$. But $(U(1))(x_0) = 0$ is impossible. The reason is as follows. Suppose that $(U(1))(x_0) = 0$. Then $|(D_2(U(1)))(m_0)| = 1$ by (48). Hence $\|D_2(U(1))\|_\infty \geq 1$. Since U is an isometry we get

$$1 = \|U(1)\|_\Sigma = \|U(1)\|_\infty + \|D_2(U(1))\|_\infty \geq 1,$$

hence $\|U(1)\|_\infty = 0$, therefore $U(1) = 0$ on K_2 and $D_2(U(1)) = 0$ follows, which is against to $\|D_2(U(1))\|_\infty \geq 1$. Thus we have that $(D_2(U(1)))(m_0) = 0$, so that $|(U(1))(x_0)| = 1$. By

$$1 \leq |(U(1))(x_0)| + \|D_2(U(1))\|_\infty \\ \leq \|U(1)\|_\infty + \|D_2(U(1))\|_\infty = \|U(1)\|_\infty + L_{U(1)} = 1,$$

we conclude that $\|D_2(U(1))\|_\infty = 0$, so $D_2(U(1)) = 0$, thus $U(1)$ is a constant function by the definition of D_2. As $|(U(1))(x_0)| = 1$, we infer that $U(1)$ is a constant function of unit modulus. In the same way, substituting $c = i$ in (47) we have that $U(i)$ is a constant function of unit modulus. Since $U(1) - U(i)$ is a constant function we have

$$\sqrt{2} = |1 - i| = \|U(1) - U(i)\|_\Sigma = \|U(1) - U(i)\|_\infty$$

as $0 = L_{U(1)-U(i)} = D_2(U(1) - U(i))$. Since $U(1)$ and $U(i)$ are constant functions we infer that $U(i) = iU(1)$ or $U(i) = -iU(1)$. Put $U_0 = \overline{U(1)}U$. Then U_0 is a surjective real-linear isometry from $\mathrm{Lip}(K_1)$ onto $\mathrm{Lip}(K_2)$ such that $U_0(1) = 1$ and $U_0(i) = i$ or $-i$. Applying [**5**, Proposition 7] we see that U_0 is also an isometry with respect to the supremum norm $\|\cdot\|_\infty$ on K_1 and K_2 respectively, hence U_0 is extended to a surjective isometry $\widetilde{U_0}$ between the uniform closure of $\mathrm{Lip}(K_j)$, which coincides with $C(K_j)$ by the Stone-Weierstrass theorem. Thus

$$\widetilde{U_0} : C(K_1) \to C(K_2)$$

is a surjective real-linear isometry with respect to the supremum norm. As K_j is the Šilov boundary for $C(K_j)$ we can apply [**3**, Theorem] to get the existence of a homeomorphism $\pi' : K_2 \to K_1$ and an open and closed subset E_2 of K_2 such that

$$\widetilde{U_0(f)} = \begin{cases} f \circ \pi', & \text{on} \quad E_2, \\ \overline{f \circ \pi'}, & \text{on} \quad K_2 \setminus E_2 \end{cases}$$

for $f \in C(K_1)$ (cf. [**6**, **18**]). As $\widetilde{U_0}(i) = U_0(i) = i$ or $-i$ we have that

$$\widetilde{U_0}(f) = f \circ \pi', \qquad f \in C(K_1)$$

if $U_0(i) = i$ and

$$\widetilde{U_0}(f) = \overline{f \circ \pi'}, \qquad f \in C(K_1)$$

if $U_0(i) = -i$. It follows that U_0 is a complex-linear map if $U_0(i) = i$ and $\overline{U_0}$ is a complex-linear map if $U_0(i) = -i$. Applying [**8**, Corollary 15] there exists a surjective isometry $\pi : K_2 \to K_1$ such that

$$U_0(f) = f \circ \pi, \qquad f \in \mathrm{Lip}(K_1)$$

if $U_0(i) = i$ and

$$\overline{U_0}(f) = f \circ \pi, \qquad f \in \mathrm{Lip}(K_1)$$

if $U_0(i) = -i$. It follows that

$$U(f) = U(1) f \circ \pi, \qquad f \in \mathrm{Lip}(K_1)$$

or

$$U(f) = U(1)\overline{f \circ \pi}, \qquad f \in \mathrm{Lip}(K_1).$$

□

Let Π be the set of all surjective isometries from K_2 onto K_1.

COROLLARY 7. $G_\Pi(\mathrm{Lip}(K_1), \mathrm{Lip}(K_2)) = \mathrm{Iso}(\mathrm{Lip}(K_1), \mathrm{Lip}(K_2))$

PROOF. Suppose that $T \in \mathrm{Iso}(\mathrm{Lip}(K_1), \mathrm{Lip}(K_2))$. By the Mazur-Ulam theorem we have that $U = T - T(0)$ is a surjective real-linear isometry from $\mathrm{Lip}(K_1)$ onto $\mathrm{Lip}(K_2)$. By Theorem 6, $U(1)$ is a constant function of unit modulus and there exists a surjective isometry $\pi \in \Pi$ such that $U(f) = U(1) f \circ \pi$ for every $f \in \mathrm{Lip}(K_1)$ or $U(f) = U(1)\overline{f \circ \pi}$ for every $f \in \mathrm{Lip}(K_1)$. It follows that $T = U + T(0) \in G_\Pi(\mathrm{Lip}(K_1), \mathrm{Lip}(K_2))$.

Suppose that $T \in G_\Pi(\mathrm{Lip}(K_1), \mathrm{Lip}(K_2))$. It is a routine work to show that T is a surjective real-linear isometry. □

5. Applications

In this section we study the problem on 2-locality for $C^1[0,1]$ and $\mathrm{Lip}[0,1]$. We first prove that $\mathrm{Iso}(\mathrm{Lip}[0,1])$ with the norm $\|\cdot\|_\Sigma$ is 2-local reflexive in $M(\mathrm{Lip}[0,1])$. The Banach algebra $\mathrm{Lip}[0,1]$ satisfies the three conditions 1), 2) and 3) for A_j in the first part of section 3.

Recall that π_0 is the identity map on the interval $[0,1]$, $\pi_1 = 1 - \pi_0$ and $\Pi_0 = \{\pi_0, \pi_1\}$.

THEOREM 8. $\mathrm{Iso}(\mathrm{Lip}[0,1])$ *is 2-local reflexive in* $M(\mathrm{Lip}[0,1])$, *where* $\mathrm{Lip}[0,1]$ *is the Banach algebra of all Lipschitz functions defined on the closed interval* $[0,1]$ *with the norm* $\|\cdot\|_\Sigma$.

PROOF. Corollary 7 asserts that $G_{\Pi_0}(\mathrm{Lip}[0,1]) = \mathrm{Iso}(\mathrm{Lip}[0,1])$, hence $\mathrm{Iso}(\mathrm{Lip}[0,1]) = G_{\Pi_0} \cap \mathrm{Iso}(\mathrm{Lip}[0,1])$. The Banach algebra $\mathrm{Lip}[0,1]$ satisfies the conditions 1), 2) and 3) for A_j in the first part of section 3. Applying Theorem 4 for $A_j = \mathrm{Lip}[0,1]$, we infer that $\mathrm{Iso}(\mathrm{Lip}[0,1])$ is 2-local reflexive in $M(\mathrm{Lip}[0,1])$. □

Next we prove that $\mathrm{Iso}(C^1[0,1])$ is 2-local reflexive in $M(C^1[0,1])$ for certain norms. Let D be a non-empty connected compact subset of $[0,1] \times [0,1]$. The norm $\|\cdot\|_{\langle D \rangle}$ on $C^1[0,1]$ is defined by

$$\|f\|_{\langle D \rangle} = \sup_{(r,s)\in D} (|f(r)| + |f'(s)|), \quad f \in C^1[0,1].$$

Let $P_j : [0,1] \times [0,1] \to [0,1]$ be the projection onto the j-th factor ($j = 1, 2$). Let D be a non-empty connected compact subset of $[0,1] \times [0,1]$. The norm $\|\cdot\|_{\langle D \rangle}$ on $C^1[0,1]$ is defined by Kawamura, Koshimizu, and Miura [**14**] as follows:

$$\|f\|_{\langle D \rangle} = \sup_{(r,s)\in D} (|f(r)| + |f'(s)|), \quad f \in C^1[0,1].$$

They studied surjective real-linear isometries between $C^1[0,1]$ onto itself for the norm $\|\cdot\|_{\langle D \rangle}$ under additional hypothesis on D. The main result of [**14**] exhibits the form of isometries on $C^1[0,1]$ for a wide class of norms and unifies the former results on isometries for several important norms such as $\|\cdot\|_\Sigma$, $\|\cdot\|_\sigma$, $\|\cdot\|_\Delta$ and so on. If $D = [0,1] \times [0,1]$, then $\|f\|_{\langle D \rangle} = \|f\|_\infty + \|f'\|_\infty$ for $f \in C^1[0,1]$. If $D = \{(t,t) : t \in [0,1]\}$, then $\|f\|_{\langle D \rangle} = \sup\{|f(t)| + |f'(t)| : t \in [0,1]\}$. We point out that applying the Mazur-Ulam theorem, their results in fact assures the forms of surjective isometries without the assumption of linearity.

THEOREM 9. *Let* D *be a non-empty connected compact subset of* $[0,1] \times [0,1]$. *Suppose that* $P_j(D) = [0,1]$ *for* $j = 1, 2$. *Then* $\mathrm{Iso}(C^1[0,1], \|\cdot\|_{\langle D \rangle})$ *is 2-local reflexive in* $M(C^1[0,1])$.

PROOF. By [**14**, Corollary] and the Mazur-Ulam theorem we infer that $G_{\Pi_0} = \mathrm{Iso}(C^1[0,1])$. The Banach space $C^1[0,1]$ satisfies the conditions 1), 2) and 3) for A_j in the first part of section 3. Applying Theorem 4 for $A_j = C^1[0,1]$, we infer that $\mathrm{Iso}(C^1[0,1])$ is 2-local reflexive in $M(\mathrm{Lip}[0,1])$. □

We point out that the 2-locality problem for surjective isometries without assuming linearity is much difficult than the 2-locality problem for surjective complex or even real linear isometries. We do not know if $\mathrm{Iso}(C[0,1], \|\cdot\|_\infty)$ is 2-local reflexive or not.

Acknowledgments. The first author was supported by JSPS KAKENHI Grant Numbers JP16K05172, JP15K04921. He would like to express his heartful thanks to Lajos Molnár for his hospitality during staying at the University of Szeged on Octorber 2018 and for mentioning the problem of 2-local reflexivity of the groups of all surjective isometries.

References

[1] H. Al-Halees and R. J. Fleming, *On 2-local isometries on continuous vector-valued function spaces*, J. Math. Anal. Appl. **354** (2009), no. 1, 70–77, DOI 10.1016/j.jmaa.2008.12.023. MR2510418

[2] F. Botelho, J. Jamison, and L. Molnár, *Algebraic reflexivity of isometry groups and automorphism groups of some operator structures*, J. Math. Anal. Appl. **408** (2013), no. 1, 177–195, DOI 10.1016/j.jmaa.2013.06.001. MR3079956

[3] A. J. Ellis, *Real characterizations of function algebras amongst function spaces*, Bull. London Math. Soc. **22** (1990), no. 4, 381–385, DOI 10.1112/blms/22.4.381. MR1058316

[4] M. Győry, *2-local isometries of $C_0(X)$*, Acta Sci. Math. (Szeged) **67** (2001), no. 3-4, 735–746. MR1876463

[5] O. Hatori, A. Jiménez-Vargas, and M. Villegas-Vallecillos, *Maps which preserve norms of non-symmetrical quotients between groups of exponentials of Lipschitz functions*, J. Math. Anal. Appl. **415** (2014), no. 2, 825–845, DOI 10.1016/j.jmaa.2014.01.088. MR3178294

[6] O. Hatori and T. Miura, *Real linear isometries between function algebras. II*, Cent. Eur. J. Math. **11** (2013), no. 10, 1838–1842, DOI 10.2478/s11533-013-0282-0. MR3080241

[7] O. Hatori, T. Miura, H. Oka, and H. Takagi, *2-local isometries and 2-local automorphisms on uniform algebras*, Int. Math. Forum **2** (2007), no. 49-52, 2491–2502, DOI 10.12988/imf.2007.07219. MR2381836

[8] O. Hatori and S. Oi, *Isometries on Banach algebras of vector-valued maps*, Acta Sci. Math. (Szeged) **84** (2018), no. 1-2, 151–183. MR3792770

[9] M. Hosseini, *Generalized 2-local isometries of spaces of continuously differentiable functions*, Quaest. Math. **40** (2017), no. 8, 1003–1014, DOI 10.2989/16073606.2017.1344889. MR3765283

[10] K. Jarosz and V. D. Pathak, *Isometries between function spaces*, Trans. Amer. Math. Soc. **305** (1988), no. 1, 193–206, DOI 10.2307/2001048. MR920154

[11] A. Jiménez-Vargas, L. Li, A. M. Peralta, L. Wang, and Y.-S. Wang, *2-local standard isometries on vector-valued Lipschitz function spaces*, J. Math. Anal. Appl. **461** (2018), no. 2, 1287–1298, DOI 10.1016/j.jmaa.2018.01.029. MR3765490

[12] A. Jiménez-Vargas and M. Villegas-Vallecillos, *2-local isometries on spaces of Lipschitz functions*, Canad. Math. Bull. **54** (2011), no. 4, 680–692, DOI 10.4153/CMB-2011-025-5. MR2894518

[13] K. Kawamura and T. Miura, *Real-linear surjective isometries between function spaces*, Topology and its Applications **226** (2017), 66–85 doi:10.1016/j.topol.2017.05.002

[14] K. Kawamura, H. Koshimizu, and T. Miura, *Norms on $C^1([0,1])$ and their isometries*, Acta Sci. Math. (Szeged) **84** (2018), no. 1-2, 239–261. MR3792775

[15] H. Koshimizu, T. Miura, H. Takagi and S.-E. Takahasi, *Real-linear isometries between subspaces of continuous functions*, J. Math. Anal. Appl. **413** (2014), 229–241 doi:10.1016/j.jmaa2013.11.050

[16] S. Kowalski and Z. Słodkowski, *A characterization of multiplicative linear functionals in Banach algebras*, Studia Math. **67** (1980), no. 3, 215–223, DOI 10.4064/sm-67-3-215-223. MR592387

[17] L. Li, A. M. Peralta, L. Wang, and Y.-S. Wang, *Weak-2-local isometries on uniform algebras and Lipschitz algebras*, Publ. Mat. **63** (2019), no. 1, 241–264, DOI 10.5565/PUBLMAT6311908. MR3908793

[18] T. Miura, *Real-linear isometries between function algebras*, Cent. Eur. J. Math. **9** (2011), no. 4, 778–788, DOI 10.2478/s11533-011-0044-9. MR2805311

[19] T. Miura, *Surjective isometries between function spaces*, Function spaces in analysis, Contemp. Math., vol. 645, Amer. Math. Soc., Providence, RI, 2015, pp. 231–239, DOI 10.1090/conm/645/12926. MR3382419

[20] T. Miura and H. Takagi, *Surjective isometries on the Banach space of continuously differentiable functions*, Problems and recent methods in operator theory, Contemp. Math., vol. 687, Amer. Math. Soc., Providence, RI, 2017, pp. 181–192. MR3635762

[21] L. Molnár, *On 2-local *-automorphisms and 2-local isometries of $\mathscr{B}(H)$* , to appear in *J. Math. Anal. Appl.*

[22] L. Molnár, *Selected preserver problems on algebraic structures of linear operators and on function spaces*, Lecture Notes in Mathematics, vol. 1895, Springer-Verlag, Berlin, 2007. MR2267033

[23] L. Molnár, *2-local isometries of some operator algebras*, Proc. Edinb. Math. Soc. (2) **45** (2002), no. 2, 349–352, DOI 10.1017/S0013091500000043. MR1912644

[24] N. V. Rao and A. K. Roy, *Linear isometries of some function spaces*, Pacific J. Math. **38** (1971), 177–192. MR0308763

[25] P. Šemrl, *Local automorphisms and derivations on $\mathcal{B}(H)$*, Proc. Amer. Math. Soc. **125** (1997), no. 9, 2677–2680, DOI 10.1090/S0002-9939-97-04073-2. MR1415338

Department of Mathematics, Faculty of Science, Niigata University, Niigata 950-2181, Japan
Email address: hatori@math.sc.niigata-u.ac.jp

Niigata Prefectural Hakkai High School, Minamiuonuma 949-6681 Japan
Email address: shiho.oi.pmfn20@gmail.com

Contemporary Mathematics
Volume **737**, 2019
https://doi.org/10.1090/conm/737/14861

When is a finite sum of box operators on a JB*-triple a Hermitian projection?

Dijana Ilišević and Lina Oliveira

ABSTRACT. We characterise the Hermitian projections which are finite sums of box operators on Cartan factors of type I, II and III, that is, on the (matrix and) operator spaces $B(\mathcal{H}, \mathcal{K})$ of bounded linear operators from a complex Hilbert space $\mathcal{H}$ to a complex Hilbert space $\mathcal{K}$, $A(\mathcal{H})$ of skew-symmetric operators on $\mathcal{H}$ and $S(\mathcal{H})$ of symmetric operators on $\mathcal{H}$.

1. Introduction

A bounded linear operator T on a complex Banach space $\mathcal{X}$ is called a *Hermitian operator* if $e^{i\varphi T}$ is an isometry for any real number φ. A *projection* on $\mathcal{X}$ is a bounded linear operator $P\colon \mathcal{X} \to \mathcal{X}$ such that $P^2 = P$. A projection P on $\mathcal{X}$ is said to be *Hermitian* if P is a Hermitian operator. The trivial projections 0 and I are Hermitian projections.

Let $\overline{P}$ be the complementary projection $I - P$. By [**5**, Lemma 2.1], a projection P is a Hermitian operator if, and only if, P is a *bicircular projection*, that is, $P + \lambda\overline{P}$ is an isometry for all modulus one complex numbers λ (cf. [**7**], [**8**]). It is easily seen that the operators $P + \lambda\overline{P}$ are always bijective, for any projection P and λ with $|\lambda| = 1$, and that P is a Hermitian projection if, and only if, $\overline{P}$ is a Hermitian projection.

A complex vector space V that possesses a triple product $(a, b, c) \mapsto \{a\, b\, c\}$ from $V \times V \times V$ to V that is symmetric and linear in the first and third variables, conjugate linear in the second variable and, for all $a, b, x, y, z \in V$, satisfies the Jordan triple identity

$$\{a\, b\, \{x\, y\, z\}\} = \{\{a\, b\, x\}\, y\, z\} - \{x\, \{b\, a\, y\}\, z\} + \{x\, y\, \{a\, b\, z\}\} \tag{1.1}$$

is called a *Jordan triple*. A *subtriple* of V is a subspace closed under the triple product.

Let V be a Jordan triple and let $a, b \in V$. Define the *box operator* $a\Box\, b : V \to V$ by

$$(a\Box\, b)(x) = \{a\, b\, x\}, \qquad x \in V.$$

A complex Banach space V is said to be a *JB*-triple* if it is a Jordan triple with a continuous triple product and, for all $a \in V$, $a\Box\, a$ is a Hermitian operator with

---

2000 *Mathematics Subject Classification.* Primary 17C65; Secondary 46L25.
Key words and phrases. JB*-triple, Cartan factor, Hermitian projection, box operator.

non-negative spectrum and $\|a\square a\| = \|a\|^2$. This implies $\|\{a\,a\,a\}\| = \|a\|^3$ (cf. [**1**, Remark 2.5.28]).

Examples of JB*-triples are C*-algebras, JB*-algebras and the Cartan factors, the latter classified in six types. Let $\mathcal{H}$ and $\mathcal{K}$ be complex Hilbert spaces. The *Cartan factors of type I, II and III* are, respectively, $B(\mathcal{H},\mathcal{K})$, $A(\mathcal{H}) = \{z \in B(\mathcal{H}) : z^t = -z\}$ and $S(\mathcal{H}) = \{z \in B(\mathcal{H}) : z^t = z\}$, with $z^t = Jz^*J$ denoting the transpose of z in the JB*-triple $B(\mathcal{H})$, where $J\colon \mathcal{H} \to \mathcal{H}$ is a conjugation, i.e., J is a conjugate linear isometric involution.

The remaining types of Cartan factors are the *spin factors* (type IV) and the *exceptional Cartan factors* $M_{1,2}(\mathbb{O})$ (type V) and $H_3(\mathbb{O})$ (type VI) of 1×2 and 3×3 Hermitian matrices over the octonions, respectively. Cartan factors of types I–IV are sometimes called the *classical Cartan factors.*

A *JC*-triple* is a closed subtriple of $B(\mathcal{H})$ for some complex Hilbert space $\mathcal{H}$. These JB*-triples are also called J*-algebras by some authors. In [**7**, Proposition 3.4], it is shown that any Hermitian projection P on a JC*-triple V satisfies

$$P(\{xyz\}) = \{P(x)yz\} - \{xP(y)z\} + \{xyP(z)\}, \qquad x, y, z \in V, \tag{1.2}$$

an identity reminiscent of the Jordan identity in JB*-triples of which JC*-triples are particular cases. In fact, it is clear from the proof of this proposition that this equality must hold for any JB*-triple V, since a bijective linear operator $T\colon V \to V$ is an isometry if, and only if,

$$T(\{xyz\}) = \{T(x)T(y)T(z)\}, \qquad x, y, z \in V$$

(cf. [**1**, Theorem 3.1.7 and Theorem 3.1.20]).

The next proposition is the converse of [**7**, Proposition 3.4] in the setting of JB*-triples.

PROPOSITION 1.1. *Let V be a JB*-triple and let $P\colon V \to V$ be a projection satisfying* (1.2). *Then P is a Hermitian projection.*

PROOF. Observe firstly that P satisfies (1.2) if, and only if, $\overline{P}$ satisfies (1.2). Since (1.2) holds and $P^2 = P$, we have, for all $x, y, z \in V$,

$$\begin{aligned} &P(\{P(x)yz\} - \{xP(y)z\} + \{xyP(z)\}) \\ = \;&\{P(x)yz\} - \{xP(y)z\} + \{xyP(z)\}, \end{aligned} \tag{1.3}$$

which firstly implies

$$\begin{aligned} &\{P(x)yz\} - \{P(x)P(y)z\} + \{P(x)yP(z)\} \\ &-\{P(x)P(y)z\} + \{xP(y)z\} - \{xP(y)P(z)\} \\ &+\{P(x)yP(z)\} - \{P(x)P(y)P(z)\} + \{xyP(z)\} \\ = \;&\{P(x)yz\} - \{xP(y)z\} + \{xyP(z)\}, \end{aligned}$$

and then

$$\begin{aligned} &2\{xP(y)z\} - 2\{P(x)P(y)z\} + 2\{P(x)yP(z)\} \\ &-\{xP(y)P(z)\} - \{P(x)P(y)P(z)\} = 0. \end{aligned} \tag{1.4}$$

Replacing y with $\overline{P}(y)$ in (1.4), we conclude

$$\{P(x)\overline{P}(y)P(z)\} = 0. \tag{1.5}$$

Replacing now P with $\overline{P}$, we obtain

$$\{\overline{P}(x)P(y)\overline{P}(z)\} = 0. \tag{1.6}$$

Furthermore, inserting $P(y)$ for y and $P(z)$ for z in (1.4), we get

$$\{\overline{P}(x)P(y)P(z)\} = 0. \tag{1.7}$$

Then, since the triple product is symmetric in the outer variables,

$$\{P(x)P(y)\overline{P}(z)\} = 0. \tag{1.8}$$

Replacing P with $\overline{P}$ in (1.7) and (1.8), we get

$$\{P(x)\overline{P}(y)\overline{P}(z)\} = \{\overline{P}(x)\overline{P}(y)P(z)\} = 0. \tag{1.9}$$

Then

$$\begin{aligned}\{P(x)yz\} &= \{P(x)P(y)P(z)\} + \{P(x)P(y)\overline{P}(z)\} \\ &\quad + \{P(x)\overline{P}(y)P(z)\} + \{P(x)\overline{P}(y)\overline{P}(z)\} \\ &= \{P(x)P(y)P(z)\}, \end{aligned} \tag{1.10}$$

where the last equality follows from (1.5), (1.8) and (1.9). Hence we have also

$$\{xyP(z)\} = \{P(x)P(y)P(z)\}. \tag{1.11}$$

Analogously,

$$\begin{aligned}\{xP(y)z\} &= \{P(x)P(y)P(z)\} + \{P(x)P(y)\overline{P}(z)\} \\ &\quad + \{\overline{P}(x)P(y)P(z)\} + \{\overline{P}(x)P(y)\overline{P}(z)\} \\ &= \{P(x)P(y)P(z)\}. \end{aligned} \tag{1.12}$$

By (1.10)–(1.12),

$$\begin{aligned} P(\{xyz\}) &= \{P(x)yz\} - \{xP(y)z\} + \{xyP(z)\} \\ &= \{P(x)P(y)P(z)\}. \end{aligned} \tag{1.13}$$

Replacing P with $\overline{P}$ we get

$$\overline{P}(\{xyz\}) = \{\overline{P}(x)\overline{P}(y)\overline{P}(z)\}. \tag{1.14}$$

Let λ be a modulus one complex number and define $T = P + \lambda\overline{P}$. Then, by (1.5)–(1.9), (1.13) and (1.14),

$$\begin{aligned}\{T(x)T(y)T(z)\} &= \{P(x)P(y)P(z)\} + \lambda\{\overline{P}(x)\overline{P}(y)\overline{P}(z)\} \\ &= P(\{xyz\}) + \lambda\overline{P}(\{xyz\}) = T(\{xyz\}), \end{aligned} \tag{1.15}$$

from which follows that T is an isometry and, therefore, P is a Hermitian projection. □

By the Jordan identity (1.1), the operators $a\Box a$ and $a\Box b + b\Box a$ satisfy the equality (1.2). Hence, whenever $a\Box a$ or $a\Box b + b\Box a$ are projections they must be Hermitian projections. Notice however that $a\Box b$ does not satisfy (1.2), in general. Namely, (1.2) holds for some $a\Box b$ if, and only if, $a\Box b = b\Box a$. In view of this, we are naturally lead to ask "Under what conditions is a finite sum of box operators a Hermitian projection?".

COROLLARY 1.2. *Let V be a JB*-triple and $a_j, b_j \in V$, $j = 1, \ldots, n$. An operator $\sum_{j=1}^{n}(a_j\Box b_j + b_j\Box a_j)$ is a projection if, and only if, it is a Hermitian projection.*

An answer to the question above is given in the next section for Cartan factors of type I, II and III. Although Cartan factors are concrete examples of JB*-triples they are also paradigmatic, since each JB*-triple can in turn be identified as a closed subtriple of an ℓ_∞-sum of Cartan factors (cf. [**3**, Theorem 3.3.19]). For more details on JB*-triples and Cartan factors, the reader is referred to [**1**, **6**].

We end this section pointing out that the motivation for the present work is, not only [**7**], but also the recent paper [**2**]. A relevant step for obtaining the matrix representation of the Tits–Kantor–Koecher Lie algebra of a JB*-triple (cf. [**2**, Theorem 4.11 and Corollary 4.13]) is the characterisation of the finite sums of box operators which coincide with (the Hermitian projection) zero, obtained in [**2**, Theorem 4.4] for the classical Cartan factors. In fact, our results can be viewed as a generalisation of [**2**, Theorem 4.4] for the Cartan factors of type I–III, upon which our proofs depend, except in the case of the type II factor.

2. Hermitian projections and box operators

Recall that, in the Cartan factors of type I, II and III, the (unique) triple product is defined, for elements $\{x\,y\,z\}$, by

$$\{xyz\} = \frac{1}{2}(xy^*z + zy^*x).$$

Hence, for a, b in the factor,

$$(a\square\, b)(x) = \frac{1}{2}(ab^*x + xb^*a).$$

In these JB*-triples, consequently, an alternative form of (1.2) is

$$P(xy^*z + zy^*x) = P(x)y^*z + P(z)y^*x - xP(y)^*z - zP(y)^*x + xy^*P(z) + zy^*P(x)$$

or, equivalently,

$$P(xy^*x) = P(x)y^*x - xP(y)^*x + xy^*P(x). \tag{2.16}$$

The structure of Hermitian projections on $S(\mathcal{H})$ and $A(\mathcal{H})$ described in [**7**] is used in the sequel to answer our question in these settings. The conjugation $J\colon \mathcal{H} \to \mathcal{H}$ considered in [**7**] is defined by

$$J(\sum \alpha_\lambda e_\lambda) = \sum \overline{\alpha}_\lambda e_\lambda,$$

where $\{e_\lambda\}_{\lambda\in\Lambda}$ is an orthonormal basis of $\mathcal{H}$ and $\overline{\alpha}_\lambda$ is the complex conjugate of $\alpha_\lambda \in \mathbb{C}$. There is no loss of generality however, since, by [**4**, Lemma 7.5.6], any two conjugations J, J' on $\mathcal{H}$ are such that $J' = uJu^*$, for some unitary operator u.

THEOREM 2.1. *Let $a_j, b_j \in S(\mathcal{H})$, $j = 1, \ldots, n$. An operator $\sum_{j=1}^n a_j\square\, b_j$ is a Hermitian projection on $S(\mathcal{H})$ if, and only if, either $\sum_{j=1}^n a_jb_j^* = 0 = \sum_{j=1}^n b_j^*a_j$ or $\sum_{j=1}^n a_jb_j^* = I_{\mathcal{H}} = \sum_{j=1}^n b_j^*a_j$. In particular, $\sum_{j=1}^n(a_j\square\, b_j + b_j\square\, a_j)$ cannot be a non-trivial projection on $S(\mathcal{H})$.*

PROOF. By [**7**, Theorem 2.3], $\sum_{j=1}^n a_j\square\, b_j$ is a Hermitian projection if, and only if, one of the following holds: $\sum_{j=1}^n a_j\square\, b_j = 0$ or $\sum_{j=1}^n a_j\square\, b_j = I_{\mathcal{H}}$.

In the first case, [**2**, Theorem 4.4] implies $\sum_{j=1}^n a_jb_j^* = 0 = \sum_{j=1}^n b_j^*a_j$. In the second case, we define $a_{n+1} = I_{\mathcal{H}}$ and $b_{n+1} = -I_{\mathcal{H}}$. Then

$$\sum_{j=1}^{n+1} a_j\square\, b_j = \sum_{j=1}^{n} a_j\square\, b_j + a_{n+1}\square\, b_{n+1} = I_{\mathcal{H}} - I_{\mathcal{H}} = 0,$$

and [**2**, Theorem 4.4] implies $\sum_{j=1}^{n+1} a_j b_j^* = 0 = \sum_{j=1}^{n+1} b_j^* a_j$. Hence $\sum_{j=1}^{n} a_j b_j^* = I_{\mathcal{H}} = \sum_{j=1}^{n} b_j^* a_j$. □

THEOREM 2.2. *Let $a_j, b_j \in A(\mathcal{H})$, $j = 1, \ldots, n$. An operator $\sum_{j=1}^{n} a_j \square b_j$ is a Hermitian projection on $A(\mathcal{H})$ if, and only if, either $\sum_{j=1}^{n} a_j b_j^* = 2p$ or $\sum_{j=1}^{n} a_j b_j^* = I_{\mathcal{H}} - 2p$, with $p = \alpha \otimes \alpha$, for some unit vector $\alpha \in \mathcal{H}$, or $p = 0$. Moreover, if $\dim \mathcal{H} = 2$, then $p = 0$.*

In particular, $\sum_{j=1}^{n}(a_j \square b_j + b_j \square a_j)$ is a projection on $A(\mathcal{H})$ if, and only if, either $\sum_{j=1}^{n}(a_j b_j^* + b_j a_j^*) = 2p$ or $\sum_{j=1}^{n}(a_j b_j^* + b_j a_j^*) = I_{\mathcal{H}} - 2p$, with $p = \alpha \otimes \alpha$, for some unit vector $\alpha \in \mathcal{H}$, or $p = 0$.

PROOF. By [**7**, Theorem 2.5], $\sum_{j=1}^{n} a_j \square b_j$ is a Hermitian projection on $A(\mathcal{H})$ if, and only if, either $\sum_{j=1}^{n} a_j \square b_j$ or $I_{\mathcal{H}} - \sum_{j=1}^{n} a_j \square b_j$ has the form $x \mapsto px + xp^t$ with $p = \alpha \otimes \alpha$ for some unit vector $\alpha \in \mathcal{H}$, or $p = 0$. In the first case, we have

$$\left(\frac{1}{2}\sum_{j=1}^{n} a_j b_j^* - p\right)x + x\left(\frac{1}{2}\sum_{j=1}^{n} b_j^* a_j - p^t\right) = 0, \qquad x \in A(\mathcal{H}), \tag{2.17}$$

while in the second case

$$\left(\frac{1}{2}\Big(I_{\mathcal{H}} - \sum_{j=1}^{n} a_j b_j^*\Big) - p\right)x + x\left(\frac{1}{2}\Big(I_{\mathcal{H}} - \sum_{j=1}^{n} b_j^* a_j\Big) - p^t\right) = 0, \qquad x \in A(\mathcal{H}). \tag{2.18}$$

Both equations have the form

$$ux + xu^t = 0, \qquad x \in A(\mathcal{H}),$$

for some $u \in B(\mathcal{H})$.

For $x, y \in \mathcal{H}$, let $x \otimes y$ denote the rank one operator on $\mathcal{H}$ defined by $(x \otimes y)(\xi) = \langle \xi, y\rangle x$. Let $\{e_\lambda\}_{\lambda \in \Lambda}$ be an orthonormal basis for $\mathcal{H}$. For all $\lambda, \mu \in \Lambda$, $\lambda \neq \mu$, we have

$$\begin{aligned} 0 &= (u(e_\lambda \otimes e_\mu - e_\mu \otimes e_\lambda) + (e_\lambda \otimes e_\mu - e_\mu \otimes e_\lambda)u^t)e_\mu \\ &= ue_\lambda + \langle u^t e_\mu, e_\mu\rangle e_\lambda - \langle u^t e_\mu, e_\lambda\rangle e_\mu \\ &= ue_\lambda + \langle ue_\mu, e_\mu\rangle e_\lambda - \langle ue_\lambda, e_\mu\rangle e_\mu, \end{aligned} \tag{2.19}$$

which implies

$$0 = \langle ue_\lambda, e_\lambda\rangle + \langle ue_\mu, e_\mu\rangle.$$

Suppose that $\dim \mathcal{H} \geq 3$. Let λ, μ, ν be distinct. Then

$$\langle ue_\lambda, e_\lambda\rangle + \langle ue_\mu, e_\mu\rangle = 0, \tag{2.20}$$

$$\langle ue_\lambda, e_\lambda\rangle + \langle ue_\nu, e_\nu\rangle = 0, \tag{2.21}$$

$$\langle ue_\mu, e_\mu\rangle + \langle ue_\nu, e_\nu\rangle = 0. \tag{2.22}$$

Then (2.20) + (2.21) - (2.22) implies $\langle ue_\lambda, e_\lambda\rangle = 0$, for every $\lambda \in \Lambda$. It follows from (2.19) that $\langle ue_\lambda, e_\nu\rangle = 0$, for every $\nu \in \Lambda$, $\nu \neq \lambda$. Finally,

$$ue_\lambda = \sum_{\nu \in \Lambda} \langle ue_\lambda, e_\nu\rangle e_\nu = 0, \qquad \lambda \in \Lambda,$$

yielding $u=0$. Hence, (2.17) implies $\sum_{j=1}^{n} a_j b_j^* = 2p$, and (2.18) implies $\sum_{j=1}^{n} a_j b_j^* = I_{\mathcal{H}} - 2p$.

Now suppose that $\dim \mathcal{H} = 2$. Then

$$\sum_{j=1}^{n} a_j b_j^* = \sum_{j=1}^{n} b_j^* a_j = \alpha I_{\mathcal{H}}$$

for some $\alpha \in \mathbb{C}$. Since $\sum_{j=1}^{n} a_j \square b_j$ is a projection, $\alpha = 0$ or $\alpha = 1$. □

THEOREM 2.3. *Let $\mathcal{H}, \mathcal{K}$ be non-null complex Hilbert spaces and let $a_j, b_j \in B(\mathcal{H}, \mathcal{K})$, for $j = 1, \ldots, n$. An operator $\sum_{j=1}^{n} a_j \square b_j$ is a Hermitian projection on $B(\mathcal{H}, \mathcal{K})$ if, and only if, one of the following holds.*

(a) *There exist a self-adjoint projection $p \in B(\mathcal{H})$ and $\alpha \in \mathbb{C}$ such that*

$$\sum_{j=1}^{n} (a_j \square b_j)(x) = xp, \qquad x \in B(\mathcal{H}, \mathcal{K}),$$

$$\sum_{j=1}^{n} a_j b_j^* = 2\alpha I_{\mathcal{K}}, \qquad \sum_{j=1}^{n} b_j^* a_j = 2p - 2\alpha I_{\mathcal{H}}.$$

(b) *There exist a self-adjoint projection $p \in B(\mathcal{K})$ and $\alpha \in \mathbb{C}$ such that*

$$\sum_{j=1}^{n} (a_j \square b_j)(x) = px, \qquad x \in B(\mathcal{H}, \mathcal{K}),$$

$$\sum_{j=1}^{n} a_j b_j^* = 2p - 2\alpha I_{\mathcal{K}}, \qquad \sum_{j=1}^{n} b_j^* a_j = 2\alpha I_{\mathcal{H}}.$$

Furthermore,

(i) *if $\mathcal{H}$ and $\mathcal{K}$ are finite dimensional, then*

$$\alpha = \frac{\dim \mathcal{R}(p)}{\dim \mathcal{H} + \dim \mathcal{K}},$$

where $\mathcal{R}(p)$ denotes the range of p;

(ii) *if either $\mathcal{H}$ or $\mathcal{K}$ is finite dimensional, then $\alpha = 0$ and $p = 0$;*

(iii) *if $\mathcal{H}$ and $\mathcal{K}$ are infinite dimensional and separable then either $I - p$ has infinite rank and $\alpha = 0$, or p has infinite rank and $\alpha = \frac{1}{2}$.*

Before proving Theorem 2.3, we consider firstly the case $n = 1$.

COROLLARY 2.4. *Let $a, b \in B(\mathcal{H}, \mathcal{K})$. An operator $a \square b$ is a Hermitian projection on $B(\mathcal{H}, \mathcal{K})$ if, and only if, $ab^* = 0 = b^*a$ or $ab^* = I_{\mathcal{K}}$, $b^*a = I_{\mathcal{H}}$. In particular, $a \square a$ cannot be a non-trivial projection on $B(\mathcal{H}, \mathcal{K})$.*

PROOF. By Theorem 2.3, one of the following holds:

(a) there exist $\alpha \in \mathbb{C}$ and a self-adjoint projection $p \in B(\mathcal{H})$ such that $ab^* = 2\alpha I_{\mathcal{K}}$ and $b^*a = 2p - 2\alpha I_{\mathcal{H}}$;

(b) there exist $\alpha \in \mathbb{C}$ and a self-adjoint projection $p \in B(\mathcal{K})$ such that $ab^* = 2p - 2\alpha I_{\mathcal{K}}$ and $b^*a = 2\alpha I_{\mathcal{H}}$.

Suppose that (a) holds. Then, on one hand,

$$\begin{aligned} b^*ab^*a &= (b^*a)^2 = (2p - 2\alpha I_{\mathcal{H}})^2 \\ &= 4p - 8\alpha p + 4\alpha^2 I_{\mathcal{H}}, \end{aligned}$$

and on the other hand

$$\begin{aligned} b^*ab^*a &= b^*(ab^*)a = 2\alpha b^*a = 2\alpha(2p - 2\alpha I_{\mathcal{H}}) \\ &= 4\alpha p - 4\alpha^2 I_{\mathcal{H}}. \end{aligned}$$

Comparing the above two equalities, we have

$$2\alpha^2 I_{\mathcal{H}} - 3\alpha p + p = 0. \tag{2.23}$$

If $p \neq I_{\mathcal{H}}$ then multiplying the above equality by $\frac{1}{2}(I_{\mathcal{H}} - p)$ we get $\alpha = 0$ and then (2.23) implies $p = 0$. If $p = I_{\mathcal{H}}$ then (2.23) implies $2\alpha^2 - 3\alpha + 1 = 0$, hence $\alpha = 1$ or $\alpha = \frac{1}{2}$. If $\alpha = 1$ and $p = I_{\mathcal{H}}$ then $ab^* = 2I_{\mathcal{K}}$ and $b^*a = 0$. Then, however, $4I_{\mathcal{K}} = (ab^*)^2 = a(b^*a)b^* = 0$, which is a contradiction. If $\alpha = \frac{1}{2}$ and $p = I_{\mathcal{H}}$ then $ab^* = I_{\mathcal{K}}$ and $b^*a = I_{\mathcal{H}}$.

The case when (b) holds can be considered analogously. □

By Corollary 2.4, if $a \square b$ is a Hermitian projection on $B(\mathcal{H},\mathcal{K})$, then it must be a trivial projection. However, for $\sum_{j=1}^n (a_j \square b_j)$ with $n \geq 2$, this does not hold necessarily, as is illustrated by the following example.

EXAMPLE 2.5. Let $\mathcal{H}$ and $\mathcal{K}$ be infinite dimensional separable Hilbert spaces. Let $p \in B(\mathcal{H})$ be a self-adjoint projection. Then p has infinite dimensional range $\mathcal{R}(p)$ or $I_{\mathcal{H}} - p$ has infinite dimensional range $\mathcal{R}(I_{\mathcal{H}} - p)$.

Suppose that $p \in B(\mathcal{H})$ has infinite dimensional range. Then there exists an isometric isomorphism, say, $u\colon \mathcal{R}(p) \to \mathcal{K}$. Let $v\colon \mathcal{H} \to \mathcal{K}$ be defined by $v(\xi) = u(p\xi)$, for every $\xi \in \mathcal{H}$. Then $u^*v = p$ and $vu^* = I_{\mathcal{K}}$. Let $t\colon \mathcal{H} \to \mathcal{K}$ be an isometric isomorphism and define $c_1 = -t$, $d_1 = t$, $c_2 = 2v$, $d_2 = u$. Then $\sum_{j=1}^2 (c_j \square d_j)(x) = xp$ for every $x \in B(\mathcal{H},\mathcal{K})$. In this case $\sum_{j=1}^2 c_j d_j^* = I_{\mathcal{K}}$, hence $\alpha = \frac{1}{2}$.

Now suppose that $I_{\mathcal{H}} - p$ has infinite dimensional range and, thus, we proceed as above. We define $v = u(I_{\mathcal{H}} - p) \in B(\mathcal{H},\mathcal{K})$, where $u\colon \mathcal{R}(I_{\mathcal{H}} - p) \to \mathcal{K}$ is an isometric isomorphism, and we also take an isometric isomorphism $t \in B(\mathcal{H},\mathcal{K})$. Then we set $c_1 = 2t$, $d_1 = t$, $c_2 = -2v$, $d_2 = u$, and we have $\sum_{j=1}^2 (c_j \square d_j)(x) = xp$, for every $x \in B(\mathcal{H},\mathcal{K})$. In this case $\sum_{j=1}^2 c_j d_j^* = 0$, hence $\alpha = 0$.

REMARK 2.6. Note that, if $\mathcal{H}$ is infinite dimensional separable, for a self-adjoint projection $p \in B(\mathcal{H})$ such that both p and $I_{\mathcal{H}} - p$ have infinite rank, the operator $x \mapsto xp$ cannot be represented as a finite sum of box operators. Namely, if we assume the opposite then, by Example 2.5, there would exist two different representations, say, $\sum_{j=1}^n c_j' \square d_j'$ and $\sum_{j=1}^n c_j'' \square d_j''$ of this operator, with $\sum_{j=1}^2 c_j'(d_j')^* = I_{\mathcal{K}}$ and $\sum_{j=1}^2 c_j''(d_j'')^* = 0$. Then, however, [**2**, Theorem 4.4] yields the contradiction.

We prove Theorem 2.3 next.

PROOF. We firstly notice that $P(x) = ux + xv$ for $u = \frac{1}{2}\sum_{j=1}^n a_j b_j^* \in B(\mathcal{K})$ and $v = \frac{1}{2}\sum_{j=1}^n b_j^* a_j \in B(\mathcal{H})$. Since $P^2 = P$ we have

$$(u - u^2)x + x(v - v^2) = 2uxv, \qquad x \in B(\mathcal{H},\mathcal{K}). \tag{2.24}$$

If we multiply (2.24) by $y \in B(\mathcal{H})$ from the right, we get

$$(u - u^2)xy + x(v - v^2)y = 2uxvy, \tag{2.25}$$

and, if we replace x with xy in (2.24), we have

$$(u - u^2)xy + xy(v - v^2) = 2uxyv. \tag{2.26}$$

As usual, we shall write $[x, y]$ for the commutator $xy - yx$. From (2.25) and (2.26), we get

$$x[v - v^2, y] = 2ux[v, y], \qquad x \in B(\mathcal{H}, \mathcal{K}), y \in B(\mathcal{H}). \tag{2.27}$$

Now we multiply (2.27) by $z \in B(\mathcal{K})$ from the left and get

$$zx[v - v^2, y] = 2zux[v, y]. \tag{2.28}$$

Replacing x with zx in (2.27) we have

$$zx[v - v^2, y] = 2uzx[v, y]. \tag{2.29}$$

Now (2.28) and (2.29) imply

$$[u, B(\mathcal{K})]B(\mathcal{H}, \mathcal{K})[v, B(\mathcal{H})] = 0. \tag{2.30}$$

For simplicity, suppose that $s \in B(\mathcal{K})$ and $t \in B(\mathcal{H})$ are such that $sB(\mathcal{H}, \mathcal{K})t = 0$. Let $\xi \in \mathcal{K}$ and $\eta \in \mathcal{H}$ be arbitrary. Then $\xi \otimes t\eta \in B(\mathcal{H}, \mathcal{K})$, thus

$$0 = s(\xi \otimes t\eta)t\eta = \langle t\eta, t\eta\rangle s\xi, \qquad \xi \in \mathcal{K}, \eta \in \mathcal{H},$$

which implies $s = 0$ or $t = 0$. Applying this to (2.30) we conclude $[u, B(\mathcal{K})] = 0$ or $[v, B(\mathcal{H})] = 0$, and therefore $u \in \mathbb{C}I_{\mathcal{K}}$ or $v \in \mathbb{C}I_{\mathcal{H}}$.

Suppose that $u = \alpha I_{\mathcal{K}}$, for some $\alpha \in \mathbb{C}$. Then $P(x) = ux + xv = x(\alpha I_{\mathcal{H}} + v)$. Let

$$p = \alpha I_{\mathcal{H}} + v \in B(\mathcal{H}). \tag{2.31}$$

Since P is a Hermitian projection, it satisfies (1.2), that is, (2.16). Inserting xp instead of $P(x)$ in (2.16), we get

$$xy^*xp = xpy^*x - xp^*y^*x + xy^*xp,$$

that is $x(p^* - p)y^*x = 0$ for all $x, y \in B(\mathcal{H}, \mathcal{K})$. In particular, for $y = x(p^* - p)$ we have $yy^*y = 0$, which firstly implies $y = 0$ and then $p^* = p$.

Assertion (b) is similarly proved.

For the converse, firstly we note that $\sum_{j=1}^n a_j \Box b_j$ is a projection on $B(\mathcal{H}, \mathcal{K})$, since it maps $x \in B(\mathcal{H}, \mathcal{K})$ to $xp \in B(\mathcal{H}, \mathcal{K})$ (or to px if (b) holds). It remains to prove that it is a Hermitian projection, that is, $\|x(p + \lambda(I_{\mathcal{H}} - p))\| = \|x\|$ for every $x \in B(\mathcal{H}, \mathcal{K})$ and every modulus one $\lambda \in \mathbb{C}$. Define $T_\lambda \colon B(\mathcal{H}, \mathcal{K}) \to B(\mathcal{H}, \mathcal{K})$ by $T_\lambda(x) = x(p + \lambda(I_{\mathcal{H}} - p))$. Since

$$\begin{aligned} T_\lambda(x)T_\lambda(y)^*T_\lambda(z) &= x(p + \lambda(I_{\mathcal{H}} - p))(p + \overline{\lambda}(I_{\mathcal{H}} - p))y^*z(p + \lambda(I_{\mathcal{H}} - p)) \\ &= xy^*z(p + \lambda(I_{\mathcal{H}} - p)) = T_\lambda(xy^*z), \qquad x, y, z \in B(\mathcal{H}, \mathcal{K}), \end{aligned}$$

T_λ is an isometry, hence $\sum_{j=1}^n a_j \Box b_j$ is a Hermitian projection.

We prove the remaining assertions only in the case when (a) holds, as the proof would be similar should we assume that (b) held.

(i) Let $\mathcal{H}, \mathcal{K}$ be finite dimensional spaces, and denote by tr the trace of an operator. We have

$$2\alpha \dim \mathcal{K} = \operatorname{tr}\,(\sum_{j=1}^n a_j b_j^*) = \operatorname{tr}\,(\sum_{j=1}^n b_j^* a_j) = 2 \dim \mathcal{R}(p) - 2\alpha \dim \mathcal{H},$$

which implies

$$\alpha = \frac{\dim \mathcal{R}(p)}{\dim \mathcal{H} + \dim \mathcal{K}}.$$

(ii) Note that, whenever $\mathcal{H}$ or $\mathcal{K}$ is finite dimensional, both $\sum_{j=1}^n a_j b_j^*$ and $\sum_{j=1}^n b_j^* a_j$ are finite rank operators.

If $p = 0$, then [**2**, Lemma 4.5] implies $\alpha = 0$. Thus in the sequel we assume $p \neq 0$.

Let $\{e_\gamma\}_{\gamma\in\Gamma}$ be an orthonormal basis for $\mathcal{H}$, containing a basis of $\mathcal{R}(p)$, and let $\{u_\lambda\}_{\lambda\in\Lambda}$ be an orthonormal basis for $\mathcal{K}$. We have

$$2\alpha = \langle \sum_{j=1}^n a_j b_j^* u_\lambda, u_\lambda\rangle_{\mathcal{K}} = \sum_{j=1}^n \sum_{\gamma\in\Gamma} \langle a_j e_\gamma, u_\lambda\rangle_{\mathcal{K}} \langle b_j^* u_\lambda, e_\gamma\rangle_{\mathcal{H}}, \qquad \lambda\in\Lambda, \tag{2.32}$$

(2.33)

$$2\langle p e_\gamma, e_\gamma\rangle_{\mathcal{H}} - 2\alpha = \langle \sum_{j-1}^n b_j^* a_j e_\gamma, e_\gamma\rangle_{\mathcal{H}} = \sum_{j=1}^n \sum_{\lambda\subset\Lambda} \langle a_j e_\gamma, u_\lambda\rangle_{\mathcal{K}} \langle b_j^* u_\lambda, e_\gamma\rangle_{\mathcal{H}}, \qquad \gamma\in\Gamma.$$

Assume firstly that $\dim \mathcal{H} < \infty$ and $\mathcal{K}$ is infinite dimensional. Since $\sum_{j=1}^n a_j b_j^*$ is a finite rank operator, by (a) of this theorem, $\alpha I_{\mathcal{K}}$ also must be a finite rank operator. However this is possible only if $\alpha = 0$, since $\mathcal{K}$ was assumed to be infinite dimensional.

But by (2.33),

$$\begin{aligned} 2\dim \mathcal{R}(p) &= 2\sum_{\gamma\in\Gamma} \langle p e_\gamma, e_\gamma\rangle_{\mathcal{H}} = \sum_{\gamma\in\Gamma} \langle \sum_{j=1}^n b_j^* a_j e_\gamma, e_\gamma\rangle_{\mathcal{H}} \\ &= \sum_{\gamma\in\Gamma} \sum_{j=1}^n \sum_{\lambda\in\Lambda} \langle a_j e_\gamma, u_\lambda\rangle_{\mathcal{K}} \langle b_j^* u_\lambda, e_\gamma\rangle_{\mathcal{H}}. \end{aligned}$$

Since Γ is finite and $\sum_{j=1}^n \sum_{\lambda\in\Lambda} \langle a_j e_\gamma, u_\lambda\rangle_{\mathcal{K}} \langle b_j^* u_\lambda, e_\gamma\rangle_{\mathcal{H}}$ converges for each $\gamma \in \Gamma$, we have

$$2\dim \mathcal{R}(p) = \sum_{\lambda\in\Lambda} \sum_{j=1}^n \sum_{\gamma\in\Gamma} \langle a_j e_\gamma, u_\lambda\rangle_{\mathcal{K}} \langle b_j^* u_\lambda, e_\gamma\rangle_{\mathcal{H}}.$$

Hence, by (2.32), $\dim \mathcal{R}(p) = 0$, i.e., $p = 0$, which contradicts our assumption.

Suppose now that $\mathcal{H}$ is infinite dimensional and $\dim \mathcal{K}$ is finite. Then, by (a) of this theorem, the operator $2p - 2\alpha I_{\mathcal{H}}$ has finite rank. If p has finite rank then $\alpha = 0$. If p has infinite rank then $\alpha = 1$ since $(2 - 2\alpha)p = (2p - 2\alpha I_{\mathcal{H}})p$ has finite rank.

In the first case, $\sum_{\gamma\in\Gamma} \langle p e_\gamma, e_\gamma\rangle_{\mathcal{H}}$ converges, since it only possesses a finite number of non-zero terms. Hence, by (2.32) and (2.33), we have

$$2\dim \mathcal{R}(p) = 2\sum_{\gamma\in\Gamma} \langle p e_\gamma, e_\gamma\rangle_{\mathcal{H}} = \sum_{\gamma\in\Gamma} \sum_{j=1}^n \sum_{\lambda\in\Lambda} \langle a_j e_\gamma, u_\lambda\rangle_{\mathcal{K}} \langle b_j^* u_\lambda, e_\gamma\rangle_{\mathcal{H}}.$$

Since Λ is finite and, by (2.32), $\sum_{j=1}^n \sum_{\gamma\in\Gamma} \langle a_j e_\gamma, u_\lambda\rangle_{\mathcal{K}} \langle b_j^* u_\lambda, e_\gamma\rangle_{\mathcal{H}}$ converges and coincides with zero for each $\lambda \in \Lambda$, it follows that

$$2\dim \mathcal{R}(p) = \sum_{\lambda\in\Lambda} \sum_{j=1}^n \sum_{\gamma\in\Gamma} \langle a_j e_\gamma, u_\lambda\rangle_{\mathcal{K}} \langle b_j^* u_\lambda, e_\gamma\rangle_{\mathcal{H}} = 0,$$

contradicting the initial hypothesis of $p \neq 0$.

In the second case, since $2p - 2\alpha I_{\mathcal{H}} = 2p - 2I_{\mathcal{H}}$ has finite rank, the set $\Gamma_0 = \{\gamma \in \Gamma : pe_\gamma = 0\}$ is finite, and, by (2.33),

$$-2|\Gamma_0| = \sum_{\gamma\in\Gamma}(2\langle pe_\gamma, e_\gamma\rangle_{\mathcal{H}} - 2) = \sum_{\gamma\in\Gamma}\sum_{j=1}^{n}\sum_{\lambda\in\Lambda}\langle a_j e_\gamma, u_\lambda\rangle_{\mathcal{K}}\langle b_j^* u_\lambda, e_\gamma\rangle_{\mathcal{H}}.$$

Since Λ is finite and, by (2.32), $\sum_{j=1}^{n}\sum_{\gamma\in\Gamma}\langle a_j e_\gamma, u_\lambda\rangle_{\mathcal{K}}\langle b_j^* u_\lambda, e_\gamma\rangle_{\mathcal{H}}$ converges, for each $\lambda \in \Lambda$, it follows that

$$-2|\Gamma_0| = \sum_{\lambda\in\Lambda}\sum_{j=1}^{n}\sum_{\gamma\in\Gamma}\langle a_j e_\gamma, u_\lambda\rangle_{\mathcal{K}}\langle b_j^* u_\lambda, e_\gamma\rangle_{\mathcal{H}} = 2|\Lambda|,$$

which is impossible.

It has thus been shown that $\alpha = 0$ and $p = 0$ is the only possibility, if either $\mathcal{H}$ or $\mathcal{K}$ is finite dimensional.

(iii) Let both $\mathcal{H}$ and $\mathcal{K}$ be infinite dimensional separable. By (a) of this theorem, we have $\sum_{j=1}^{n}(a_j\Box\, b_j)(x) = xp$ for every $x \in B(\mathcal{H},\mathcal{K})$. By Remark 2.6, either p has infinite rank, or $I_{\mathcal{H}} - p$ has infinite rank. But then $xp = \sum_{j=1}^{2}(c_j\Box\, d_j)(x)$ for every $x \in B(\mathcal{H},\mathcal{K})$, where c_1, c_2, d_1, d_2 are as in Example 2.5. Then $\sum_{j=1}^{n} a_j\Box\, b_j - \sum_{j=1}^{2} c_j\Box\, d_j = 0$. By [**2**, Theorem 4.4], $\sum_{j=1}^{n} a_j b_j^* - \sum_{j=1}^{2} c_j d_j^* = 0$. Hence, $\alpha = 0$, if $I_{\mathcal{H}} - p$ has infinite rank, and $\alpha = \frac{1}{2}$, if p has infinite rank. □

COROLLARY 2.7. *Let $\mathcal{H}$ and $\mathcal{K}$ be complex Hilbert spaces such that either $\mathcal{H}$ or $\mathcal{K}$ is finite dimensional. Let $a_j, b_j \in B(\mathcal{H},\mathcal{K})$, $j = 1,\dots,n$. Then $\sum_{j=1}^{n}(a_j\Box\, b_j + b_j\Box\, a_j)$ is a projection if and only if $\sum_{j=1}^{n}(a_j b_j^* + b_j a_j^*) = 0 = \sum_{j=1}^{n}(b_j^* a_j + a_j^* b_j)$. In particular, $\sum_{j=1}^{n} a_j\Box\, a_j$ is a projection if and only if $a_j = 0$, for every $j = 1,\dots,n$.*

REMARK 2.8. The operators of the form $x \mapsto xp$ (resp., $x \mapsto px$) for self-adjoint projections $p \in B(\mathcal{H})$ (resp., $p \in B(\mathcal{K})$) are always Hermitian projections on $B(\mathcal{H},\mathcal{K})$, but by Theorem 2.3 they are not always representable as finite sums of box operators.

REMARK 2.9. By the proof of Theorem 2.3 (iii), if both $\mathcal{H}$ and $\mathcal{K}$ are infinite dimensional separable and $\sum_{j=1}^{n} a_j\Box\, b_j$ is a Hermitian projection, then it can be represented as a sum of two box operators, that is, $\sum_{j=1}^{n} a_j\Box\, b_j = \sum_{j=1}^{2} c_j\Box\, d_j$ for some $c_1, c_2, d_1, d_2 \in B(\mathcal{H},\mathcal{K})$. Moreover, by Corollary 2.4, if it is a nontrivial projection then it cannot be represented as a box operator, that is, as $c\Box\, d$ for some $c, d \in B(\mathcal{H},\mathcal{K})$.

REMARK 2.10. If $\mathcal{H}$ and $\mathcal{K}$ are of the same finite dimension or both infinite dimensional separable, then $P\colon B(\mathcal{H},\mathcal{K}) \to B(\mathcal{H},\mathcal{K})$ is representable as a finite sum of box operators if, and only if, $I - P$ is representable as a finite sum of box operators. This follows from the fact that $P + (I - P) = I$ is a box operator $t\Box\, t$, where $t \in B(\mathcal{H},\mathcal{K})$ is an isometric isomorphism. However, it is not the case in general: by Theorem 2.3 (ii) it is sufficient to take a finite dimensional $\mathcal{H}$ and an infinite dimensional $\mathcal{K}$ (or vice versa) and a trivial projection P.

REMARK 2.11. By Theorem 2.3 (i), if $\mathcal{K} = \mathcal{H}$ and $\dim \mathcal{H}$ is finite and non-zero, then

$$\alpha = \frac{\dim \mathcal{R}(p)}{2\dim \mathcal{H}},$$

hence $0 \leq \alpha \leq \frac{1}{2}$. Thus $\alpha = 0$ if, and only if, $p = 0$, and $\alpha = \frac{1}{2}$ if, and only if, $p = I_{\mathcal{H}}$. By Example 2.5, this is no longer true if $\mathcal{K} = \mathcal{H}$ is infinite dimensional.

We end this paper with an example showing that in a finite dimensional case one can indeed have α different from 0 and $\frac{1}{2}$.

EXAMPLE 2.12. Let $\mathcal{H}$ be a $2k$-dimensional complex Hilbert space and let $p \in B(\mathcal{H})$ be a selfadjoint projection of rank k, more precisely, $p = I_k \oplus 0_k$. Let

$$a_1 = \begin{bmatrix} 0 & \frac{1}{2}I_k \\ \frac{1}{2}I_k & 0 \end{bmatrix}, \quad b_1 = \begin{bmatrix} 0 & -I_k \\ I_k & 0 \end{bmatrix}, \quad a_2 = b_2 = \begin{bmatrix} I_k & 0 \\ 0 & 0 \end{bmatrix}.$$

Then $\sum_{j=1}^{2}(a_j \square b_j)(x) = xp$. In this case $\alpha = \frac{1}{2}\sum_{j=1}^{2} a_j b_j^* = \frac{1}{4}$.

Acknowledgments. The first author has been fully supported by the Croatian Science Foundation [project number IP-2016-06-1046]. The second author was partially supported by the FCT/Portugal grant UID/MAT/04459/2013.

References

[1] C.-H. Chu, *Jordan structures in geometry and analysis*, Cambridge Tracts in Mathematics, vol. 190, Cambridge University Press, Cambridge, 2012. MR2885059

[2] C.-H. Chu and L. Oliveira, *Tits-Kantor-Koecher Lie algebras of JB*-triples*, J. Algebra **512** (2018), 465–492, DOI 10.1016/j.jalgebra.2018.07.013. MR3841531

[3] Y. Friedman and B. Russo, *The Gel′fand-Naĭmark theorem for* JB*-*triples*, Duke Math. J. **53** (1986), no. 1, 139–148, DOI 10.1215/S0012-7094-86-05308-1. MR835800

[4] H. Hanche-Olsen and E. Størmer, *Jordan operator algebras*, Monographs and Studies in Mathematics, vol. 21, Pitman (Advanced Publishing Program), Boston, MA, 1984. MR755003

[5] J. E. Jamison, *Bicircular projections on some Banach spaces*, Linear Algebra Appl. **420** (2007), no. 1, 29–33, DOI 10.1016/j.laa.2006.05.009. MR2277626

[6] O. Loos, Bounded symmetric domains and Jordan pairs, Mathematical Lectures, University of California, Irvine, 1977.

[7] L. L. Stachó and B. Zalar, *Bicircular projections on some matrix and operator spaces*, Linear Algebra Appl. **384** (2004), 9–20, DOI 10.1016/j.laa.2003.11.014. MR2055340

[8] L. L. Stachó and B. Zalar, *Bicircular projections and characterization of Hilbert spaces*, Proc. Amer. Math. Soc. **132** (2004), no. 10, 3019–3025, DOI 10.1090/S0002-9939-04-07333-2. MR2063123

Department of Mathematics, Faculty of Science, University of Zagreb, Croatia
Email address: ilisevic@math.hr

Center for Mathematical Analysis, Geometry and Dynamical Systems *and* Department of Mathematics, Instituto Superior Técnico, Universidade de Lisboa, Av. Rovisco Pais 1049-001 Lisboa, Portugal
Email address: lina.oliveira@tecnico.ulisboa.pt

Contemporary Mathematics
Volume **737**, 2019
https://doi.org/10.1090/conm/737/14862

2-local isometries on $C^{(n)}([0,1])$

Kazuhiro Kawamura, Hironao Koshimizu, and Takeshi Miura

Abstract. Let $C^{(n)}([0,1])$ be the Banach space of all n-times continuously differentiable functions on $[0,1]$ with the norm

$$\|f\|_C = \sup_{t\in[0,1]} \sum_{k=0}^{n} |f^{(k)}(t)|/k!.$$

We prove that every 2-local isometry on $(C^{(n)}([0,1]), \|\cdot\|_C)$ is a surjective complex linear isometry. Two proofs, hopefully illuminating different aspects of the operator on the space, are presented.

1. Introduction

A mapping $T: N \to N$ on a normed linear space $(N, \|\cdot\|_N)$ over the complex number field $\mathbb{C}$ is called an *isometry* if $\|T(f) - T(g)\|_N = \|f - g\|_N$ for all $f, g \in N$ (neither the linearity nor the surjectivity of the mapping is assumed). Motivated by the notion of 2-local automorphisms and derivations due to Šemrl [**9**], Molnár introduced the notion of 2-local isometry in [**6**]. A mapping $S: N \to N$ is called a *2-local isometry* if for each $f, g \in N$ there exists a surjective complex linear isometry $T_{f,g}: N \to N$, depending on f and g, such that $S(f) = T_{f,g}(f)$ and $S(g) = T_{f,g}(g)$. Again neither the surjectivity nor the linearity of the mapping S is assumed. Characterizing 2-local isometries on various function spaces has been studied by several authors. For example, Győry [**1**] gave the characterization of 2-local isometries on $C_0(X)$, the Banach space of all continuous complex-valued functions vanishing at infinity defined on a first countable σ-compact locally compact Hausdorff space X.

This paper deals with the space $C^{(n)}([0,1])$ $(n \geq 1)$, the space of all n-times continuously differentiable functions on $[0,1]$. Several norms are known which makes the space a Banach space. Hosseini [**2**] investigated 2-local isometries on $C^{(n)}([0,1])$, assuming the surjectivity and the real-linearity of such mappings, with respect to the norm $\|f\|_n$ defined by $\|f\|_n = \max\{|f(0)|, |f'(0)|, \cdots, |f^{(n-1)}(0)|, \|f^{(n)}\|_\infty\}$, where $\|\cdot\|_\infty$ is the supremum norm on $[0,1]$. We study 2-local isometries, without the surjectivity/linearity assumption, on the space $C^{(n)}([0,1])$ with respect to the

2010 *Mathematics Subject Classification.* Primary: 46J10.
Key words and phrases. isometry, local isometry, n-times continuously differentiable function.
The first author is supported by JSPS KAKENHI Grant Number 17K05241. The third author was supported by JSPS KAKENHI Grant Number 15K04921 and 16K05172.

norms $\|\cdot\|_C$ and $\|\cdot\|_\Sigma$ defined by:

$$\|f\|_C = \sup_{t\in[0,1]} \sum_{k=0}^{n} \frac{|f^{(k)}(t)|}{k!} \quad \text{and} \quad \|f\|_\Sigma = \sum_{k=0}^{n} \|f^{(k)}\|_\infty$$

where $f^{(0)}(t) = f(t)$. We prove that every 2-local isometry on the space $(C^{(n)}([0,1]), \|\cdot\|_C)$ is a surjective complex linear isometry on the space. The same holds for 2-local isometries on $(C^{(n)}([0,1]), \|\cdot\|_\Sigma)$.

Two proofs of our theorem are presented: one is based on the idea due to Jiménez-Vargas and Villegas-Vallecillos in [**4**, Proof of Theorem 2.1] and the other is based on the Bartle-Graves theorem. We hope that these proofs with different viewpoints would give us a new perspective to the 2-local-reflexivity problem. It is most likely that the present method could be applied to study 2-local isometries on some other function spaces. Its full development is a subject of future study. It should be mentioned that the 2-local reflexivity of various function spaces including $C^{(1)}([0,1])$ has been studied in a general framework by Hatori and Oi in [**3**]. Studying the connection of our method with theirs is also a subject of further study.

Before proceeding, let us observe that every 2-local isometry $S : N \to N$ is an isometry; for each $f, g \in N$ there exists an isometry $T : N \to N$ such that $S(f) = T(f), S(g) = T(g)$ and we have

$$\|S(f) - S(g)\|_N = \|T(f) - T(g)\|_N = \|f - g\|_N.$$

In particular every 2-local isometry is a continuous injection.

2. Main result and its First Proof

Let $\mathbb{T} = \{z \in \mathbb{C} : |z| = 1\}$. The following is our main theorem.

THEOREM 2.1. *Let $n \geq 1$ be an integer and let S be a 2-local isometry of $C^{(n)}([0,1])$ with respect to the norm $\|f\|_C = \sup_{t\in[0,1]} \sum_{k=0}^{n} |f^{(k)}(t)|/k!$. There exists a constant $c \in \mathbb{T}$ such that $S(f)(t) = cf(t)$ for all $f \in C^{(n)}([0,1])$ and $t \in [0,1]$, or $S(f)(t) = cf(1-t)$ for all $f \in C^{(n)}([0,1])$ and $t \in [0,1]$.*

Both of our proofs depend on the characterization of complex linear $\|\cdot\|_C$-isometries on $C^{(n)}([0,1])$ due to Pathak:

THEOREM 2.2. [**7**, Theorem 2.5] *Let $n \geq 1$ be an integer. For each surjective complex linear isometry T on $(C^{(n)}([0,1]), \|\cdot\|_C)$, there exists $c \in \mathbb{T}$ such that $T(f)(t) = cf(t)$ for all $f \in C^{(n)}([0,1])$ and $t \in [0,1]$, or $T(f)(t) = cf(1-t)$ for all $f \in C^{(n)}([0,1])$ and $t \in [0,1]$.*

This section gives the first proof of Theorem 2.1 on the basis of the idea due to Jiménez-Vargas and Villegas-Vallecillos in [**4**, Proof of Theorem 2.1]. For later use, let id be the identity function on $[0,1]$, $r : [0,1] \to [0,1]$ be the map given by $r(t) = 1-t$, $t \in [0,1]$, and let $\mathcal{H} = \{\mathrm{id}, r\}$, the only isometries on the unit interval $[0,1]$. Note that

$$\varphi \circ \varphi = \mathrm{id}, \quad \varphi \in \mathcal{H}. \tag{2.1}$$

PROOF OF THEOREM 2.1. Let S be a 2-local isometry on $C^{(n)}([0,1]), \|\cdot\|_C$. We denote by $\mathbf{1}$ the constant function which takes the value 1. For each $g \in C^{(n)}([0,1])$ there exists a surjective complex linear isometry T_g on $C^{(n)}([0,1])$ such that $S(\mathbf{1}) = T_g(\mathbf{1})$ and $S(g) = T_g(g)$. By Theorem 2.2, there exist $c_g \in \mathbb{T}$ and $\phi_g \in \mathcal{H}$ such that $T_g(f)(t) = c_g f(\phi_g(t))$ for all $f \in C^{(n)}([0,1])$ and $t \in [0,1]$. Observe that $c_g = T_g(\mathbf{1}) = S(\mathbf{1})$ and in particular c_g does not depend on g. Define $S_0 \colon C^{(n)}([0,1]) \to C^{(n)}([0,1])$ by $S_0 = \overline{S(\mathbf{1})}S$. For each $g \in C^{(n)}([0,1])$, we have

$$S_0(g) = \overline{S(\mathbf{1})}S(g) = \overline{S(\mathbf{1})}T_g(g) = \overline{S(\mathbf{1})}c_g(g \circ \phi_g) = g \circ \phi_g.$$

Hence, we get

$$S_0(g) = g \circ \phi_g, \quad g \in C^{(n)}([0,1]). \tag{2.2}$$

For an arbitrary $t_0 \in [0,1]$, choose a function $h_0 \in C^{(n)}([0,1])$ such that

$$h_0(t_0) = 1 \qquad \text{and} \qquad 0 \le h_0(t) < 1 \quad \text{for all } t \in [0,1] \setminus \{t_0\}. \tag{2.3}$$

For a function $f \in C^{(n)}([0,1])$, we define the set $E_{t_0,f}$ by

$$E_{t_0,f} = \{t \in [0,1] : S_0(f)(t) = f(t_0)\} = \{t \in [0,1] : f(\phi_f(t)) = f(t_0)\}.$$

In what follows we prove that the set $\cap_{f \in C^{(n)}([0,1])} E_{t_0,f}$ is a singleton. More strongly we show the following equality for an arbitrary function h_0 satisfying (2.3):

$$\cap_{f \in C^{(n)}([0,1])} E_{t_0,f} = \{\phi_{h_0}(t_0)\}. \tag{2.4}$$

First observe from (2.3) that

$$\begin{aligned} E_{t_0,h_0} &= \{t \in [0,1] : h_0(\phi_{h_0}(t)) = h_0(t_0) = 1\} \\ &= \{t \in [0,1] : \phi_{h_0}(t) = t_0\} = \{\phi_{h_0}(t_0)\}. \end{aligned}$$

Thus for the proof of (2.4) it suffices to show that $\phi_{h_0}(t_0) \in E_{t_0,f}$ for each $f \in C^{(n)}([0,1])$. Take an arbitrary function $f \in C^{(n)}([0,1])$ and take a surjective complex linear isometry T_{f,h_0} on $C^{(n)}([0,1])$ such that $S(f) = T_{f,h_0}(f)$ and $S(h_0) = T_{f,h_0}(h_0)$. Theorem 2.2 implies $S(f) = c_{f,h_0}(f \circ \varphi_{f,h_0})$ and $S(h_0) = c_{f,h_0}(h_0 \circ \varphi_{f,h_0})$ for some $c_{f,h_0} \in \mathbb{T}$ and $\varphi_{f,h_0} \in \mathcal{H}$. Using (2.2), we obtain

$$\begin{aligned} f \circ \phi_f &= S_0(f) = \overline{S(\mathbf{1})}S(f) = \overline{S(\mathbf{1})}\, c_{f,h_0}(f \circ \varphi_{f,h_0}), \\ h_0 \circ \phi_{h_0} &= S_0(h_0) = \overline{S(\mathbf{1})}\, c_{f,h_0}(h_0 \circ \varphi_{f,h_0}). \end{aligned}$$

Recalling (2.1) and using the second equality we obtain

$$1 = h_0(t_0) = \overline{S(\mathbf{1})}\, c_{f,h_0} h_0(\varphi_{f,h_0}(\phi_{h_0}(t_0))).$$

Hence $S(\mathbf{1})\overline{c_{f,h_0}} = h_0(\varphi_{f,h_0}(\phi_{h_0}(t_0))) \ge 0$. This and $S(\mathbf{1})\overline{c_{f,h_0}} \in \mathbb{T}$ imply $S(\mathbf{1})\overline{c_{f,h_0}} = 1$. Thus

$$f \circ \phi_f = f \circ \varphi_{f,h_0} \qquad \text{and} \qquad h_0 \circ \phi_{h_0} = h_0 \circ \varphi_{f,h_0}. \tag{2.5}$$

Then (2.1) and the second equality of (2.5) show:

$$1 = h_0(t_0) = h_0(\phi_{h_0}(\phi_{h_0}(t_0))) = h_0(\varphi_{f,h_0}(\phi_{h_0}(t_0))),$$

and, by (2.3), we have $\varphi_{f,h_0}(\phi_{h_0}(t_0)) = t_0$. Combining (2.2) and (2.5), we obtain

$$S_0(f)(\phi_{h_0}(t_0)) = f(\phi_f(\phi_{h_0}(t_0))) = f(\varphi_{f,h_0}(\phi_{h_0}(t_0))) = f(t_0).$$

Consequently, $\phi_{h_0}(t_0) \in E_{t_0,f}$ for each $f \in C^{(n)}([0,1])$ which proves the equality (2.4). Since $\phi_{h_0}(t_0) = t_0$ or $1 - t_0$, what we have actually shown is:

$$\cap_{f \in C^{(n)}([0,1])} E_{t_0,f} = \{t_0\} \text{ or } \{1 - t_0\} \text{ for each } t_0 \in [0,1]. \tag{2.6}$$

The condition (2.4) allows us to define a map $\psi\colon [0,1] \to [0,1]$ with the property that

$$\cap_{f\in C^{(n)}([0,1])} E_{t,f} = \{\psi(t)\}.$$

By the definition of $E_{t,f}$, we have

$$S_0(f)(\psi(t)) = f(t), \quad f \in C^{(n)}([0,1]), t \in [0,1] \tag{2.7}$$

and by (2.6), $\psi(t) \in \{t, 1-t\}, t \in [0,1]$.

To finish the proof we need to show that $\psi \in \mathcal{H}$. For the proof let $E_1 = \{t \in [0,1] : \psi(t) = t\}$ and $E_{-1} = \{t \in [0,1] : \psi(t) = 1-t\}$. We see $[0,1] = E_1 \cup E_{-1}$ and $E_1 \cap E_{-1} = \{1/2\}$. We prove that

(∗) $E_1 \cap [0,1/2)$ is closed in $[0,1/2)$.

For the proof, let (t_n) be a sequence in $E_1 \cap [0,1/2)$ such that $t_n \to t_0 \in [0,1/2)$ and suppose that $t_0 \in E_{-1}$. Then

$$S_0(f)(t_0) = \lim_n S_0(f)(t_n) = \lim_n S_0(f)(\psi(t_n)) = \lim_n f(t_n) = f(t_0)$$

and also

$$S_0(f)(1-t_0) = S_0(f)(\psi(t_0)) = f(t_0),$$

hence $S_0(f)(t_0) = S_0(f)(1-t_0)$ for all $f \in C^{(n)}([0,1])$. Recalling (2.2) it follows that $f(\phi_f(t_0)) = f(\phi_f(1-t_0))$ with $\phi_f \in \mathcal{H}$. This implies $f(t_0) = f(1-t_0)$ for all $f \in C^{(n)}([0,1])$, a contradiction because $t_0 \neq 1-t_0$. Thus we have $t_0 \in E_1$ and (∗) is proved.

Similarly $E_{-1} \cap [0,1/2)$ is closed in $[0,1/2)$ and thus

$$E_1 \supset [0,1/2] \quad \text{or} \quad E_{-1} \supset [0,1/2].$$

Likewise we have $E_1 \supset [1/2,1]$ or $E_{-1} \supset [1/2,1]$. Suppose that both inclusions $E_1 \supset [0,1/2]$ and $E_{-1} \supset [1/2,1]$ hold. Then we see $\psi(t) = t$ for $t \in [0,1/2]$ and $\psi(t) = 1-t$ for $t \in [1/2,1]$. Hence $\psi([0,1]) = [0,1/2]$. However by (2.2) and (2.7), we have

$$f(\phi_f(\psi(t))) = S_0(f)(\psi(t)) = f(t), \quad f \in C^{(n)}([0,1]), t \in [0,1],$$

hence we obtain $f(\phi_f([0,1/2])) = f([0,1])$ for each $f \in C^{(n)}([0,1])$, which is impossible. Similarly both of the inclusions "$E_{-1} \supset [0,1/2], E_1 \supset [1/2,1]$" do not hold and what remains is either of the equalities:

$$E_1 = [0,1] \quad \text{or} \quad E_{-1} = [0,1]$$

which means that $\psi \in \mathcal{H}$. Using (2.1) and (2.7) we have $S_0(f)(t) = S_0(f)(\psi(\psi(t))) = f(\psi(t))$ for $f \in C^{(n)}([0,1]), t \in [0,1]$. This completes the proof. □

By examining the above proof, we see readily that the role of the norm $\|\cdot\|_C$ in the theorem is just to ensure the characterization theorem, Theorem 2.2. Such a theorem has been known also for the norm $\|\cdot\|_\Sigma$ on the space $C^{(1)}([0,1])$ [**8**, Theorem 4.1]. Therefore we have

COROLLARY 2.3. *Every 2-local isometry on $C^{(1)}([0,1])$ with the norm $\|f\|_\Sigma = \|f\|_\infty + \|f'\|_\infty$ for $f \in C^{(1)}([0,1])$ is a surjective complex linear isometry on $(C^{(1)}([0,1]), \|\cdot\|_\Sigma)$, where $\|g\|_\infty = \sup_{t\in[0,1]} |g(t)|$.*

3. The second proof

This section gives an alternative proof of Theorem 2.1 applying the following form of the Bartle-Graves theorem [**5**].

THEOREM 3.1. [**5**, Example 1.3, Example 1.3*, Corollary on p.364, and Theorem 3.2"] *Let $u : E \to B$ be a bounded linear surjection between Banach spaces E and B. Then*

(1) *there exists a continuous map $s : B \to E$ such that $u \circ s = \mathrm{id}_B$ and $s(0) = 0$.*
(2) *The map $\Phi : \mathrm{Ker}\, u \times B \to E$ defined by*

$$\Phi(e,b) = e + s(b), \quad (e,b) \in \mathrm{Ker}\, u \times B$$

is a homeomorphism.

In particular the restrictions $\Phi|_{\mathrm{Ker}\, u\times\{0\}} : \mathrm{Ker}\, u\times\{0\} \to \mathrm{Ker}\, u$ and $\Phi|_{\mathrm{Ker}\, u\times(B\setminus\{0\})} : \mathrm{Ker}\, u \times (B \setminus \{0\}) \to E \setminus \mathrm{Ker}\, u$ are homeomorphisms. The inverse homeomorphism $\Phi^{-1} : E \to \mathrm{Ker}\, u \times B$ is given by

$$\Phi^{-1}(e) = (e - (s \circ u)(e), u(e)), \quad e \in E.$$

THE SECOND PROOF OF THEOREM 2.1. Let $S : C^{(n)}([0,1]) \to C^{(n)}([0,1])$ be a 2-local isometry. As is noticed in Section 1, S is a continuous operator. For each $g \in C^{(n)}([0,1])$, choose a complex linear $\|\cdot\|_C$-isometry $T_g : C^{(n)}([0,1]) \to C^{(n)}([0,1])$ such that $S(g) = T_g(g), S(\mathbf{1}) = T_g(\mathbf{1})$. Again by Theorem 2.2, there exist $c_g \in \mathbb{C}$ and $\phi_g : [0,1] \to [0,1]$ such that $T_g f(t) = c_g f(\phi_g(t))$, $f \in C^{(n)}([0,1])$, where $c_g \in \mathbb{T}$ and $\phi_g \in \mathcal{H}$. As in the first proof, we have $c_g = S(\mathbf{1})$ and thus $S(g)(t) = (S(\mathbf{1}))g(\phi_g(t))$ for $g \in C^{(n)}([0,1])$ and $t \in [0,1]$. Recall $\mathcal{H} = \{\mathrm{id}, r\}$ where $r(t) = 1-t, t \in [0,1]$. As before, let $S_0(f) = \overline{S(\mathbf{1})}S(f) = f \circ \phi_f$, $f \in C^{(n)}([0,1])$ (see (2.2)). It is a continuous mapping.

For $g \in C^{(n)}([0,1])$ with $g(t_0) \neq g(1-t_0)$ for some $t_0 \in [0,1]$, the map ϕ_g is uniquely determined, because $(S(\mathbf{1}))g(t) = (S(\mathbf{1}))g(1-t)$ for each $t \in [0,1]$ forces $g(t) = g(1-t)$ for each $t \in [0,1]$. Let

$$Z = \{g \in C^{(n)}([-1,1]) : g \circ r \neq g\}.$$

By the above remark, we have a well-defined map $Q\colon Z \to \{\pm 1\}$ given by

$$Q(g) = \begin{cases} 1 & \text{if } \phi_g(t) = t, \\ -1 & \text{if } \phi_g(t) = 1-t \end{cases}.$$

We show that Q is continuous. It suffices to prove that $Q^{-1}(1)$ and $Q^{-1}(-1)$ are closed in Z. Suppose that a sequence (g_n) in $Q^{-1}(1)$ converges to $g \in Z$. By the continuity of S_0, we have $S_0(g_n) \to S_0(g)$ and thus $g_n \circ \phi_{g_n} \to g \circ \phi_g$. Since $Q(g_n) = 1$ for each n, we have $\phi_{g_n} = \mathrm{id}$ and thus $g = g \circ \phi_g$. Since $g \in Z$, we have $\phi_g = \mathrm{id}$ and $g \in Q^{-1}(1)$. The same argument shows that $Q^{-1}(-1)$ is closed in Z. Next we prove

(**) Z is connected.

For the proof, let $R : C^{(n)}([0,1]) \to C^{(n)}([0,1])$ be the linear operator defined by

$$Rf = f \circ r - f, \quad f \in C^{(n)}([0,1]).$$

Then $Z = C^{(n)}([0,1]) \setminus \mathrm{Ker}\, R$. We have $\mathrm{Im}\, R = A := \{f \mid f \circ r = -f\}$: in fact, the inclusion $\mathrm{Im}\, R \subset A$ is readily verified. If $h \circ r = -h$, then $R(-\frac{1}{2}h) = \frac{1}{2}(-h\circ r + h) =$

h and $h \in \operatorname{Im} R$. Thus we see the above equality. The subspace A is a Banach space as a closed subspace of $C^{(n)}([0,1])$. Applying Theorem 3.1 and the remark after the theorem, we see

$$Z = C^{(n)}([0,1]) \setminus \operatorname{Ker} R \text{ is homeomorphic to } \operatorname{Ker} R \times (A \setminus \{0\}).$$

It is easy to see that $A \setminus \{0\}$ is connected because A is an infinite dimensional Banach space, and so is the product $\operatorname{Ker} R \times (A \setminus \{0\})$ and hence Z is connected which proves $(**)$.

The continuity of the map Q and $(**)$ imply that Q is a constant map. Therefore, there exists $\phi \in \mathcal{H}$ such that ,

$$S(f)(t) = (S(\mathbf{1}))f(\phi(t)), \; f \in Z, \; t \in [0,1]. \tag{3.1}$$

Furthermore the space Z is dense in $C^{(n)}([0,1])$. In fact, it is easy to find a non-zero function h_0 with $\|h_0\|_C$ being arbitrarily small such that $h_0 \neq h_0 \circ r$. Then for each $g \in C^{(n)}([0,1])$ with $g = g \circ r$, the function $g + h_0$ is an approximation of g with $g + h_0 \in Z$. By the continuity of S, we see that (3.1) holds for each $f \in C^{(n)}([0,1])$. This proves the theorem. □

References

[1] M. Győry, *2-local isometries of* $C_0(X)$, Acta Sci. Math. (Szeged) **67** (2001), no. 3-4, 735–746. MR1876463

[2] M. Hosseini, *Generalized 2-local isometries of spaces of continuously differentiable functions*, Quaest. Math. **40** (2017), no. 8, 1003–1014, DOI 10.2989/16073606.2017.1344889. MR3765283

[3] O. Hatori and S. Oi, *2-local isometries on function spaces,* preprint, arXiv:1812.10342v2.

[4] A. Jiménez-Vargas and M. Villegas-Vallecillos, *2-local isometries on spaces of Lipschitz functions*, Canad. Math. Bull. **54** (2011), no. 4, 680–692, DOI 10.4153/CMB-2011-025-5. MR2894518

[5] E. Michael, *Continuous selections. I*, Ann. of Math. (2) **63** (1956), 361–382, DOI 10.2307/1969615. MR0077107

[6] L. Molnár, *2-local isometries of some operator algebras*, Proc. Edinb. Math. Soc. (2) **45** (2002), no. 2, 349–352, DOI 10.1017/S0013091500000043. MR1912644

[7] V. D. Pathak, *Isometries of* $C^{(n)}[0,1]$, Pacific J. Math. **94** (1981), no. 1, 211–222. MR625820

[8] N. V. Rao and A. K. Roy, *Linear isometries of some function spaces*, Pacific J. Math. **38** (1971), 177–192. MR0308763

[9] P. Šemrl, *Local automorphisms and derivations on* $\mathcal{B}(H)$, Proc. Amer. Math. Soc. **125** (1997), no. 9, 2677–2680, DOI 10.1090/S0002-9939-97-04073-2. MR1415338

Institute of Mathematics, University of Tsukuba, Tsukuba, Ibaraki 305-8571, Japan
Email address: kawamura@math.tsukuba.ac.jp

National Institute of Technology, Yonago College, Yonago 683-8502, Japan
Email address: koshimizu@yonago-k.ac.jp

Department of Mathematics, Faculty of Science, Niigata University, Niigata 950-2181, Japan
Email address: miura@math.sc.niigata-u.ac.jp

Contemporary Mathematics
Volume **737**, 2019
https://doi.org/10.1090/conm/737/14863

Quotients of tensor product spaces

Monika and T. S. S. R. K. Rao

Abstract. Let X, Y, Z be Banach spaces. For the identity operator I on X and a quotient operator $Q : Z \to Y$, in this paper we investigate conditions under which $I \otimes Q$ and $Q \otimes I$ are again quotient operators on the respective tensor product spaces. Let $J \subset X$ be a closed subspace. For the quotient injective tensor product space $(X\hat{\otimes}_\epsilon Y)/(J\hat{\otimes}_\epsilon Y)$, we consider several geometric properties of Banach space X and J under which this quotient space is isometric to $(X/J)\hat{\otimes}_\epsilon Y$. We show that for a L^1-predual space X and for a special type of M-ideal $J \subset X$, the spaces under consideration are isometric.

1. Introduction

Let X, Y be real Banach spaces and let $J \subset Y$ be a closed subspace. Let $Q : Y \to Y/J$ be the quotient map. Let $I : X \to X$ be the identity map. In this paper we investigate conditions under which the tensor operator $I \otimes Q$ or $Q \otimes I$ is again a quotient operator on the appropriate tensor product space. Most often we assume the condition that the kernel of the tensor operator is again a tensor product space. We refer to the monographs [**10**] and [**2**] Chapter VIII, for basic concepts and notations.

In Section 2 we start a review of properties preserved under taking projective tensor products. Since in general the kernel of the operator $I\hat{\otimes}_\pi Q$ need not be a subspace of $X\hat{\otimes}_\pi Y$ we consider special situations, like the case of space of Bochner integrable functions, when the kernel of the quotient operator is again a projective tensor product space.

Using the notion of an ideal due to [**7**] (see Section 2) we show that a closed subspace J of X is an ideal if and only if $X\hat{\otimes}_\pi J$ is a subspace of $X\hat{\otimes}_\pi Y$ for every Banach space Y.

Since injective tensor products preserve subspace operation, an interesting question in injective tensor product theory is to determine conditions on X or Y so that the operator $Q\hat{\otimes}_\epsilon I : X\hat{\otimes}_\epsilon Y \to (X/J)\hat{\otimes}_\epsilon Y$ is a quotient map whose kernel is $J\hat{\otimes}_\epsilon Y$. Such questions for $I\hat{\otimes}_\epsilon Q$ were considered when

2010 *Mathematics Subject Classification.* Primary 47L05, 46B28, 46B25.
Key words and phrases. Injective and projective tensor products, quotients of tensor products, L^1-preduals, M-ideals.

X is a $\mathcal{L}_{\infty,1+\epsilon}$-space, for all $\epsilon > 0$, see [**10**] Theorem 3.6 on page 51 . Let Ω be a compact Hausdorff space. The space of continuous functions $C(\Omega)$ is a $\mathcal{L}_{\infty,1+\epsilon}$-space, for all $\epsilon > 0$. We also have the identification of $C(\Omega)\hat{\otimes}_\epsilon Y$ as the space of vector-valued continuous functions $C(\Omega, Y)$ (see [**2**] Example VIII.6). For a closed subspace $Z \subset Y$, let $Q : Y \to Y/Z$ be the quotient map. It is easy to see that $I\hat{\otimes}_\epsilon Q : C(\Omega, Y) \to C(\Omega, Y/Z)$ has kernel $C(\Omega, Z)$. Thus one has the isometry $C(\Omega, Y/Z) = C(\Omega, Y)/C(\Omega, Z)$.

In section 3 we study conditions under which $Q\hat{\otimes}_\epsilon I$ is a quotient map on $X\hat{\otimes}_\epsilon Y$. Unlike the results in [**10**] we need a condition on the kernel of the quotient map. Clearly we have $J\hat{\otimes}_\epsilon Y$ is a subspace of $ker(Q\hat{\otimes}_\epsilon I)$. We assume that $ker(Q\hat{\otimes}_\epsilon I) = J\hat{\otimes}_\epsilon Y$.

Similar to the classical situation of [**10**, Theorem 3.6], we consider Banach spaces X for which X^* is isometric to $L^1(\mu)$ for a positive measure μ, the so called L^1-predual spaces. See [**8**] and [**6**] for structure theory of these spaces. The spaces $C(\Omega)$ are L^1-predual spaces.

In order to prove the isometry of quotient spaces, we need a stronger notion of ideal, called M-ideal (see Section 3). Here we establish the isometry of quotient spaces by showing that for a L^1-predual space X and a special type of M-ideal $J \subset X$, for any Banach space Y, $X/J\hat{\otimes}_\epsilon Y$ is isometric to $(X\hat{\otimes}_\epsilon Y)/(J\hat{\otimes}_\epsilon Y)$. Our proof involves vector-valued integral representation theory from [**12**] and a characterization of L^1-preduals in terms of boundary measures from the work of Effros, [**3**]. Our ideas lead to a vector-valued version of an extension Theorem due to Alfsen and Effros.

It is an open question that for a separable Banach space X in which for every extreme point x^* of the dual unit ball, $\ker(x^*)$ is an M-ideal, the validity of the quotient isometry result for all M-ideals $J \subset X$ and for all Banach spaces Y, implies that X is a L^1-predual space?

2. Projective Tensors

Let X, Y and Z be Banach spaces. $\mathcal{B}(X \times Y, Z)$ denotes the Banach space of bounded bilinear mappings from $X \times Y$ into Z and $\mathcal{L}(X, Y)$ denotes the Banach space of bounded linear functions from X into Y. We can define a norm on the tensor product $X \otimes Y$, the projective norm as follows

$$\pi(u) = \inf\left\{ \sum_{i=1}^n \|x_i\|\|y_i\| : u = \sum_{i=1}^n x_i \otimes y_i \right\}$$

π defined this way is a norm on $X \otimes Y$ and $\pi(x \otimes y) = \|x\|\|y\|$. We shall denote by $X \otimes_\pi Y$ the tensor product $X \otimes Y$ endowed with projective norm π and its completion by $X\hat{\otimes}_\pi Y$. We will call the Banach space $X\hat{\otimes}_\pi Y$ as the projective tensor product of X and Y. It can be shown that $\mathcal{B}(X \times Y) = (X\hat{\otimes}_\pi Y)^*$, which yields a new formula for the projective norm

$$\pi(u) = \sup\{|\langle u, B\rangle| : B \in \mathcal{B}(X \times Y), \|B\| \leq 1\}.$$

Similarly the identification $(X\hat{\otimes}_\pi Y)^* = \mathcal{L}(X, Y^*)$ also gives us the following formula for the projective norm

$$\begin{aligned}\pi(u) &= \sup\{|\langle u, S\rangle| : S \in \mathcal{L}(X, Y^*), \|S\| \leq 1\} \\ &= \sup\{|\langle u, T\rangle| : T \in \mathcal{L}(Y, X^*), \|T\| \leq 1\}\end{aligned}$$

In general the projective tensor product does not respect subspaces. In other words, if Z is a subspace of X, then the induced norm on $Z\hat{\otimes}_\pi Y$ by $X\hat{\otimes}_\pi Y$ is not, in general the projective norm, since if Z is a subspace of X then for any $u \in Z\hat{\otimes}_\pi Y$, $\pi(u; X\hat{\otimes}_\pi Y) \leq \pi(u; Z\hat{\otimes}_\pi Y)$. However the projective tensor product respect quotients i.e., if Z is a closed subspace of X then $(X/Z)\hat{\otimes}_\pi Y$ is a quotient of $X\hat{\otimes}_\pi Y$ and $Y\hat{\otimes}_\pi (X/Z)$ is a quotient of $Y\hat{\otimes}_\pi X$.

The following Proposition from [**10**] shows that the projective tensors behave nicely with quotient subspaces. We recall from [**10**] an operator $Q : Z \to Y$ is a quotient operator if Q is surjective and $\|y\| = \inf\{\|z\| : z \in Z, Q(z) = y\}$ for every $y \in Y$, or, equivalently, Q maps the open unit ball of Z onto the open unit ball of Y which simply means Y is isometrically isomorphic to the quotient space $Z/kerQ$.

PROPOSITION 1. *Let J be a closed subspace of X and $Q : X \to X/J$ be the natural quotient operator. If I denotes the identity operator on Y, then $Q\hat{\otimes}_\pi I : X\hat{\otimes}_\pi Y \to (X/J)\hat{\otimes}_\pi Y$ (and similarly $I\hat{\otimes}_\pi Q$) is a quotient operator for each Banach space Y.*

DEFINITION 2. *Let (Ω, Σ, μ) be a measure space. A μ-measurable function $f : \Omega \to X$ is Bochner integrable if*

$$\int_\Omega \|f(t)\| d\mu(t) < \infty$$

$L^1(\mu, X)$ denotes the space of all Bochner integrable functions.

PROPOSITION 3. *For any Banach space X and a subspace Y of X, we have $L^1(\mu, X/Y) \cong L^1(\mu, X)/L^1(\mu, Y)$.*

PROOF. Let $Q : X \to X/Y$ be the quotient map. By [**10**], $L^1(\mu) \hat{\otimes}_\pi X = L^1(\mu, X)$. With this identification we claim that the tensor operator $I\hat{\otimes}_\pi Q : L^1(\mu, X) \to L^1(\mu, X/Y)$ is just $f \to Q \circ f$. Since both the maps are continuous, we need to verify this only at simple functions. For $s = \sum_1^n \chi_{A_i} x_i$, $(I\hat{\otimes}_\pi Q)(s) = \sum_1^n \chi_{A_i} Q(x_i)$. On the other hand $Q \circ s = \sum_1^n \chi_{A_i} Q(x_i)$. Hence the claim. Therefore $(I\hat{\otimes}_\pi Q)(f) = Q \circ f$. Hence kernel of this map is precisely $L^1(\mu, Y)$. From Proposition 1, we know that $I\hat{\otimes}_\pi Q$ is a quotient map, we have the relation $L^1(\mu, X)/L^1(\mu, Y) = L^1(\mu, X/Y)$. □

Let us recall a definition from [**10**].

DEFINITION 4. *Let $\lambda > 1$. A Banach space X is said to be an $\mathcal{L}_{1,\lambda}$-space if every finite dimensional subspace M of X contained in a finite dimensional subspace N, of dimension k, say, whose Banach-Mazur distance from the space l_1^k is at most λ. If X is a $\mathcal{L}_{1,\lambda}$-space for some $\lambda > 1$ then X is said to be an $\mathcal{L}_1$-space.*

In order to state the next Theorem let us introduce a terminology [**10**]. We shall say that *projective tensor products* with X respects subspaces *isomorphically* if, for every Banach space Y, and every subspace Z of Y, the projective norm on $X\hat{\otimes}_\pi Z$ is equivalent to the norm induced by the projective norm on $X\hat{\otimes}_\pi Y$. The following result is from [**10**].

PROPOSITION 5. *If X is a $\mathcal{L}_1$ space then the projective tensor products with X respect subspaces isomorphically.*

The following Theorem (from [**2**, Page 230]) is one more instance where the projective tensors respect subspaces.

THEOREM 6. *Let U and V be subspaces of X and Y respectively. Then $U\hat{\otimes}_\pi V$ is a subspace of $X\hat{\otimes}_\pi Y$ if and only if every bounded bilinear form on $U \times V$ extends to a bounded bilinear form on $X \times Y$ with the same norm.*

PROOF. 'Only if' part follows from Hahn Banach theorem and the fact that $\mathcal{B}(U \times V) = (U\hat{\otimes}_\pi V)^*$. For the 'if' part let us assume that every bounded bilinear function from $U \times V$ extends to a bounded bilinear function on $X \times Y$ with same norm. Let $u \in U\hat{\otimes}_\pi V$, then

$$\begin{aligned} \pi(u; U\hat{\otimes}_\pi V) &= \sup\{|B(u)| : B \in \mathcal{B}(U \times V), ||B|| = 1\} \\ &\leq \sup\{|B(u)| : B \in \mathcal{B}(X \times Y), ||B|| = 1\} \\ &= \pi(u; X\hat{\otimes}_\pi Y) \end{aligned}$$

□

REMARK 7. *Let $(\Omega, \mathcal{A}, \mu)$ be a finite measure space. Let $\mathcal{B} \subset \mathcal{A}$ be a sub σ-algebra. Let $L^1(\mathcal{B}, \mu)$ denote the space of $\mathcal{B}$- measurable integrable functions. Let $E : L^1(\mathcal{A}, \mu) \to L^1(\mathcal{B}, \mu)$ be the conditional expectation operator. We recall that E is a linear contraction and is a projection. For any Banach space X, as an interesting application of Theorem 6 we can immediately conclude that the canonical inclusion operator from $L^1(\mathcal{B}, \mu)\hat{\otimes}_\pi X$ to $L^1(\mathcal{A}, \mu)\hat{\otimes}_\pi X$ is an isometry.*

COROLLARY 8. *Let U be a subspace of X. Then $U\hat{\otimes}_\pi Y$ is a subspace of $X\hat{\otimes}_\pi Y$ if and only if every bounded linear operator from U into Y^* extends to an operator of the same norm X into Y^*.*

PROOF. Follows from the observation that $\mathcal{B}(X \times Y) = \mathcal{L}(X, Y^*)$ □

We recall from [**10**] that a Banach space X is said to be an injective space, if for every Banach space Z and every subspace W of Z, every operator from W into X extends to an operator from Z into X of the same norm. The next example follows from the previous corollary.

EXAMPLE 9. *Let X be a Banach space such that X^* is not injective (for example ℓ^2). Then by above corollary for some Banach space Z, there exists a closed linear subspace W of Z such that $W\hat{\otimes}_\pi X$ is not a subspace of $Z\hat{\otimes}_\pi X$.*

We now will look at some more cases when projective tensor products respect subspaces. Let us first recall a definition from [**7**].

DEFINITION 10. *A closed subspace J of X is an ideal, if there exists a linear projection $P : X^* \to X^*$ such that $ker(P) = J^\perp$ and $\|P\| = 1$.*

PROPOSITION 11. *If U is an ideal of X then for any Banach space Y, $U\hat{\otimes}_\pi Y$ is a subspace of $X\hat{\otimes}_\pi Y$.*

PROOF. Let $T : U \to Y^*$ then $T^{**} : U^{**} = U^{\perp\perp} \to Y^{***}$ is a norm preserving extension. Let $Q : Y^{***} \to Y^*$ be the canonical contractive projection. Now $Q \circ T^{**} : U^{\perp\perp} \to Y^*$ is a again a norm-preserving extension. Suppose we assume that there is contractive projection $R : X^{**} \to U^{\perp\perp}$ Then $Q \circ T^{**} \circ R|_X : X \to Y^*$ is the required norm preserving extension of T. Hence by Corollary 8, the proposition follows. □

In what follows for a Banach space X, we denote by $X^{(IV)}$ the fourth dual of X. We continue to embed a Banach space canonically in its bidual. We also implicitly use the canonical isometry of X^{**} and $X^{\perp\perp} \subset X^{(IV)}$. We make a one time exception and denote by $Q : X^{(IV)} \to X^{**}$ the canonical projection $Q(\Lambda) = \Lambda|_{X^*}$ of norm one. We note that this operator is also continuous when the domain and the range are equipped with the weak*-topology. It is also routine to verify that for a closed subspace $U \subset X$, $Q(U^{(IV)}) = U^{\perp\perp}$. We first recall a Proposition from [**7**].

PROPOSITION 12. *If M is a closed subspace of a Banach space X, then the following statements are equivalent:*

(1) *$M^\perp$ is the kernel of a norm-one projection in X^*.*
(2) *$M^{\perp\perp}$ is the image of a norm-one projection in X^{**}.*

The following theorem along with Proposition 11 gives us a necessary and sufficient condition for a closed subspace to be an ideal.

THEOREM 13. *Let $U \subset X$ be a closed subspace. Suppose for every Banach space Y , every operator $T : U \to Y^*$ has a norm preserving extension to X. Then U is an ideal in X.*

PROOF. We exhibit an onto projection $R : X^{**} \to U^{\perp\perp}$ such that $\|R\| = 1$. Consider the canonical embedding $i : U \to U^{**}$. As the range space is a dual space, by hypothesis there exists $T : X \to U^{**}$ such that $T = i$ on U. We have $T^{**} : X^{**} \to U^{(IV)}$. Let $R = Q \circ T^{**}$. Clearly $\|R\| = 1$. Since the operators are also weak*-continuous, to see that R is a projection, we will show that it is identity on the weak*-dense subspace U of $U^{\perp\perp}$.

For $u \in U$, $R(u) = Q(T(u)) = Q(u) = u$. Therefore R is the required projection of norm one. □

REMARK 14. *In the case of $L^1(\mu)$, since $Y = L^1(\mu)^*$ is an injective space, clearly $U\hat{\otimes}_\pi Y$ is a subspace of $X\hat{\otimes}_\pi Y$. Since $U\hat{\otimes}_\pi Y$ is isometric to $Y\hat{\otimes}_\pi U$, we get the inclusion relation for projective tensors.*

REMARK 15. *Consider ℓ^2. Since this is not an injective space, so by Corollary 8, there is a Banach space X and a closed subspace Y such that $Y\hat{\otimes}_\pi \ell^2$ is not a subspace of $X\hat{\otimes}_\pi \ell^2$. If $Q : X \to X/Y$ is the quotient map then the tensor operator $Q\hat{\otimes}_\pi I : X\hat{\otimes}_\pi \ell^2 \to (X/Y)\hat{\otimes}_\pi \ell^2$ by proposition 1 is a quotient operator. Note however that the kernel of this operator is $Y\hat{\otimes}_\pi \ell^2$, and the latter space is not a subspace of $X\hat{\otimes}_\pi \ell^2$. So we can conclude that kernel of a projective tensor map need not be a projective tensor space.*

3. Injective tensor products

Let X_1 denote the closed unit ball of X and for a dual space, let $\partial_e X_1^*$ denote the set of extreme points of X_1^* and these sets are equipped with the weak*-topology. *Throughout this section we assume that quotient operators are of norm one.*

In the next Theorem we give a generic way of proving quotient space results on injective tensor product spaces. We use a result that if $Q : Z \to Y$ is a quotient operator then $I\hat{\otimes}_\epsilon Q : X\hat{\otimes}_\epsilon Z \to X \otimes_\epsilon Y$ is a quotient operator on $X\hat{\otimes}_\epsilon Z$ if for every compact operator $S : X^* \to Y$ and for $\epsilon > 0$, there exists an operator $T : X^* \to Z$ such that $Q \circ T = S$ and $\|T\| \leq \|S\| + \epsilon$. See Exercise 3.3 in [**10**].

Let $\mathcal{L}(X, Y)$ and $\mathcal{K}(X, Y)$ denote space of bounded operators and compact operators respectively. Let $Q : Z \to Y$ be a quotient operator. Thus if $\Phi : \mathcal{L}(X^*, Z) \to \mathcal{L}(X^*, Y)$ is defined by $\Phi(S) = Q \circ S$, then the condition in Exercise 3.3 states that Φ is a quotient operator.

We next consider a partial converse of the result in Exercise 3.3.

Suppose X^{**} has the AP (approximation property), then the space of compact operators $\mathcal{K}(X^*, Z) = X^{**}\hat{\otimes}_\epsilon Z$ (see Corollary 4.13 of [**10**]).

The following proposition is now easy to see.

PROPOSITION 16. *Let X be a Banach space such that X^{**} has the A.P. Let $Q : Z \to Y$ be a quotient operator then $\Phi : \mathcal{K}(X^*, Z) \to \mathcal{K}(X^*, Y)$ defined by $\Phi(T) = Q \circ T$ is a quotient operator if and only if $I\hat{\otimes}_\epsilon Q$ is a quotient operator on $X^{**}\hat{\otimes}_\epsilon Z$.*

In the following theorem we exhibit a situation when quotient operators behave well when restricted to ideals.

THEOREM 17. *Let W be a Banach space such that W^{**} has the A.P. Let $X \subset W$ be an ideal. Let $Q : Z \to Y$ be a quotient operator. Suppose $I\hat{\otimes}_\epsilon Q$ on $W^{**}\hat{\otimes}_\epsilon Z$ is a quotient operator. Then $I\hat{\otimes}_\epsilon Q$ on $X\hat{\otimes}_\epsilon Z$ is a quotient operator.*

PROOF. Since X is an ideal in W, there exists a projection $P : W^* \to W^*$ of norm one such that $ker(P) = X^\perp$. It is easy to see that $P(W^*)$ is isometric to X^* via the operator $P(w^*) \to w^*|_X$. In what follows we ignore this embedding and consider $X^* \subset W^*$ and write $P(x^*) = x^*$ for all $x^* \in X^*$. Now suppose $S : X^* \to Y$ be a compact operator and let $\epsilon > 0$.

Consider $S \circ P : W^* \to Y$. Since W^{**} has the A.P, by our hypothesis and the above Proposition, there exists a compact operator $T' : W^* \to Z$ such that $Q \circ T' = S \circ P$ and $\|T'\| \leq \|S\| + \epsilon$. Now let $T = T'|_{X^*} \to Z$ and for $x^* \in X^*$,

$$Q(T(x^*)) = Q(T'(x^*)) = S(P(x^*)) = S(x^*).$$

Also $\|T\| \leq \|T'\| \leq \|S\| + \epsilon$. Therefore the canonical quotient operator Φ on $\mathcal{K}(X^*, Z)$ is a quotient operator.

Thus we get from Exercise 3.3 that $I \hat{\otimes}_\epsilon Q : X \hat{\otimes}_\epsilon Z \to X \hat{\otimes}_\epsilon Y$ is a quotient operator. □

The following corollary extends Proposition 3.5 of [**10**] to L^1-predual spaces.

COROLLARY 18. *Let X be a L^1-predual space. Let $Q : Z \to Y$ be a quotient operator. Then $I \hat{\otimes}_\epsilon Q$ on $X \hat{\otimes}_\epsilon Z$ is a quotient operator.*

PROOF. Since $X^* = L^1(\mu)$, we have, X^{**} is isometric to a $C(K)$-space. It is easy to see that X and all of its duals have the approximation property. See Example 4.2 and Corollary 4.7 in [**10**] . Since X under the canonical embedding is an ideal in X^{**}, by Proposition 3.5 of [**10**] and the above theorem we get the conclusion. □

We recall from [**4**] Chapter I that a closed subspace $J \subset X$ is an M-ideal, if there is a linear projection $P : X^* \to X^*$ such that $ker(P) = J^\perp$ and $\|x^*\| = \|P(x^*)\| + \|x^* - P(x^*)\|$ for all $x^* \in X^*$. In this case $X^* = J^* \oplus_1 J^\perp$ as a ℓ^1-direct sum and hence $\partial_e X_1^* = \partial_e J_1^* \cup \partial_e (J^\perp)_1$.

In what follows we use the fact $\partial_e (X \hat{\otimes}_\epsilon Y)_1^* = \{x^* \otimes y^* : x^* \in \partial_e X_1^*,\ y^* \in \partial_e Y_1^*\}$. See Theorem VI.1.3 in [**4**].

In order to give a positive answer to the quotient question, we consider subspaces $J \subset X$ that are M-ideals. Our first Lemma is well-known and included here for the sake of completeness we include its simple proof. As before let Q denote the quotient map.

LEMMA 19. *Let $J \subset X$ be an M-ideal and let $E = \partial_e J_1^\perp$. Then $E \subset \partial_e X_1^*$ and $J = \{x \in X : e^*(x) = 0 \ for \ all \ e^* \in E\}$.*

PROOF. Clearly J is contained in the set on the right hand side. Suppose $x \in X$ and $e^*(x) = 0$ for all $e^* \in E$. If $x \notin J$ then since by an application of the Krein-Milman theorem $\|Q(x)\| = e^*(x) \neq 0$ for some $e^* \in E$. This contradiction shows that $x \in J$. Hence the claim. □

We next consider classes of Banach spaces X where certain weak*-subsets of $\partial_e X_1^*$ determine M-ideals. Let X be a real L^1-predual space and let $E \subset \partial_e X_1^*$ be a weak*-compact set such that $E \cap -E = \emptyset$. Let $J = \{x \in X : e^*(x) = 0 \ for \ all \ e^* \in E\}$. It is known J is an M-ideal in X (see Corollary 1.3 of [**9**] applied to the real scalar field) and that X/J is isometric to $C(E)$. Thus we have $(X/J) \hat{\otimes}_\epsilon Y$ is isometric to $C(E, Y)$ (see [**2**] Example VIII.6).

In the following theorem we use integral representations for subspaces of vector-valued continuous functions. See [**12**], [**11**] for an exposition on this. We refrain from introducing the notation and terminology of vector-valued boundary measures. As in [**3**] we embed X in a canonical way as a point separating closed subspace of $C(X_1^*)$. In what follows the role of a Choquet simplex in [**11**] is being replaced by the dual unit ball of a L^1-predual space and towards the uniqueness of representing measures we use Theorem 3.2 [**3**]. We will also be using Singer's representation of elements of $C(E,Y)^*$ as Y^*-valued vector measures. See [**2**] page 181 as well as Theorem VIII.5.

THEOREM 20. *Let X be a L^1-predual space and let $E \subset \partial_e X_1^*$ be a weak*-compact set such that $E \cap -E = \emptyset$. Let $J = \{x \in X : e^*(x) = 0 \ \forall e^* \in E\}$. For any Banach space Y, the quotient space $(X\hat{\otimes}_\epsilon Y)/(J\hat{\otimes}_\epsilon Y)$ is isometric to $(X/J)\hat{\otimes}_\epsilon Y = C(E,Y)$.*

PROOF. We recall that by Proposition VI.3.1 the hypothesis implies that $J\hat{\otimes}_\epsilon Y$ is an M-ideal in $X\hat{\otimes}_\epsilon Y$.

We next show that $J\hat{\otimes}_\epsilon Y = \{\tau \in X\hat{\otimes}_\epsilon Y : (e^* \otimes y^*)(\tau) = 0 \ for \ all \ e^* \in E, \ y^* \in \partial_e Y_1^*\}$.

Let $\tau = \sum_1^n x_i \otimes y_i$, for $x_i \in J$, $y_i \in Y$.

Let $e^* \in E$ and $y^* \in \partial_e Y_1^*$. $(e^* \otimes y^*)(\tau) = \sum_1^n e^*(x_i)y^*(y_i) = 0$. To prove the converse we invoke the fact that $J\hat{\otimes}_\epsilon Y$ is an M-ideal in $X\hat{\otimes}_\epsilon Y$. Thus $\partial_e(X\hat{\otimes}_\epsilon Y)_1^* = \partial_e(J\hat{\otimes}_\epsilon Y)_1^* \cup \partial_e(J\hat{\otimes}_\epsilon Y)_1^\perp$. We further note that as J is an M-ideal in X, $\partial_e X_1^* = \partial_e J_1^* \cup \partial_e J_1^\perp$

We will show that $(J\hat{\otimes}_\epsilon Y)^\perp \subset \{\tau \in X\hat{\otimes}_\epsilon Y : (e^* \otimes y^*)(\tau) = 0 \ for \ all \ e^* \in E, \ y^* \in \partial_e Y_1^*\}^\perp$. Since we have weak*-closed subspaces, by an application of the Krein-Milman theorem, it is enough to show that extreme points of the set on the LHS is in the RHS set.

Let $\Lambda \in \partial_e(J\hat{\otimes}_\epsilon Y)^\perp$. From our remarks before, $\Lambda \in \partial_e(X\hat{\otimes}_\epsilon Y)_1^*$, since J is an M-ideal in X. $\Lambda = x^* \otimes y^*$ where $x^* \in E$ and $y^* \in \partial_e Y_1^*$. For $\tau \in \{\tau \in X\hat{\otimes}_\epsilon Y : (e^* \otimes y^*)(\tau) = 0 \ for \ all \ e^* \in E, \ y^* \in \partial_e Y_1^*\}$, clearly $\Lambda(\tau) = 0$. Hence the claim.

To define $\Psi : (X\hat{\otimes}_\epsilon Y)/(J\hat{\otimes}_\epsilon Y) \to C(E,Y)$, we first define it on a dense set. Let $\tau \in X\hat{\otimes}_\epsilon Y$, define $\Psi([\tau])(e^*) = (\sum_1^n x_i \otimes y_i)(e^*) = \sum_1^n e^*(x_i)y_i$, where $\tau = \sum_1^n x_i \otimes y_i$ for $x_i \in X$ and $y_i \in Y$, $e^* \in E$. To show that Ψ is a well defined map, suppose $\tau \in J \otimes Y$. As before let $\tau = \sum_1^n x_i \otimes y_i$ for $x_i \in X$ and $y_i \in Y$. We may assume without loss of generality that y_i's are linearly independent. Let $e^* \in E$ and $y^* \in \partial_e Y_1^*$. $(e^* \otimes y^*)(\sum_1^n x_i \otimes y_i) = 0 = \sum_1^n e^*(x_i)y^*(y_i)$. Thus for all $y^* \in \partial_e Y_1^*$, $y^*(\sum_1^n e^*(x_i)y_i) = 0$, so that $\sum_1^n e_i^*(x_i)y_i = 0$. By the assumption of linear independence of y_i's $e^*(x_i) = 0$ for all i. Since this holds for all $e^* \in E$, we get by the Lemma 1 that $x_i \in J$ for all i. This shows that Ψ is well defined.

Since $\partial_e C(E,Y)_1^* = \{e^* \otimes y^* : e^* \in E,\ y^* \in \partial_e Y_1^*\}$,

$$\|\Psi([\tau])\| = sup\{|\sum_1^n e^*(x_i)y^*(y_i)| : e^* \in E,\ y^* \in \partial_e Y_1^*\}$$

Since $J\hat{\otimes}_\epsilon Y$ is an M-ideal in $X\hat{\otimes}_\epsilon Y$, we have, $\partial_e(J\hat{\otimes}_\epsilon Y)_1^\perp = \{e^* \otimes y^* : e^* \in E,\ y^* \in \partial_e Y_1^*\}$.

Thus $\|[\tau]\| = sup\{|\sum_1^n e^*(x_i)y^*(y_i)| : e^* \in E,\ y^* \in \partial_e Y_1^*\}$.

Therefore Ψ is an isometry on the dense set, the algebraic tensor $(X \otimes Y)/(J \otimes Y)$ and hence extends to an into isometry of the injective tensor product.

To see that Ψ is onto, suppose there exists a non-zero $F \in C(E,Y)^*$ such that $\int_E \tau dF = 0$ for all $\tau \in X\hat{\otimes}_\epsilon Y$. We extend F to a boundary regular Borel Y^*-valued measure on X_1^*, supported by E, still denoted by F. We still have $\int_{X_1^*} \tau dF = 0$ for all $\tau \in X \otimes Y$. Since X is a L^1-predual space, by Theorem 1.3 of [**11**] and Theorem 3.2 [**3**], as $E \cap -E = \emptyset$, by uniqueness of representing boundary measures we get that $F = 0$. Therefore Ψ is an onto isometry.

Now composing Ψ with the canonical isometry $\Phi : C(E,Y) \to (X/J)\hat{\otimes}_\epsilon Y$ we get that $(X\hat{\otimes}_\epsilon Y)/(J\hat{\otimes}_\epsilon Y)$ is isometric to $(X/J)\hat{\otimes}_\epsilon Y$.

□

Our next result can be considered as a special case of vector-valued version of the celebrated result of Alfsen and Effors [**1**], Corollary 5.5, which says that for a M-ideal $J \subset X$ and a weak*-continuous linear functional f on $J^\perp$ extends to an element of X with the same norm.

COROLLARY 21. *Let X be a L^1-predual space, $E \subset \partial_e X_1^*$ be a weak*-compact set such that $E \cap -E = \emptyset$. Let Y be a Banach space and let $f : E \to Y$ be a continuous function. There exists a $\tau' \in X\hat{\otimes}_\epsilon Y$ such that $\|\tau'\| = \|f\|$ and for any $e^* \in E$ and $y^* \in \partial_e Y_1^*$, $(e^* \otimes y^*)(\tau') = y^*(f(e^*))$.*

PROOF. Let $f \in C(E,Y)$ and let $J = \{x \in X : e^*(x) = 0 \ for\ all\ e^* \in E\}$. By above theorem we have that there exists $\tau \in X\hat{\otimes}_\epsilon Y$ such that $\Psi(\tau) = f$ and $d(\tau, J\hat{\otimes}_\epsilon Y) = \|f\|$. $J\hat{\otimes}_\epsilon Y$ being an M-ideal is a proximinal subspace (see Proposition II.1.1 [**4**]), so there is a $\Lambda \in J\hat{\otimes}_\epsilon Y$ such that $d(\tau, J\hat{\otimes}_\epsilon Y) = \|\tau - \Lambda\| = \|f\|$. Let $\tau' = \tau - \Lambda$. Now for $e^* \in E$ and $y^* \in \partial_e Y_1^*$, $(e^* \otimes y^*)(\tau') = (e^* \otimes y^*)(\tau) = y^*(f(e^*))$.

□

REMARK 22. *It may be noted that since the appearance of* [**1**] *there have been several geometric proof of the proximinality of M-ideals by various authors (see* [**5**]*) that are independent of the circle of ideas from* [**1**]. *Thus there is no circularity in the proof of the corollary.*

References

[1] E. M. Alfsen and E. G. Effros, *Structure in real Banach spaces. I, II*, Ann. of Math. (2) **96** (1972), 98–128; ibid. (2) 96 (1972), 129–173, DOI 10.2307/1970895. MR0352946

[2] J. Diestel and J. J. Uhl Jr., *Vector measures*, American Mathematical Society, Providence, R.I., 1977. With a foreword by B. J. Pettis; Mathematical Surveys, No. 15. MR0453964

[3] E. G. Effros, *On a class of real Banach spaces*, Israel J. Math. **9** (1971), 430–458, DOI 10.1007/BF02771459. MR0296658

[4] P. Harmand, D. Werner, and W. Werner, *M-ideals in Banach spaces and Banach algebras*, Lecture Notes in Mathematics, vol. 1547, Springer-Verlag, Berlin, 1993. MR1238713

[5] V. Indumathi and S. Lalithambigai, *A new proof of proximinality for M-ideals*, Proc. Amer. Math. Soc. **135** (2007), no. 4, 1159–1162, DOI 10.1090/S0002-9939-06-08701-6. MR2262920

[6] H. E. Lacey, *The isometric theory of classical Banach spaces*, Springer-Verlag, New York-Heidelberg, 1974. Die Grundlehren der mathematischen Wissenschaften, Band 208. MR0493279

[7] Å. Lima, *The metric approximation property, norm-one projections and intersection properties of balls*, Israel J. Math. **84** (1993), no. 3, 451–475, DOI 10.1007/BF02760953. MR1244680

[8] J. Lindenstrauss, *Extension of compact operators*, Mem. Amer. Math. Soc. No. **48** (1964), 112. MR0179580

[9] T. S. S. R. K. Rao, *Characterizations of some classes of L^1-preduals by the Alfsen-Effros structure topology*, Israel J. Math. **42** (1982), no. 1-2, 20–32, DOI 10.1007/BF02765007. MR687931

[10] R. A. Ryan, *Introduction to tensor products of Banach spaces*, Springer Monographs in Mathematics, Springer-Verlag London, Ltd., London, 2002. MR1888309

[11] P. Saab, *The Choquet integral representation in the affine vector-valued case*, Aequationes Math. **20** (1980), no. 2-3, 252–262, DOI 10.1007/BF02190517. MR577491

[12] P. Saab, *Integral representation by boundary vector measures*, Canad. Math. Bull. **25** (1982), no. 2, 164–168, DOI 10.4153/CMB-1982-022-4. MR663609

Department of Mathematical Sciences, University of Memphis, Memphis, Tennessee 38152

Email address: myadav@memphis.edu

Department of Mathematics, Ashoka University, Rai, Sonepat Haryana-131029, India

Email address: srin@fulbrightmail.org

Contemporary Mathematics
Volume **737**, 2019
https://doi.org/10.1090/conm/737/14864

Into isometries of Banach spaces

T. S. S. R. K. Rao

ABSTRACT. In this article we consider several geometric properties of a Banach space X which are preserved under every isometric embedding of X into spaces of functions on discrete sets.

1. Introduction

Let X be a real Banach space and let $Y \subset X$ be a closed subspace. An important question in the isometric theory of Banach spaces is to study relative properties of Y in X, that are preserved by the range of every into isometry (called an embedding) of the space Y into X . Such an investigation of the finite dimensional structure of the range space of into isometries was done in [**19**].

Here we consider the following three relative properties of Y in X.

(1) For any $y_1, y_2, ..., y_n \in Y$ and $x \in X$ there is a $y_0 \in Y$ such that $\|y_i - y_0\| \leq \|y_i - x\|$ for $1 \leq i \leq n$.
(2) For any $x \in X$ there is a $y_0 \in Y$ such that $\|y - y_0\| \leq \|y - x\|$ for all $y \in Y$.
(3) There is a surjective linear projection $P : X \to Y$ such that $\|P\| = 1$.

Subspaces satisfying property 1) were called cental subspaces in [**3**] and spaces satisfying property 2) were called almost constrained subspaces in [**5**] and existence sets in [**11**]. See also [**9**].

By taking $y_0 = P(x)$ it is easy to see that 3) $\Rightarrow$ 2). Clearly 2) $\Rightarrow$ 1). Lindenstrauss ([**16**]) gave an example to show that 2) need not imply 3) by constructing a Banach space X and a closed subspace $Y \subset X$ such that for any $x \in X$ there is a projection of norm one from $span\{x, Y\}$ onto Y and there exists $x_1, x_2 \in X$ for which there is no projection of norm one from $span\{x_1, x_2, Y\}$ onto Y.

2000 *Mathematics Subject Classification.* Primary 47L05, 46B28, 46B25.
Key words and phrases. Into isometries, L^1-preduals, almost constrained subspaces.

Let c_0 and ℓ^∞ denote spaces of sequences of real numbers converging to 0 and the space of bounded sequences respectively, with both the spaces equipped with the supremum norm. Let $n > 2$, $y_1, y_2, ..., y_n \in c_0$, $x \in \ell^\infty$ and let $\epsilon = min_{1 \leq 1 \leq n}\{\|y_i - x\|\}$. There exists a $N > 0$ such that $|y_i(k)| \leq \epsilon$ for all $k > N$ and $1 \leq i \leq n$. Since for $n > 2$, any n-pair-wise intersecting intervals have a point in common, choose real numbers α_j such that $|y_i(j) - \alpha_j| \leq |y_i(j) - x(j)|$ for $1 \leq i \leq n$ and $1 \leq j \leq N$. Let $y_0 \in c_0$, be such that $y_0(j) = 0$ for $j > N$ and $y_0(j) = \alpha_j$ for $1 \leq j \leq N$. Now $\|y_0 - y_i\| \leq \|y_i - x\|$ for $1 \leq i \leq n$. Thus $c_0 \subset \ell^\infty$ is a central subspace.

On the other hand for the constant sequence 1 in ℓ^∞, if there is a $y_0 \in c_0$ such that $sup_{i \geq 1}|y(i) - y_0(i)| \leq sup_{i \geq 1}|y(i) - 1|$ for all $y \in c_0$, by taking $y_n = 2e_n$, where e_n is the sequence with 1 in the nth place and 0's elsewhere, we have $|2 - y_0(i)| \leq 1$ for all i, which is a contradiction since $y_0 \in c_0$. So c_0 is not an almost constrained subspace of ℓ^∞. See also Proposition 8.

The third property, $Y \subset X$ being the range of a contractive projection has been well studied in the literature. In this case one also says Y is one-complemented in X. We refer to the survey article [**20**] for several interesting results in this direction. We have given only few references that have appeared after [**20**]. We note that if $P : X \to X$ is a contractive projection, then $P(X)^*$ is isometric to $P^*(X^*)$. We note that all the three properties are transitive and for $Z \subset Y \subset X$, if the stronger of the properties is assumed between Y and X, the weaker property is transitive from Z to X.

We investigate the question when do these properties gets preserved under every embedding of Y in X? Let $\Delta \subset [0, 1]$ denote the Cantor set. It is known that (see [**10**] Theorem 4 in section 22) for the space of continuous functions, $C(\Delta)$ has an isometric embedding in $C([0, 1])$ which is the range of a projection of norm one and another embedding in $C([0, 1])$ where it is not even a complemented subspace (see [**1**], page 91 and Proposition 4.4.6). I am grateful to Professor Gilles Godefroy for bringing the results from [**1**] to my attention. To quote from [**20**] 'it seems that the one-complementability of a subspace $Y \subset X$ in fact depends on the way that Y is embedded in X'. For a discrete set Γ, let $\ell^\infty(\Gamma)$ and $c_0(\Gamma)$ denote the space of bounded functions and functions vanishing at ∞ on Γ respectively, equipped with the supremum norm and let $\ell^1(\Gamma)$ denote the space of countably supported and absolutely summable functions on Γ equipped with the ℓ^1-norm.

We show that for any infinite discrete set Γ and for a closed subspace $Y \subset c_0(\Gamma)$ all the three properties are equivalent. Moreover if Y has any of the properties, then it has them under every embedding of Y in $c_0(\Gamma)$. This uses a result of Ando and Douglas ([**2**], [**7**]), that the range of a projection of norm one in a L^1-space is again a L^1-space. These results extend the results from [**5**] and [**11**] from countable to uncountable discrete sets under a weaker assumption, by completely different methods. We also show that any almost constrained subspace of $\ell^\infty(\Gamma)$ is the range of a projection of norm one, improving a result from [**11**]. As an application of these results we give a new proof of a result of Lima [**12**] that distinguishes spaces in which any collection of 3-pair-wise intersecting closed balls having non-empty intersection from spaces where this holds for 4 closed balls.

We recall that a Banach space X is said to be a L^1-predual space if X^* is isometric to $L^1(\mu)$ for some positive measure μ. For a compact set K, the space $C(K)$ and for a discrete set Γ, since $\ell^\infty(\Gamma)$ is isometric to the space $C(\beta(\Gamma))$, where $\beta(\Gamma)$ is the Stone-$\hat{C}$ech compactification of Γ, and the spaces $c_0(\Gamma)$, are all L^1-predual spaces. See [**15**] and Chapter 7 of [**10**] for properties of these spaces and more examples. If Y is a L^1-predual space such that whenever $Z \subset Y$ is a L^1-predual then it is an almost constrained subspace, we show that Y is isometric to $c_0(\Gamma)$ for a discrete set Γ.

Let K be a compact set and let $\sigma : K \to K$ be a homeomorphism such that $\sigma(\sigma(k)) = k$ for all $k \in K$. Let $Y = \{f \in C(K) : f \circ \sigma = -f\}$. It is easy to see that $P : C(K) \to Y$ defined by $P(f) = \frac{f - f\circ\sigma}{2}$ is a contractive projection. Since $C(K)^*$ is a L^1-space and P^* is a contractive projection onto Y^*, by the Ando-Douglas' theorem Y is also a L^1-predual space. Such spaces are called C_σ-spaces. See Chapter 3, Section 10 of [**10**]. We show that if X a separable L^1-predual space with a non-separable dual such that the range of every self isometry is an almost constrained subspace, then X is isometric to a C_σ-space.

QUESTION 1. *For an uncountable compact metric space K, for the space $C(K)$, is there always a self embedding of $C(K)$ which is not an almost constrained subspace of $C(K)$?*

For an uncountable discrete set Γ we show that any finite dimensional central subspace of $\ell^1(\Gamma)$ is the range of a projection of norm one.

I thank the referee for the suggestions that improved the readability of the paper.

2. $c_0(\Gamma)$ and $\ell^\infty(\Gamma)$ spaces

We need a well known Lemma on the structure of projections of norm one. See [**14**] . The proofs are included for the sake of completeness. In what follows we canonically embed a Banach space X in its bidual X^{**}. We recall that under this identification we have $Y \subset Y^{\perp\perp} \subset X^{**}$ and Y is a weak*-dense subspace of $Y^{\perp\perp}$. We also note that Y^{**} is canonically isometric to $Y^{\perp\perp}$. We denote by $Q : X^{***} \to X^* \subset X^{***}$ the canonical surjective projection, $\Lambda \to \Lambda|X$ and note that, $\|Q\| = 1$, $\ker(Q) = X^\perp$. Also Q is a weak*-weak* continuous map.

LEMMA 2. *Let $Y \subset X$ be a closed subspace. Suppose $P : X^{**} \to Y^{\perp\perp}$ be a surjective projection with $\|P\| = 1$. There exists a contractive projection $R : X^* \to X^*$ such that $\ker(R) = Y^\perp$ and thus $R(X^*)$ is isometric to Y^*.*

PROOF. Let $Q : X^{***} \to X^{***}$ be the canonical projection. Let $R = Q \circ P^*|X^*$. It is easy to see that R is a contractive projection. Let x^* be such that $R(x^*) = 0$. For any $y \in Y$, by the definition of Q, $0 = R(x^*)(y) = P^*(x^*)(y) = P(y)(x^*) = x^*(y)$, since $y \in Y^{\perp\perp}$. Thus $x^* \in Y^\perp$. On the other hand for $x^* \in Y^\perp$, $\Lambda \in X^{**}$, $P^*(x^*)(\Lambda) = x^*(P(\Lambda)) = 0$. Thus $R(x^*) = Q(P^*(x^*)) = Q(0) = 0$. Now the quotient space $X^*/Ker(R) = X^*/Y^\perp$ is isometric to Y^*, therefore $R(X^*)$ is isometric to Y^*. □

In what follows we consider the duality, $c_0(\Gamma)^* = \ell^1(\Gamma)$, $c_0(\Gamma)^{**} = \ell^\infty(\Gamma)$. Our next lemma deals with the structure of projections of norm one in $\ell^1(\Gamma)$.

LEMMA 3. *Let $P : \ell^1(\Gamma) \to \ell^1(\Gamma)$ be a projection of norm one. Then range of P is a weak*-closed subspace. In particular if $Y \subset c_0(\Gamma)$ is such that $Y^{\perp\perp}$ is the range of a projection of norm one in $\ell^\infty(\Gamma)$, then Y is the range of a projection of norm one in $c_0(\Gamma)$.*

PROOF. Let $P : \ell^1(\Gamma) \to \ell^1(\Gamma)$ be a projection of norm one. It is well known that $P(\ell^1(\Gamma)) = \ell^1(\Gamma')$ for some discrete set Γ'. We also recall that under the canonical embedding,

$$\ell^\infty(\Gamma)^* = \ell^1(\Gamma) \oplus_1 (c_0(\Gamma))^\perp$$

(an ℓ^1-direct sum). Similarly

$$\ell^1(\Gamma')^{\perp\perp} = \ell^\infty(\Gamma')^* = \ell^1(\Gamma') \oplus_1 (c_0(\Gamma'))^\perp.$$

It now follows from the proof of Proposition IV.1.10 in **[6]** that $\ell^1(\Gamma')$ is a weak*-closed subspace of $\ell^1(\Gamma)$. If $Y \subset c_0(\Gamma)$ is such that $Y^{\perp\perp}$ is the range of a projection of norm one in $\ell^\infty(\Gamma)$, by Lemma 1, there is a projection $R : \ell^1(\Gamma) \to \ell^1(\Gamma)$ of norm one such that $ker(R) = Y^\perp$. Now the range and null space of R are weak*-closed subspaces. Let $M \subset c_0$ be a closed subspace such that $M^\perp = range(R)$. Thus by an application of the separation theorem and the open mapping theorem we have $c_0 = M \bigoplus Y$. If P is the projection associated with this decomposition with range as Y, we get that $P^* = R$ is a weak*-continuous map. Thus there exists a projection $P : c_0(\Gamma) \to c_0(\Gamma)$ of norm one such that $range(P) = Y$. □

Let X be a Banach space suppose $Y \subset X$ is isometric to $\ell^\infty(\Gamma)$ for some discrete set Γ. Then ignoring the embedding let $\ell^\infty(\Gamma) \subset X$. For $\gamma \in \Gamma$, let $e_\gamma : \ell^\infty(\Gamma) \to \mathbb{R}$ be defined by $e_\gamma(x) = x(\gamma)$ for $x \in \ell^\infty(\Gamma)$. Let $e'_\gamma \in X^*$ be a norm preserving extension of e_γ, for $\gamma \in \Gamma$. Let $P : X \to \ell^\infty(\Gamma)$ be defined by $P(x)(\gamma) = e'_\gamma(x)$. It is easy to see that P is a projection of norm one with range $\ell^\infty(\Gamma)$.

We recall that a closed subspace $M \subset X$ is said to be a M-ideal if there exists a linear projection $P : X^* \to X^*$ such that $ker(P) = M^\perp$

and $\|x^*\| = \|P(x^*)\| + \|x^* - P(x^*)\|$ for all $x^* \in X^*$. Such a projection P is called an L-projection. See Chapter I of [**6**] for several examples and properties of these spaces. In particular if X is a L^1-predual space, and $J \subset X$ is a M-ideal then both J and the quotient space X/J are L^1-predual spaces. If P is a projection such that $\|x\| = max\{\|P(x)\|, \|x - P(x)\|\}$ for all $x \in X$, we say that P is a M-projection and the range of P is called a M-summand of X. It is also known that P is an L-projection if and only if P^* is a M-projection. See Chapter 1 of [**6**]. If $Y \subset X$ is a M-ideal then $P^{**} : X^{**} \to X^{**}$ is a M-projection whose range is $Y^{\perp\perp} = Y^{**}$.

THEOREM 4. *Let $Y \subset c_0(\Gamma)$ be an infinite dimensional closed subspace. The following statements are equivalent.*

(1) *Y is isometric to $c_0(\Gamma')$ for some discrete set Γ'.*
(2) *Let $\Psi : Y \to c_0(\Gamma)$ be any into isometry. $\Psi(Y)$ is the range of a projection of norm one in $c_0(\Gamma)$*
(3) *Let $\Psi : Y \to c_0(\Gamma)$ be any into isometry. Then $\Psi(Y)$ is a central subspace of $c_0(\Gamma)$.*

PROOF. 1) $\Rightarrow$ 2): Suppose Y is isometric to $c_0(\Gamma')$ for some discrete set Γ'. It is enough to show that Y is the range of a projection of norm one in $c_0(\Gamma)$. Now $Y^{\perp\perp} = Y^{**} = \ell^\infty(\Gamma')$. So by our observation before the theorem, it follows that $Y^{\perp\perp}$ is the range of a projection of norm one in $\ell^\infty(\Gamma)$. The conclusion now follows from Lemma 3. 2) $\Rightarrow$ 3): Easy to see. 3) $\Rightarrow$ 1): We may assume without loss of generality that $Y \subset c_0(\Gamma)$ is a central subspace. We will first show that for any four pairwise intersecting balls $\{B(a_i, r_i)\}_{1 \leq i \leq 4}$ in Y, $\cap_1^4 B(a_i, r_i) \neq \emptyset$. Since $a_i \in c_0(\Gamma)$ and $|a_i(\alpha) - a_j(\alpha)| \leq r_i + r_j$ for all $\alpha \in \Gamma$ and for all i, j. By arguments similar to the ones given in the Introduction (in the case of c_0 and ℓ^∞), it is easy to see that there exists $a \in c_0(\Gamma)$ such that $\|a - a_i\| \leq r_i$ for $1 \leq i \leq 4$. Now by hypothesis there exists $a_0 \in Y$ such that $\|a_i - a_0\| \leq \|a_i - a\| \leq r_i$ for $1 \leq i \leq 4$. Hence the claim. Therefore by a well known characterization of L^1-predual spaces in terms of 4-ball intersection property ([**10**] Theorem 6 in Section 21 and [**15**]), Y^* is isometric to $L^1(\mu)$ for some positive measure μ. Since Γ is a discrete set, it is easy to see that μ is a purely atomic measure. Thus Y^* isometric to $\ell^1(\Gamma')$ for some discrete set Γ' (see [**10**] Theorem 6 in Section 22). So by arguments similar to the ones given during the proof of Lemma 3, we get that Y^* is a weak*-closed subspace of $\ell^1(\Gamma)$.

We next show that any M-ideal J in Y is a M-summand. It would then follow from the results in [**18**] that Y is isometric to $c_0(\Gamma')$ for some discrete set Γ'. Let $J \subset Y$ be a M-ideal. It follows from the remarks made above that $J^* \subset Y^* \subset \ell^1(\Gamma)$, J^* is isometric to $\ell^1(A)$ for some discrete set A. Thus J^* is a weak* closed subspace of $\ell^1(\Gamma)$ and hence of Y^*. Therefore the L-projection corresponding to $J^\perp$ is weak*-continuous in Y^*, so that J is the range of a M-projection. □

PROPOSITION 5. *Let Γ be an infinite discrete set and let X be a L^1-predual space. Let $\Phi : X \to c_0(\Gamma)$ be an into isometry. $Y = \Phi(X)$ is the range of a projection of norm one in $c_0(\Gamma)$.*

PROOF. We note that $Y^{\perp\perp} = Y^{**} = \Phi((X))^{**} = \Phi^{**}(X^{**}) \subset \ell^\infty(\Gamma) = c_0(\Gamma)^{**}$. Since X^{**} is a $\mathcal{P}_1$-space (see [**15**]), there is a projection of norm one $R : \ell^\infty(\Gamma) \to \ell^\infty(\Gamma)$ with $range(R) = Y^{\perp\perp}$. By Lemma 3, we get that there is a projection of norm one $R : c_0(\Gamma) \to Y$ such that $range(R) = Y$. □

Our next proposition shows that a weaker form of this property determines isometrically spaces of the form $c_0(\Gamma)$ among L^1-predual spaces.

PROPOSITION 6. *Let Y be a L^1-predual space such that for any L^1-predual space X and for any isometry $\Phi : X \to Y$, $\Phi(X)$ is an almost constrained subspace of Y, then Y is isometric to $c_0(\Gamma)$ for a discrete set Γ.*

PROOF. We will show that any M-ideal in Y is a M-summand. It would then follow that Y is isometric to $c_0(\Gamma)$ for some discrete set Γ, Let $Z \subset Y$ be any M-ideal. Since Y is a L^1-predual space, so is Z. Applying the hypothesis for the inclusion map, we get that Z is an almost constrained subspace of Y. Therefore by Proposition 3.17 of [**4**] Z is a M-summand in Y. □

It is easy to see that if $Y \subset X$ is the range of a projection of norm one, then so is $Y^{\perp\perp}$ in X^{**}. Our analysis shows that the converse is true when $X = c_0(\Gamma)$.

COROLLARY 7. *Let Y be a L^1-predual space. Suppose for every closed subspace $X \subset Y$ such that $X^{\perp\perp}$ is one-complemented subspace of Y^{**}, X is one-complemented in Y. Then Y is isometric to $c_0(\Gamma)$ for some discrete set Γ.*

PROOF. In view of Proposition 6, ignoring the isometric embedding, let $X \subset Y$ be a L^1-predual space. Since X^{**} is a $\mathcal{P}_1$-space (see [**15**]), $X^{\perp\perp}$ is the range of a projection of norm one on Y^{**}. Therefore by hypothesis, X is a one complemented subspace of Y and hence is an almost constrained subspace. Now the conclusion follows from Proposition 6. □

We next note that any almost constrained subspace of $\ell^\infty(\Gamma)$ is the range of a projection of norm one. The authors of [**11**] (see page 126) use strong geometric techniques to deduce the same when Γ is a finite set.

PROPOSITION 8. *Let $Y \subset \ell^\infty(\Gamma)$ be an almost constrained subspace. Then Y is the range of a projection of norm one on $\ell^\infty(\Gamma)$.*

PROOF. Let $I : Y \to Y$ be identity map. We first claim that any collection $\{B(y_i, r_i)\}_{i \in I}$ of pair-wise intersecting closed balls in Y have non-empty intersection. To see this consider the family of larger balls in $\ell^\infty(\Gamma)$ with the same centers and radii. It is easy to see that any finitely many of them intersect in $\ell^\infty(\Gamma)$. Since ℓ^∞ is a dual space, by weak*-compactness of closed balls there exists a $\alpha \in \ell^\infty$ such that $\|y_i - \alpha\| \le r_i$ for all $i \in I$. Since Y is an almost constrained subspace, there exists a $y_0 \in Y$ such that $\|y_i - y_0\| \le r_i$ for all $i \in I$.

Now consider the identity map $I : Y \to Y$. For any $\tau \in \ell^\infty(\Gamma)$, since the closed balls $\{B(y, \|y - \tau\|)\}_{y \in Y}$ pair-wise intersect, let $y_0 \in Y$ be such that $\|y_0 - y\| \le \|y - \tau\|$ for all $y \in Y$, it is easy to see that $I' : span\{\tau, Y\} \to Y$ defined by $I'(y + a\tau) = y + ay_0$ for any real number a, is a norm preserving extension of I and is a projection onto Y.

An application of Zorn's Lemma based exhaustion argument can be used now to get a contractive projection $P : \ell^\infty(\Gamma) \to Y$. □

As an application we give a new proof of a weaker form of Theorem 4.3 from [**12**]. Our analysis is intricately related to the structure of finite dimensional subspaces.

We note that if X is a finite dimensional space such that X^* contains an isometric copy of $\ell^\infty(k)$ for some $k > 1$, then since there is a contractive projection $P : X^* \to \ell^\infty(k)$, we get that X contains an isometric copy of $\ell^1(k)$ as the range of a projection of norm one. By our observations on ranges of projections of norm one, if a Banach space has an isometric copy $\ell^1(k)$ as the range of a projection of norm one, then X^* has a copy of $\ell^\infty(k)$ as the range of a projection of norm one. We also note that $(x, y) \to (\frac{x+y}{2}, \frac{x-y}{2})$ is an isometry of $\ell^\infty(2)$ and $\ell^1(2)$. For $k > 2$, since the unit ball of $\ell^\infty(k)$ has 2^k extreme points with coordinates coming from $\{\pm 1\}$ where as the unit ball of $\ell^1(k)$ has $2k$ extreme points with only one coordinate as ± 1 and the rest 0 , we see that these spaces are not isometric.

THEOREM 9. *Let X be a real Banach space. If X^* contains an isometric copy of $\ell^\infty(3)$ then there exists 4 pair-wise intersecting closed balls in X having empty intersection. Suppose X is finite dimensional and for any 3 pair-wise intersecting balls in X intersect and there exists 4 pair-wise intersecting closed balls in X having empty intersection. Then X^* has an isometric copy of $\ell^\infty(3)$.*

PROOF. Suppose X^* contains an isometric copy of $\ell^\infty(3)$ and every collection of 4 pair-wise intersecting closed balls in X has non-empty intersection. Then by Theorem 6.1 in [**16**], X^* is isometric to $L^1(\mu)$. Therefore from our remarks made earlier any isometric copy of $\ell^\infty(3)$ in X^* will be a range of a projection of norm one in $L^1(\mu)$. Hence by Ando-Douglas' theorem we get $\ell^\infty(3)$ is isometric to $\ell^1(3)$. A contradiction. So X has a set of 4 pair-wise intersecting balls whose intersection is empty.

Conversely suppose that X is a finite dimensional space and in X any set of 3 pair-wise intersecting balls intersect and has a set of 4 pair-wise intersecting balls whose intersection is empty. It follows from Corollary

7.4 of [**8**] that X is a combination of ℓ^1 and ℓ^∞ sums of one dimensional subspaces of X. If X has no copy of $\ell^1(3)$ then in Hansen and Lima's decomposition ([**8**]) of X only finite dimensional $\ell^\infty(k)$ spaces appear. It is easy to see that in this case any set of 4 pair-wise intersecting balls have non-empty intersection. So X has an isometric copy of $\ell^1(3)$. Since X is a combination of ℓ^1 and ℓ^∞ sums of one dimensional subspaces of X, there is an isometric copy of $\ell^1(3)$ which is the range of a projection of norm one. Thus X^* has an isometric copy of $\ell^\infty(3)$. □

3. Large separable L^1-predual spaces

Conjecture: If X is a separable L^1-predual space such that X^* is not separable, then there is a self isometry of X whose range is not almost constrained in X?

We only have the following 'positive' result. We recall that for a set $C \subset X$, $c_0 \in C$ is said to be a centre of symmetry if for all $c \in C$, $2c_0 - c \in C$. In what follows we will use a result of Lima and Uttersrud, [**13**], that a Banach space X is isometric to a C_σ space if and only if any three intersecting balls in X have a centre of symmetry.

THEOREM 10. *Let X be a separable L^1-predual space with a non-separable dual. Suppose the range of every isometric embedding of X is an almost constrained subspace of X. Then X is a C_σ space.*

PROOF. Let Δ denote the Cantor set. Since X is a separable L^1-predual space with X^* non-separable, there is an isometry $\Psi : C(\Delta) \to X$ (see [**10**] Theorem 4 in section 22). Since X is a separable Banach space, let $\Phi : X \to C(\Delta)$ be the canonical embedding. Now $\Psi \circ \Phi : X \to X$ is an isometric embedding, so that by hypothesis, $(\Psi \circ \Phi)(X)$ is an almost constrained subspace of X and hence is an almost constrained subspace of $\Psi(C(\Delta))$. Since $\Psi(C(\Delta))$ is a $C(K)$ space for some compact set K, we next show that an almost constrained subspace Z of a $C(K)$ space is a C_σ space. This completes the proof.

Let $\{B(z_i, r_i)\}_{1 \leq i \leq 3}$ be closed balls in Z with $\|z_0 - z_i\| \leq r_i$ for $1 \leq i \leq 3$. Since $C(K)$ is a C_σ space, let f_0 be a centre of symmetry for the intersection of these 3 closed balls in $C(K)$ with centres at z_i. Since Z is an almost constrained subspace, let $P : span\{Z, f_o\} \to Z$ be a projection of norm one. We claim that $P(f_0)$ is a centre of symmetry for the intersecting balls in Z.

Let $z \in Z$ and $\|z - z_i\| \leq r_i$ for $1 \leq i \leq 3$. $\|2P(f_0) - z - z_i\| = \|P(2f_0 - z - z_i)\| \leq r_i$, for all i, as f_0 is a centre of symmetry. Thus $2P(f_0) - z$ is in the intersection of the 3 balls. Hence $P(f_0)$ is a centre of symmetry. Therefore Z is isometric to a C_σ space. □

4. Subspaces of $\ell^1(\Gamma)$

In this section we consider the equivalence of the 3 properties considered here for $\ell^1(\Gamma)$ for an infinite discrete set Γ. We continue to use the duality $c_0(\Gamma)^* = \ell^1(\Gamma)$. Since Γ is possibly uncountable, our arguments differ from traditional arguments based on basic sequences, see [**17**], Proposition 2.a.1.

THEOREM 11. *Let Γ be an infinite discrete set. Let $Y \subset \ell^1(\Gamma)$ be a finite dimensional central subspace. Then Y is the range of a projection of norm one. In particular Y is isometric to $\ell^1(k)$ where $k = dim(Y)$.*

PROOF. We will first indicate a separable reduction procedure. Let $y_1, ..., y_k$ be a basis for Y. Since every element of $\ell^1(\Gamma)$ has countable support, there exists a countable set $A \subset \Gamma$ such that $Y \subset \ell^1(A)$ and $\ell^1(\Gamma) = \ell^1(A) \bigoplus_1 \ell^1(\Gamma - A)$. Let $P : \ell^1(\Gamma) \to \ell^1(A)$ be the contractive projection associated with this decomposition. We next note that Y is an almost constrained subspace of $\ell^1(A)$. It would then follow from Theorem 2.4 in [**11**] that there exists a surjective contractive projection $R : \ell^1(A) \to Y$. Therefore $R \circ P$ is the required contractive projection onto Y. Thus by Ando-Douglas' theorem again, Y is isometric to $\ell^1(k)$ where $k = dim(Y)$.

Since P is a contractive projection, clearly $Y \subset \ell^1(A)$ is a central subspace. Let $x \in \ell^1(A)$. Consider the family of closed balls $\{B(y, \|y-x\|)\}_{y\in Y}$ in Y. For any $y_1, .., y_n \in Y$, since Y is a central subspace of $\ell^1(A)$, there exists a $y_0 \in Y$ such that $\|y_i - y_0\| \leq \|y_i - x\|$ for $1 \leq i \leq n$. Thus any finite collection of sets from $\{B(y, \|y-x\|)\}_{y\in Y}$ intersect. Therefore by compactness we get a $y_0 \in Y$ such that $\|y - y_0\| \leq \|y - x\|$ for all $y \in Y$. Hence Y is an almost constrained subspace of $\ell^1(A)$. □

REMARK 12. *Arguments similar to the ones given above can be used to show that if $Y \subset \ell^1(\Gamma)$ is a separable almost constrained subspace then Y is the range of a projection of norm one on $\ell^1(\Gamma)$.*

QUESTION 13. *If $Y \subset \ell^1(\Gamma)$ is an almost constrained subspace, is it a weak*-closed subspace, w. r. t the duality $c_0(\Gamma)^* = \ell^1(\Gamma)$?*

References

[1] F. Albiac and N. J. Kalton, *Topics in Banach space theory*, 2nd ed., Graduate Texts in Mathematics, vol. 233, Springer, [Cham], 2016. With a foreword by Gilles Godefory. MR3526021

[2] T. Andô, *Contractive projections in L_p spaces*, Pacific J. Math. **17** (1966), 391–405. MR0192340

[3] P. Bandyopadhyay and T. S. S. R. K. Rao, *Central subspaces of Banach spaces*, J. Approx. Theory **103** (2000), no. 2, 206–222, DOI 10.1006/jath.1999.3420. MR1749962

[4] P. Bandyopadhyay and S. Dutta, *Almost constrained subspaces of Banach spaces*, Proc. Amer. Math. Soc. **132** (2004), no. 1, 107–115, DOI 10.1090/S0002-9939-03-07146-6. MR2021253

[5] P. Bandyopadhyay and S. Dutta, *Almost constrained subspaces of Banach spaces. II*, Houston J. Math. **35** (2009), no. 3, 945–957. MR2534290

[6] P. Harmand, D. Werner, and W. Werner, *M-ideals in Banach spaces and Banach algebras*, Lecture Notes in Mathematics, vol. 1547, Springer-Verlag, Berlin, 1993. MR1238713

[7] R. G. Douglas, *Contractive projections on an $\mathfrak{L}_1$ space*, Pacific J. Math. **15** (1965), 443–462. MR0187087

[8] A. B. Hansen and Å. Lima, *The structure of finite-dimensional Banach spaces with the 3.2. intersection property*, Acta Math. **146** (1981), no. 1-2, 1–23, DOI 10.1007/BF02392457. MR594626

[9] A. Kamińska, H. J. Lee, and G. Lewicki, *Extreme and smooth points in Lorentz and Marcinkiewicz spaces with applications to contractive projections*, Rocky Mountain J. Math. **39** (2009), no. 5, 1533–1572, DOI 10.1216/RMJ-2009-39-5-1533. MR2546654

[10] H. E. Lacey, *The isometric theory of classical Banach spaces*, Springer-Verlag, New York-Heidelberg, 1974. Die Grundlehren der mathematischen Wissenschaften, Band 208. MR0493279

[11] G. Lewicki and G. Trombetta, *Optimal and one-complemented subspaces*, Monatsh. Math. **153** (2008), no. 2, 115–132, DOI 10.1007/s00605-007-0510-4. MR2373365

[12] Å. Lima, *Intersection properties of balls and subspaces in Banach spaces*, Trans. Amer. Math. Soc. **227** (1977), 1–62, DOI 10.2307/1997452. MR0430747

[13] Å. Lima and U. Uttersrud, *Centers of symmetry in finite intersections of balls in Banach spaces*, Israel J. Math. **44** (1983), no. 3, 189–200, DOI 10.1007/BF02760970. MR693658

[14] Å. Lima, *The metric approximation property, norm-one projections and intersection properties of balls*, Israel J. Math. **84** (1993), no. 3, 451–475, DOI 10.1007/BF02760953. MR1244680

[15] J. Lindenstrauss, *Extension of compact operators*, Mem. Amer. Math. Soc. No. **48** (1964), 112. MR0179580

[16] J. Lindenstrauss, *On projections with norm* $1 - --$*an example*, Proc. Amer. Math. Soc. **15** (1964), 403–406, DOI 10.2307/2034513. MR0161126

[17] J. Lindenstrauss and L. Tzafriri, *Classical Banach spaces. I*, Springer-Verlag, Berlin-New York, 1977. Sequence spaces; Ergebnisse der Mathematik und ihrer Grenzgebiete, Vol. 92. MR0500056

[18] T. S. S. R. K. Rao, *Characterizations of some classes of* L^1*-preduals by the Alfsen-Effros structure topology*, Israel J. Math. **42** (1982), no. 1-2, 20–32, DOI 10.1007/BF02765007. MR687931

[19] T. S. S. R. K. Rao, *Into isometries that preserve finite dimensional structure of the range*, Problems and recent methods in operator theory, Contemp. Math., vol. 687, Amer. Math. Soc., Providence, RI, 2017, pp. 219–224. MR3635765

[20] B. Randrianantoanina, *Norm-one projections in Banach spaces*, Taiwanese J. Math. **5** (2001), no. 1, 35–95, DOI 10.11650/twjm/1500574888. International Conference on Mathematical Analysis and its Applications (Kaohsiung, 2000). MR1816130

Department of Mathematics, Ashoka University, Rai, Sonepat Haryana-131029, India

Email address: srin@fulbrightmail.org

Contemporary Mathematics
Volume **737**, 2019
https://doi.org/10.1090/conm/737/14865

Results on topological properties of operations on function spaces

Holly Renaud

Abstract. In this paper, we establish openness of the maximum and minimum operations on spaces of scalar-valued integrable functions, $L^p(\Omega,\ \mathcal{A},\ \lambda)$, $1 \leq p \leq \infty$, with Ω a topological space and λ a finite measure on a σ-algebra $\mathcal{A}$ of subsets of Ω. We consider the openness property of the dense-defined standard multiplication on $L^p(\Omega)$, $1 \leq p < \infty$; we observe that multiplication on these settings is only defined on a dense set. We adapt the definition of openness for the multiplication to include dense-defined products and then prove that the multiplication on $L^p(\Omega)$, $1 \leq p < \infty$, restricted to its domain, is uniformly open. We also establish the openness for the multiplication on $L^\infty(\Omega)$. The last section of the paper deals with the connection between openness of the multiplication on spaces of continuous functions and topological properties of the domain of those functions.

1. Introduction

The interest on openness properties of maps may be traced to the fact that the open mapping theorem does not hold for bilinear maps (see [**14**]). This leads to questions on when certain collections of maps are open or which elements in the domain of those maps are points of local openness (see Definition 1.1). Several operations on classical Banach spaces have been studied from this point of view; for example, see [**1**], [**2**], [**3**] and references therein.

In [**6**], we investigate conditions for openness of multiplication in a variety of spaces of continuous and differentiable functions. In this paper, we study openness properties of certain binary operations on spaces of integrable functions. We start by recalling the definitions of open and uniformly open functions between two normed spaces.

Definition 1.1. *(cf. [**6**] and [**12**]) A function f between two normed spaces, $f : E \to F$, is said to be open, provided that for every point $x \in E$ and every neighborhood of x, V, $f(V)$ contains a neighborhood of $f(x)$. We rephrase this statement as follows:*

$$\forall\, x \in E,\ \forall \epsilon > 0\ \exists \delta > 0\ \ s.t.\ \ f(\mathcal{B}(x,\epsilon)) \supset \mathcal{B}(f(x),\delta). \tag{1.1}$$

2010 *Mathematics Subject Classification.* Primary 47H07; Secondary 46B42, 54C35.
Key words and phrases. topological dimension, topological stable rank, multiplication-almost openness property.

If δ in (1.1) can be chosen independently of the point x, then we say that f is uniformly open. The points x in E, where the function f satisfies the condition formulated in (1.1), are called points of local openness of f.

For simplicity of exposition, we shall use the following terminology.

DEFINITION 1.2. *Given a space X and a binary operation on X, say $T : X \times X \to X$, we say that X has the "T-openness property" (T-Op) if T is an open map. Furthermore, X has the "T-uniform openness property" (T-UOp) if T is uniformly open.*

For the particular case of a multiplication on X, we say that X has the M-Op or M-UOp respectively.

A topic of our interest is to decide whether multiplication on L^p spaces ($1 \leq p < \infty$) is open. We observe that this operation is not defined for every pair of functions (i.e. the product of two p-integrable functions is not necessarily p-integrable). We explain this obstruction with an example:

Define $P : L^1([0,1]) \times L^1([0,1]) \longrightarrow L^1([0,1])$, where $(f,g) \mapsto f \cdot g$, and consider $f(x) = g(x) = x^{-(1/2)} \in L^1([0,1])$. Note that $f(x) \cdot g(x) = x^{-1}$. Thus,

$$\|f \cdot g\|_1 = \left(\int_{[0,1]} x^{-1}\, dx \right) = \infty$$

but $x^{-1} \notin L^1([0,1])$. Thus, the standard product is not necessarily defined on all of $L^p([0,1]) \times L^p([0,1])$. Similar examples can be constructed for $L^p([0,1])$ with $1 \leq p < \infty$.

Certain classical spaces may have a multiplication that is not well defined for every pair of elements but is, however, defined on a dense subset of pairs. As explained before, the domain of the product $P(f,g) = f \cdot g$ on $L^p([0,1])$, with $1 \leq p < \infty$, is not $L^p([0,1]) \times L^p([0,1])$, but it clearly contains $L^p([0,1]) \times L^\infty([0,1]) \cup L^\infty([0,1]) \times L^p([0,1])$, where $L^\infty([0,1])$ denotes the set of all measurable and essentially bounded functions on $[0,1]$. This motivates the following definition:

DEFINITION 1.3. *A topological space X is said to have the multiplication-almost (or uniform) openness property M-aOp (or M-aUOp, respectively) if and only if $X \times X$ contains a dense subset A such that multiplication, restricted to A, is well defined and open (or uniformly, respectively).*

In Section 2, we establish the openness of the maximum and minimum operations in L^p-spaces.

In Section 3, we prove that $L^p(\Omega)$, $1 \leq p < \infty$, has the M-aUOp. We also show that the multiplication on $L^\infty(\Omega)$ is uniformly open.

In Section 4, we recall the definition of stable topological dimension of a space. We start by giving a brief overview of existing results on spaces of continuous functions defined on a topological space with zero topological dimension. Then, we apply these results to derive a result for $C(\Omega, E)$, with Ω a compact Hausdorff space and E an algebra. This is extended to other classical Banach spaces.

2. The $\vee$ and $\wedge$ operations

Let Ω be a topological space and λ a finite measure on a σ-algebra of subsets of Ω. The space $L^p(\Omega)$ ($1 \leq p < \infty$) consists of all real valued p-integrable functions defined on Ω. We define the operations maximum and minimum, denoted by $\vee$

($\wedge$, respectively). Given f and g in $L^p(\Omega)$, $(f \vee g)(x) = \max\{f(x), g(x)\}$, for λ-almost every $x \in \Omega$. Similarly $(f \wedge g)(x) = \min\{f(x), g(x)\}$, or λ-almost every $x \in \Omega$. If $\vee : L^p(\Omega) \times L^p(\Omega) \to L^p(\Omega)$ (or $\wedge : L^p(\Omega) \times L^p(\Omega) \to L^p(\Omega)$) is open ($\wedge : L^p(\Omega) \times L^p(\Omega) \to L^p(\Omega)$ is open) then we say that $L^p(\Omega)$ has the $\vee$-open property ($L^p(\Omega)$ has the $\wedge$-open property, respectively).

PROPOSITION 2.1. *Let $1 \le p < \infty$. $L^p(\Omega)$ has the $\vee$-uniform openness and the $\wedge$-uniform openness properties.*

PROOF. Given f and g in L^p and $\epsilon > 0$, we show that there exists δ such that

$$\mathcal{B}(f \vee g, \delta) \subset \mathcal{B}(f, \epsilon) \vee \mathcal{B}(g, \epsilon).$$

Let $A = \{x \in \Omega : f(x) > g(x) + \frac{\epsilon}{6}\}$ and $B = \{x \in \Omega : g(x) > f(x) + \frac{\epsilon}{6}\}$. Then let $f \vee g = f \cdot \chi_A + g \cdot \chi_B + f \vee g \cdot \chi_C$, with $C = (A \cup B)^c$. Let $h \in \mathcal{B}(f \vee g, \delta)$ with $0 < \delta < \frac{\epsilon}{6}$. Since

$$\|(h - f \vee g) \cdot \chi_A\|_p = \|(h - f) \cdot \chi_A\|_p < \delta < \frac{\epsilon}{6},$$

$$\|(h - f \vee g) \cdot \chi_B\|_p = \|(h - g) \cdot \chi_B\|_p < \delta < \frac{\epsilon}{6},$$

$$\|(h - f) \cdot \chi_C\|_p \le \|(h - f \vee g) \cdot \chi_C\|_p + \|(f \vee g - f) \cdot \chi_C\|_p < 2\delta < \frac{\epsilon}{3},$$

and

$$\|(h - g) \cdot \chi_C\|_p \le \|(h - f \vee g) \cdot \chi_C\|_p + \|(f \vee g - g) \cdot \chi_C\|_p < \frac{\epsilon}{3},$$

we set $f_1 = h \cdot \chi_{A \cup C} + f \cdot \chi_B$ and $g_1 = h \cdot \chi_{B \cup C} + g \cdot \chi_A$. Then we have $\|f - f_1\|_p = \|(f - h) \cdot \chi_C \ + (f - h) \cdot \chi_A\| < \dfrac{\epsilon}{3} + \dfrac{\epsilon}{6} < \dfrac{\epsilon}{2}$. Similarly, $\|g - g_1\|_p < \dfrac{\epsilon}{2}$. It is left to check that $h = f_1 \vee g_1$. We see that

$$\begin{aligned} f_1 \vee g_1\,(x) &= (h \cdot \chi_{A \cup C} + f \cdot \chi_B) \vee (h \cdot \chi_{B \cup C} + g \cdot \chi_A)\,(x) \\ &= \begin{cases} h(x) \vee g(x) & \text{if} \quad x \in A \\ f(x) \vee h(x) & \text{if} \quad x \in B \\ h(x) & \text{if} \quad x \in C \end{cases} \\ &= h(x), \end{aligned}$$

since $g(x) > f(x) + \dfrac{\epsilon}{6}$ on B and $f(x) > g(x) + \dfrac{\epsilon}{6}$ on A. A similar argument is used to show that $L^p(\Omega)$ has the $\wedge$-open property. $\square$

We recall that $L^\infty(\Omega)$ consists of all measurable functions $f : \Omega \to \mathbb{R}$, for which there exists $M > 0$ such that $\lambda\{x \in \Omega : |f(x)| \ge M\} = 0$. Each function is a representative of an equivalence class of all functions that coincide with the given one, except possibly on a set of λ-measure zero. This space is endowed with $\|f\|_\infty = \inf\{M > 0 : \ \lambda\{x \in \Omega : |f(x)| \ge M\} = 0\}$. We observe that the proof given for Proposition 2.1 also shows the next corollary.

COROLLARY 2.2. *$L^\infty(\Omega)$ has the $\vee$- and $\wedge$-uniform openness properties.*

The same results have been proved for spaces of continuous functions: for details see [**12**].

3. Openness of Multiplication on spaces of integrable functions

In this section, we study the openness of multiplication on spaces of integrable functions. We consider a product of p-integrable functions on the interval [0, 1]. We consider the dense-defined product $P(f,g) = f \cdot g$. The domain of P contains $L^p([0,1]) \times L^\infty([0,1]) \cup L^\infty([0,1]) \times L^p([0,1])$, where $L^\infty([0,1])$ denotes the space of all essentially bounded functions. Recall Definition 1.3 from Section 1. We stated the following result in [**6**] but did not provide the details of the proof; we do so here.

THEOREM 3.1. *Let $1 \le p < \infty$ and P be the multiplication on $L^p(\Omega, \mathcal{A}, \lambda)$ with domain $Dom(P)$. Then P has the M-aUOp.*

PROOF. Without loss of generality, we may assume that $\lambda(\Omega) = 1$. It is clear that for $t > 0$, the identity, $Id : L^p(\Omega, \mathcal{A}, \lambda) \to L^p(\Omega, \mathcal{A}, t\lambda)$, is a t-isometry (i.e. $\|f\|_p^{t\lambda} = t\|f\|_p^{\lambda}$).

We use a similar strategy to the one given in [**2**]. Specifically, we show that, for $\delta = \dfrac{\epsilon^2}{4^2}$, $B(f \cdot g, \delta) \subseteq B(f, \epsilon) \cdot B(g, \epsilon)$ for every $f, g \in L^p(\Omega)$. We assume, without loss of generality, that $\epsilon < 1$. We subdivide Ω into disjoint sets as follows: $A_g = \{x : |g(x)| > \frac{\epsilon}{4}\}$, $A_f = \{x \notin A_g : |f(x)| > \frac{\epsilon}{4}\}$, and $B = (A_f \cup A_g)^c = \{x : |f(x)| \text{ and } |g(x)| \le \frac{\epsilon}{4}\}$. Let $h \in B(f \cdot g, \delta)$. We set $f_1 = \frac{h}{g} \cdot \chi_{A_g} + f \cdot \chi_{A_f} + \sqrt{|h|} \cdot \chi_B$ and $g_1 = g \cdot \chi_{A_g} + \frac{h}{f} \cdot \chi_{A_f} + \sqrt{|h|} \cdot e^{i\theta_h(x)} \cdot \chi_B$. We observe that

$$f_1 \cdot g_1 = h \cdot \chi_{A_g} + h \cdot \chi_{A_f} + h \cdot \chi_B = h.$$

We show that $g_1 \in B(g, \epsilon)$; the argument that $f_1 \in B(f, \epsilon)$ is similar. We have

$$\begin{aligned} \|g - g_1\|_p^p &= \int |g - g_1|^p d\lambda \\ &= \int_{A_f} \left| \frac{h}{f} - g \right|^p d\lambda + \int_{A_g} |g - g|^p d\lambda + \int_B \left| \sqrt{|h|} \cdot e^{i\theta_h(x)} - g \right|^p d\lambda \\ &= \int_{A_f} \left| \frac{h}{f} - g \right|^p d\lambda + \int_B \left| \sqrt{|h|} \cdot e^{i\theta_h(x)} - g \right|^p d\lambda \qquad (a) \end{aligned}$$

and

$$\begin{aligned} \|h - f \cdot g\|_p^p &= \int_{A_f} |h - f \cdot g|^p d\lambda + \int_{A_g} |h - f \cdot g|^p d\lambda + \int_B |h - f \cdot g|^p d\lambda \\ &\ge \int_{A_f} |h - f \cdot g|^p d\lambda + \int_B |h - f \cdot g|^p d\lambda. \end{aligned}$$

Since $h \in B(f \cdot g, \delta)$, we have that $\|h - f \cdot g\|_p^p < \delta^p$. Thus, by the above inequality, we see that both $\int_{A_f} |h - f \cdot g|^p d\lambda < \delta^p$ and $\int_B |h - f \cdot g|^p d\lambda < \delta^p$. Thus,

$$\delta^p > \int_{A_f} |h - f \cdot g|^p d\lambda = \int_{A_f} |f|^p \left| \frac{h}{f} - g \right|^p d\lambda \ge \left(\frac{\epsilon}{4}\right)^p \int_{A_f} \left| \frac{h}{f} - g \right|^p d\lambda$$

and

$$\left(\frac{4\delta}{\epsilon}\right)^p > \int_{A_f} \left| \frac{h}{f} - g \right|^p d\lambda \qquad (b).$$

We also have that

$$\left(\int_B |h - f\cdot g|^p d\lambda\right)^{1/p} \geq \left(\int_B |h|^p d\lambda\right)^{1/p} - \left(\int_B |f\cdot g|^p d\lambda\right)^{1/p}.$$

Now observe that

$$\left(\int_B |h|^p d\lambda\right)^{1/p} \leq \left(\int_B |h - f\cdot g|^p d\lambda\right)^{1/p} + \left(\int_B |f\cdot g|^p d\lambda\right)^{1/p} \leq \delta + \left(\frac{\epsilon}{4}\right)^2.$$

Using that $\left(\sqrt{|h|}\cdot e^{i\theta_h(x)}\right)^p \in L^2([0,1])$, an application of Holder's Inequality yields

$$\begin{aligned}\left(\int_B \left(\sqrt{|h|}\cdot e^{i\theta_h(x)}\right)^p d\lambda\right)^{1/p} &\leq \left[\left(\int_B |h|^p d\lambda\right)^{1/2} \cdot \left(\int_B 1 d\lambda\right)^{1/2}\right]^{1/p} \\ &\leq \left(\int_B |h|^p d\lambda\right)^{1/2p} \qquad (c).\end{aligned}$$

Hence, the inequality displayed in (c) implies

$$\left(\int_B \left(\sqrt{|h|}\cdot e^{i\theta_h(x)}\right)^p d\lambda\right)^{1/p} \leq \left(\int_B |h|^p d\lambda\right)^{1/(2p)} = \sqrt{\|h\cdot\chi_B\|_p} \leq \sqrt{\delta + \left(\frac{\epsilon}{4}\right)^2}.$$

By (a), we have that

$$\begin{aligned}\|g-g_1\|_p^p &= \left(\left(\left(\int_B \left|\sqrt{|h|}\cdot e^{i\theta_h(x)} - g\right|^p d\lambda\right)^{1/p}\right)^p + \left(\int_{A_f} \left|\frac{h}{f} - g\right|^p d\lambda\right)\right) \\ &\leq \left[\left(\int_B \left|\sqrt{|h|}\cdot e^{i\theta_h(x)}\right|^p d\lambda\right)^{1/p} + \left(\int_B |g|^p d\lambda\right)^{1/p}\right]^p \\ &\quad + \left(\int_{A_f} \left|\frac{h}{f} - g\right|^p d\lambda\right) \\ &\leq \left(\sqrt{\delta + \left(\frac{\epsilon}{4}\right)^2} + \frac{\epsilon}{4}\right)^p + \left(\frac{4\delta}{\epsilon}\right)^p, \quad \text{by } (b) \text{ and } (c).\end{aligned}$$

Therefore $\|g-g_1\|_p \leq \frac{4\delta}{\epsilon} + \sqrt{\delta + \left(\frac{\epsilon}{4}\right)^2} + \frac{\epsilon}{4} < \epsilon$. This completes the proof. □

COROLLARY 3.2. *$L^\infty(\Omega, \mathcal{A}, \lambda)$ has the multiplication-uniform openness property (M-UOp).*

PROOF. The proof follows the one given for Theorem 3.1. We just remark that $\|g-g_1\|_\infty = max\{\|g\cdot\chi_{A_f} - g_1\cdot\chi_{A_f}\|_\infty, \|g\cdot\chi_{A_g} - g_1\cdot\chi_{A_g}\|_\infty, \|g\cdot\chi_B - g_1\cdot\chi_B\|_\infty\}$. □

4. Openness and Dimension

In this section, we include results that show an interconnection between openness of the multiplication on $C(\Omega, E)$, with Ω a compact Hausdorff space and E a unital algebra, and the topological structure of Ω. The motivation for this topic comes from a paper by Draga and Kania (cf. [**7**]). We start with the definition of dimension of a topological space.

We follow Lebesgue's approach to dimension. Given two open coverings of a topological space X, $\mathcal{U}$ and $\mathcal{V}$, we say that $\mathcal{U}$ refines $\mathcal{V}$, denoted by $\mathcal{U} \prec \mathcal{V}$, if for every open set $U \in \mathcal{U}$ there exists at least one element $V \in \mathcal{V}$ such that $U \subset V$. The order of a finite open covering $\mathcal{U}$ of X is equal to the maximal number of elements in $\mathcal{U}$ that contain a point in X. More precisely, given $x \in X$, $ord_x\mathcal{U}$ is the number of elements in $\mathcal{U}$ that contain x. The order of $\mathcal{U}$ is equal to the maximum of $ord_x\mathcal{U}$, for every $x \in X$.

DEFINITION 4.1. *(see [**11**]) If, for every finite open cover of X, there exists a finite open refinement of order less than or equal to $n+1$, then the dimension of X is less or equal to n, and we write $dim(X) \leq n$. If $dim(X) \leq n$ but $dim(X) \leq n-1$ does not hold, then $dim(X) = n$. If $dim(X) \leq n$ does not hold for every n, then $dim(X) = \infty$. By convention, $dim\emptyset = -1$. This number, $ordX$, is called the covering dimension or stable topological dimension of X.*

A similar definition to the one given above, but using cozero sets instead of coverings by open sets, yields to the strong covering dimension of X, and this is denoted by $Dim(X)$. We formulate some results about spaces of topological dimension zero.

PROPOSITION 4.2. *(cf. [**11**], p.348-350 and p. 361)*
(1) A topological space X has zero topological dimension if and only if every finite open covering has a refinement that is a partition (i.e. a covering by pairwise disjoint clopen sets).
(2) In a normal topological space X, $dim(X) = 0$ if and only if $Dim(X) = 0$.
(3) If X is a compact space, then $dim(X) = 0$ if and only if X is totally disconnected.
(4) A completely regular space X is strongly zero dimensional if and only if its Stone-Čech compactification (βX) is totally disconnected (e.g. $\beta\mathbb{N}$).

We shall invoke a proposition from [**7**], which we recall next, for an easier reading.

PROPOSITION 4.3. *(see Proposition 4.6 in [**7**]) Let X be a zero dimensional compact Hausdorff space. Then $C(X)$ (this refers to either the real or complex algebra) has the M-UOp.*

We now state a proposition that concerns the openness property for spaces of algebra-valued continuous functions.

PROPOSITION 4.4. *Let Ω be a compact Hausdorff topological space and E a Banach algebra. The following holds:*

1. *If $C(\Omega, E)$ has the M-Op (or M-UOp), then E has the M-Op (or M-UOp, repectively), and*
2. *If $C(\Omega, E)$ has the M-Op (or M-UOp) and E is unital, then $C(\Omega)$ has the M-Op (or M-UOp, respectively).*

PROOF. We just notice that E is embedded as a subalgebra of $C(\Omega, E)$ via the constant functions (i.e. $T : E \longrightarrow C(\Omega, E)$ such that Tu is the constant function equal to u). Let e be the unit in E. We define $S : C(\Omega) \longrightarrow C(\Omega, E)$, given by $Sf = f \cdot e$. It is clear that S is an isometric algebra isomorphism. This implies the statement and completes the proof. □

In [**13**], M. Rieffel introduced the left (right) topological stable rank of a unital Banach algebra A to be the least integer n such that the set of n-tuples of elements of A, which generate A as a left (right) ideal, is dense in the product A^n. If no such integer exists, then we say the respective rank is infinite. We shall consider commutative algebras, so we talk about topological stable rank. The next definition concerns the topological stable rank for $n = 1$.

DEFINITION 4.5. *(cf. Definition 2.3 in [**7**]) A unital Banach algebra has topological stable rank 1 ($tsr A = 1$), when the group of all the invertible elements in A, $GL(A)$, is dense in A.*

It is easy to see that the algebra of all n square matrices with complex entries, $M_n(\mathbb{C})$, has topological rank 1. Further, if A is a unital C^* algebra, then the invertible elements of A are dense in A if and only if the invertible elements of $M_n(A)$ are dense in $M_n(A)$ (cf. Theorem 3.3 in [**13**]).

In a unital Banach algebra A, we say that $a \in A$ is a topological zero divisor if and only if

$$\inf\{\|x \cdot a\| + \|a \cdot x\| : x \in A, \|x\| = 1\} = 0.$$

This condition is equivalent to saying that there exists a sequence of norm 1 elements in A, $\{x_n\}$, such that $\lim_{n\to\infty} \|x_n \cdot a\| = 0$ and $\lim_{n\to\infty} \|a \cdot x_n\| = 0$. The next result can be found in [**4**], and we also include its proof for completeness of exposition.

PROPOSITION 4.6. *Let A be a unital Banach algebra. Then the boundary of $GL(A)$ consists of topological zero divisors.*

PROOF. We denote by $\partial GL(A)$ the boundary of $GL(A)$. Let $a \in \partial GL(A)$. Then $a \notin GL(A)$, since $GL(A)$ is open. Hence, there exists a sequence of invertible elements in A that converges to a, say $a_n \to a$. Then $\|a_n^{-1}\|$ is an unbounded sequence. Otherwise, if $M > 0$ is such that $\|a_n^{-1}\| \leq M$, for every $n \in \mathbb{N}$, we have

$$\|a_n^{-1} - a_m^{-1}\| \leq M^2 \cdot \|a_n - a_m\|,$$

for every n and $m \in \mathbb{N}$.

The sequence $\{a_n^{-1}\}_n$ satisfies the Cauchy condition, so it converges. If b denotes the limit of $\{a_n^{-1}\}_n$, we have that $a \cdot b = b \cdot a = 1$. This is impossible. Therefore $\|a_n^{-1}\|$ is unbounded, and we select a subsequence, $\{a_{n_k}\}$, such that, for every $k \in \mathbb{N}$, $\|a_{n_k}^{-1}\| \geq k$. We set $x_k = \|a_{n_k}^{-1}\|^{-1} \cdot a_{n_k}^{-1}$, and then we have

$$a \cdot x_k = (a - a_{n_k}) \cdot x_n + \|a_{n_k}^{-1}\|^{-1} \cdot e \to 0.$$

Similarly, we conclude that $\{x_k \cdot a\}_k$ converges to zero. This implies that a is a topological zero divisor. This completes the proof. □

The next proposition appears in [**7**] for unital Banach algebras. We formulate its statement and show a proof that gives a scheme to generalize the statement for a larger collection of spaces.

PROPOSITION 4.7. *If Ω is a compact Hausdorff space with positive topological dimension, E is a unital commutative C^*-algebra, and $C(\Omega, E)$ has the M-Op, then the set of invertible elements of E is dense in E (i.e. the stable topological rank of E is equal to 1 ($tsr(E) = 1$)).*

PROOF. Since $dim(\Omega) > 0$, there exists S connected with at least two points x_1 and x_2 (see proposition 4.2 (3), and also [**9**]). We denote the unit in E by

e, and, without loss of generality, we may assume that $\|e\| = 1$. We define f such that $f(x_1) = e$ and $f(x_2) \notin GL(E)$. We select $0 < \epsilon < 1$ so that we have $\mathcal{B}(f(x_2), \epsilon) \cap GL(E) = \emptyset$. We recall that the condition $tsr(E) \neq 1$ implies that the group of invertible elements in E is not dense. We apply a generalization of Tietze's extension theorem in [**5**] (see Dugundji's extension theorem for compact metric spaces in [**8**]) to ensure the existence of a continuous mapping F that extends f. Since $F(S)$ is connected, it intersects $\partial GL(E)$, which consists of topological zero divisors. Hence, there exists $s \in S$ such that $F(s)$ is a topological zero divisor of E. We notice that, if an element $a \in E$ is a topological zero divisor, then a^* is also a topological zero divisor. Moreover, the product of a topological zero divisor and an element in E is also a topological zero divisor. We consider the product $\mathcal{B}(F^*, \epsilon) \cdot \mathcal{B}(F, \epsilon)$. For every $g \in \mathcal{B}(F^*, \epsilon)$ and $h \in \mathcal{B}(F, \epsilon)$, we have that $g(x_1)$ and $h(x_1)$ are invertible elements in E. Moreover, $g(x_2)$ and $h(x_2)$ are not invertible, as $g(x_2)$ and $h(x_2) \in \mathcal{B}(f(x_2), \epsilon)$ and $\mathcal{B}(f(x_2), \epsilon) \cap GL(E) = \emptyset$. Then there exists s_0 such that $g(s_0)$ is a topological zero divisor and $g(s_0) \cdot h(s_0)$ is also a topological zero divisor. Since E is a $C^\star$-algebra, there exists a modulus 1 element in E, $\{z_n\}$, such that $g(s_0) \cdot z_n \to 0$. Then $\|g(s_0) \cdot h(s_0) \cdot z_n\| \leq \|g(s_0) \cdot z_n\| \cdot \|h(s_0)\| \to 0$, and $\|z_n \cdot g(s_0) \cdot h(s_0)\| \leq \|z_n \cdot g(s_0)\| \cdot \|h(s_0)\| \to 0$. Every function in $\mathcal{B}(F^*, \epsilon) \cdot \mathcal{B}(F, \epsilon)$ has a topological zero divisor in the range. Let $F^* \cdot F + \frac{1}{n} \cdot e$, for n large enough, be such that $\frac{1}{n} < \epsilon$. Then the spectrum of $F^* \cdot F + \frac{1}{n} \cdot e$, denoted $\sigma(F^* \cdot F + \frac{1}{n} \cdot e)$, is such that $\sigma(F^* \cdot F + \frac{1}{n} \cdot e) \subset [\frac{1}{n}, \infty)$. Since $0 \notin \left[\frac{1}{n}, \infty\right)$, we have that $F^* \cdot F + \frac{1}{n} \cdot e$ is invertible. Thus, the range does not contain any topological zero divisors. Therefore, $F^* \cdot F \notin int\,(\mathcal{B}(F^*, \epsilon) \cdot \mathcal{B}(F, \epsilon))$. Thus, $C(\Omega, E)$ does not have the M-Op. This completes the proof. □

We observe that for range spaces of $C(\Omega, E)$ with E a unital C^*-algebra, we have that $C(\Omega)$ has the M-Op, if $C(\Omega, E)$ has the M-Op. This implies that $tsr(E) = 1$. The converse is not true. Just consider $E = \mathbb{R}$. $\mathbb{R}$ has the M-UOp, but $C(\Omega, \mathbb{R})$ does not have the M-Op.

The proof given for Proposition 4.7 works for several other classes of spaces of vector-valued continuous functions. First, we consider spaces of Lipschitz functions, $Lip(\Omega, E)$, with Ω a compact metric space and E a unital commutative C^* algebra. This space consists of all Lipschitz functions

$$Lip(\Omega, E) = \{f : \Omega \to E : \ f \text{ is Lipschitz}\}$$

endowed with any of the norms:

(1) $\|f\|_1 = \|f\|_\infty + L(f)$;
(2) $\|f\|_m = \max\{\|f\|_\infty, L(f)\}$.

These two norms are equivalent:

$$\|f\|_m \leq \|f\|_1 \leq 2\|f\|_m.$$

We recall that $L(f) = \sup_{x \neq y} \frac{\|f(x)-f(y)\|_E}{d(x,y)}$ with d denoting the metric on Ω.

We first notice that spaces of Lipschitz functions $(Lip(\Omega, E), \|\cdot\|_1)$ are Banach algebras with continuous $*$-operation. The existence of Lipschitz extensions is assured in the next theorem.

THEOREM 4.8. *(cf.* [**16**] *p. 16) Let X be a metric space and X_0 be a subset of X.*

(1) *For any* $f_0 : X_0 \to \mathbb{R}$ *there exists* $f : X \to \mathbb{R}$ *such that* $f|_{X_0} = f_0$, $L(f) = L(f_0)$ *and* $\|f\|_\infty = \|f_0\|_\infty$.
(2) *For any* $f_0 : X_0 \to \mathbb{C}$ *there exists* $f : X \to \mathbb{C}$ *such that* $f|_{X_0} = f_0$, $L(f) \leq \sqrt{2}L(f_0)$ *and* $\|f\|_\infty = \|f_0\|_\infty$.

4.1. Sequence Spaces. We now consider multiplication in sequence spaces, particularly the spaces of convergent sequences (c) and ℓ^∞.

COROLLARY 4.9. *The Banach algebra of all convergent sequences* c *has the M-UOp, and so does the subalgebra* c_0.

PROOF. We show that the one-point compactification of $\mathbb{N}$, $\mathbb{N} \cup \{p\}$, has topological dimension zero. Let $\mathcal{U}$ be a finite open covering of $\mathbb{N} \cup \{p\}$. We denote by U an open set in $\mathcal{U}$ that contains p. Then U^c is compact in $\mathbb{N}$, and, as compact subsets of $\mathbb{N}$ must be compact, we have that $\mathbb{N} \setminus U$ is finite (i.e. $\{n_1, \ldots, n_k\}$). Then we just consider the partition $\{\{n_1\}, \ldots, \{n_k\}, U\}$. This is a refinement of $\mathcal{U}$, consisting of clopen sets. Hence $dim(\mathbb{N} \cup \{p\}) = 0$ by (1) of Proposition 4.2. We also have that c is isometric to the space $C(\mathbb{N} \cup \{p\})$. By Proposition 4.3 in [**7**], we conclude that c has the M-UOp. Since c_0 is a closed subalgebra of c, we have that c_0 also has the M-UOp. □

We employ similar reasoning to prove that ℓ^∞ also has the M-UOp.

COROLLARY 4.10. *The Banach algebra of all bounded sequences* ℓ^∞ *has the M-UOp.*

PROOF. We observe that ℓ^∞ is isometric to $C(\beta\mathbb{N})$, the space of all continuous functions defined on the Stone Čech compactification of $\mathbb{N}$ ($\beta\mathbb{N}$). Proposition 3.9 in [**15**] formulates that $\beta\mathbb{N}$ is totally disconnected. Part (4) of Proposition 4.2 implies that $dim\beta\mathbb{N} = 0$. This completes the proof. □

REMARK 4.11. *Applying Theorem 4.4 gives that* $c(C([0,1], \mathbb{R}))$ *does not have the M-Op. We recall that* $c(C([0,1], \mathbb{R}))$ *consists of all uniformly convergent sequences of functions in* $C([0,1], \mathbb{R})$.

References

[1] M. Balcerzak, A. Majchrzycki, and P. Strobin, *Uniform openness of multiplication in Banach spaces* L^p, Studia Mathematica 170 (2005), no. 2, 203–209.

[2] M. Balcerzak, A. Majchrzycki, and A. Wachowicz, *Openness of multiplication in some function spaces*, Taiwanese J. Math. **17** (2013), no. 3, 1115–1126, DOI 10.11650/tjm.17.2013.2521. MR3072279

[3] E. Behrends, *Where is matrix multiplication locally open?*, Linear Algebra Appl. **517** (2017), 167–176, DOI 10.1016/j.laa.2016.12.014. MR3592017

[4] F. F. Bonsall and J. Duncan, *Complete normed algebras*, Springer-Verlag, New York-Heidelberg, 1973. Ergebnisse der Mathematik und ihrer Grenzgebiete, Band 80. MR0423029

[5] F. Botelho and T. S. S. R. K. Rao, *A note on bi-contractive projections on spaces of vector valued continuous functions*, Concr. Oper. **5** (2018), no. 1, 42–49, DOI 10.1515/conop-2018-0005. MR3903602

[6] F. Botelho and H. Renaud, *Topological properties of operations on spaces of differentiable functions*, Adv. Oper. Theory **4** (2019), no. 1, 305–320, DOI 10.15352/aot.1804-1351. MR3862623

[7] S. Draga and T. Kania, *When is multiplication in a Banach algebra open?*, Linear Algebra Appl. **538** (2018), 149–165, DOI 10.1016/j.laa.2017.10.007. MR3722833

[8] J. Dugundji, *An extension of Tietze's theorem*, Pacific J. Math. **1** (1951), 353–367. MR0044116

[9] R. Engelking, *Theory of dimensions finite and infinite*, Sigma Series in Pure Mathematics, vol. 10, Heldermann Verlag, Lemgo, 1995. MR1363947

[10] A. Komisarski, *A connection between multiplication in $C(X)$ and the dimension of X*, Fund. Math. **189** (2006), no. 2, 149–154, DOI 10.4064/fm189-2-4. MR2214575

[11] G. Naber, *Set theoretic topology, with emphasis on problems from the theory of coverings, zero dimensionality and cardinal invariants*, University Microfilms International, Ann Arbor, Michigan, 1977.

[12] H. Renaud, *Topological properties of the standard operations on spaces of continuous functions and integrable functions*, Dissertation Prospectus, University of Memphis, 2018.

[13] M. A. Rieffel, *Dimension and stable rank in the K-theory of C^*-algebras*, Proc. London Math. Soc. (3) **46** (1983), no. 2, 301–333, DOI 10.1112/plms/s3-46.2.301. MR693043

[14] W. Rudin, *Functional analysis*, McGraw-Hill Book Co., New York-Düsseldorf-Johannesburg, 1973. McGraw-Hill Series in Higher Mathematics. MR0365062

[15] R. C. Walker, *The Stone-Čech compactification*, Springer-Verlag, New York-Berlin, 1974. Ergebnisse der Mathematik und ihrer Grenzgebiete, Band 83. MR0380698

[16] N. Weaver, *Lipschitz algebras*, World Scientific Publishing Co., Inc., River Edge, NJ, 1999. MR1832645

Department of Mathematical Sciences, The University of Memphis, Memphis, Tenessee 38152

Email address: hrenaud@memphis.edu

Contemporary Mathematics
Volume **737**, 2019
https://doi.org/10.1090/conm/737/14866

Support sets of nonlinear functionals

Jessica E. Stovall and William A. Feldman

Abstract. Any Dedekind complete Banach lattice E with a quasi-interior point e is lattice isomorphic to a space of continuous, extended real-valued functions defined on a compact Hausdorff space X. An orthogonally additive, continuous, monotonic, and subhomogeneous nonlinear functional $T : E \to \mathbb{R}$ is studied. In this case, the concept of integration is no longer valid. However, a measure $\overline{\mu}$ related to the nonlinear operator T can be constructed as well as an associated linear operator. This measure and linear operator can be used to study nonlinear functionals. It is shown that this associated linear operator is unique and results regarding support sets will be presented.

1. Preliminaries

Let E be a Dedekind complete Banach lattice with order continuous norm and quasi-interior point e. Then there exists an extremally disconnected compact topological space X with the property that E is Riesz isomorphic to an ideal in $C^\infty(X)$, where $C^\infty(X)$ is the continuous extended real-valued functions, each finite on a dense subset of X. The order ideal $I(e)$ generated by e is identified with $C(X)$, the continuous real-valued functions on X, and is dense in E. Here the space X can be viewed as the lattice homomorphisms on the order ideal generated by e together with the product topology. For details regarding the representation theory see [**7**]. As needed, the elements in E will be identified with their representation in $C^\infty(X)$ [**6, 7**].

An operator $T : E \to F$ between two Banach lattices is called *monotonic* if $T(f) \leq T(g)$ whenever $0 \leq f \leq g$. The operator T is called *orthogonally additive* if $T(f+g) = T(f) + T(g)$ for $f \geq 0$, $g \geq 0$, and $f \wedge g = 0$. Furthermore, T is said to be *subhomogeneous* if for $f \geq 0$ and $\alpha > 0$, there exist positive constants $m(\alpha)$ and $M(\alpha)$ with $m(\alpha)$ a monotone function of α and unbounded so that $m(\alpha)T(f) \leq T(\alpha f) \leq M(\alpha)T(f)$ and $M(\alpha)$ goes to zero as α goes to zero.

Let T be an orthogonally additive, continuous, monotonic, and subhomogeneous operator mapping a Banach lattice E with an order continuous norm and quasi-interior point to $\mathbb{R}$. Note that since T is orthogonally additive, then $T(0) = 0$. Furthermore, T is monotonic, thus T is positive. The nonlinear operator $T : C(X) \to \mathbb{R}$ defined by $T(f) = L(f^2)$ for $f \in C(X)$, where L is a positive, continuous linear functional is an example of an operator satisfying all of the conditions of the previous paragraph. More generally, for appropriate functions ϕ and λ, the

2010 *Mathematics Subject Classification.* Primary 47H07; Secondary 46B42, 54C35.
Key words and phrases. Banach Lattice, Nonlinear Operator.

map $T(f) = \phi(L(\lambda f))$ will also satisfy these conditions. For further details about these type of non-linear operators, see [**1**, **3**–**5**, **8**].

Let $\mathcal{K}$ be the set of all clopen subsets of X and define a set function $\mu : \mathcal{K} \to [0, \infty]$ by $\mu(K) = T(\chi_K)$ for any $K \in \mathcal{K}$. It was shown in [**1**] that μ is a premeasure. Furthermore, we denote by $\overline{\mu}$ the Carathéodory extension of μ. So every clopen subset K of X is measurable with respect to the outer measure μ^* induced by μ and $\mu(K) = \mu^*(K) = \overline{\mu}(K)$. Additionally, $\overline{\mu}$ is a Baire measure.

In this setting, the support set of T is defined by

$$K_T = X \setminus \bigcup_{\{h \in I(e)^+ : T(h) = 0\}} \{x \in X : h(x) > 0\},$$

where $I(e)^+$ denotes the positive elements in the ideal generated by e. That is, $I(e)^+$ is the bounded non-negative functions on X. Notice in that if T is linear, then K_T is equivalent to the support of the corresponding measure. For further details on support sets for these types of operators, see [**1**, **3**, **4**, **8**].

Define $L(f) = \int f d\overline{\mu}$ for any $f \geq 0$ in E. Then L is a positive operator from E^+ to $[0, \infty]$. Restricting to $I(e)^+$, the bounded nonnegative functions on X, then L is a positive linear functional. L and T agree on the characteristic functions and the support of L is equal to the support of T. See [**1**] for more details on these properties.

An order continuous map, $J : E^+ \to [0, \infty]$, which is a linear functional restricted to the positive elements in a dense principal ideal, is *associated* with T if there exists a quasi-interior point, u of E such that for every decomposition $u = u_1 + u_2$ of u where $u_1 \wedge u_2 = 0$ and u_1 and u_2 are greater than or equal to zero, then $J(u_1) = T(u_1)$ and $J(u_2) = T(u_2)$. Since $u = u + 0$ meets the conditions for the above decomposition, it follows that if J is associated with T, then $J(u) = T(u)$ for the quasi-interior point u as described above. Since this is true for every decomposition of u as described above, J and T still agree on the characteristic functions with respect to the corresponding spectrum from u, where the spectrum is defined to be the set of all real homomorphisms $\phi \in E^*$ such that $\phi(u) = 1$. Additionally, if J is associated with T, then the support of J is equal to the support of T. Thus, the conditions observed for L are also satisfied for other operators associated with T. For more information about these associated operators, see [**1**]. The associated linear operators can now be utilized in studying this class of nonlinear operators.

2. Associated Linear Operators

For this section, a Banach lattice E, with a quasi-interior point e is considered. The Banach lattice E is identified with a subspace of functions in $C^\infty(X)$. Recall $C(X)$ is dense in E, which is embedded in $C^\infty(X)$ and $I(e) = C(X)$. Additionally, this implies that X is a compact, Hausdorff space.

THEOREM 2.1. *Let E be a Banach lattice with an order continuous norm and a quasi-interior point and let $T : E \to \mathbb{R}$ be nonlinear, orthogonally additive, continuous, monotonic, and subhomogeneous. Then there exists a unique linear operator associated with T.*

PROOF. From [**1**] it is known that for nonlinear operators T that are orthogonally additive, continuous, monotonic, and subhomogeneous that associated linear operators exists. In particular, both L and any operator J as discussed in the preliminaries are associated linear operators.

Let T be as described and let J be an associated linear operator. Then for any clopen set K, the operators T and J agree on χ_K. So if K_i are clopen, then on any linear combination $\sum_{i=1}^{N} \alpha_i \chi_{K_i}$, the linearity of J gives $J(\sum_{i=1}^{N} \alpha_i \chi_{K_i}) = \sum_{i=1}^{N} \alpha_i J(\chi_{K_i}) = \sum_{i=1}^{N} \alpha_i T(\chi_{K_i})$. This is the only choice for J.

Now let C_E be the sublattice of E consisting of the span of characteristic functions of clopen sets in X, that is C_E consists of all functions that can be expressed as a finite sum $\sum_{i=1}^{N} \alpha_i \chi_{K_i}$ where the K_i are clopen. Let ${C_E}^+$ is the cone of the sublattice C_E. Thus, J is unique on ${C_E}^+$. By Proposition 4 from [**2**], ${C_E}^+$ is order dense in E^+. Now, $J = L$ on ${C_E}^+$, thus $J = L$ on E^+. It follows that the associated linear operator L found in [**1**] is unique. □

3. Measures of Points

For this section, Banach lattices of the type in the preceding section are considered. Thus, for a positive, continuous, linear functional $L : E \to \mathbb{R}$ and for any positive f in $C(X)$, the Riesz Representation Theorem gives L a Baire measure, m, such that $L(f) = \int f \, dm$. Furthermore, the definition of the support of L is $K_L = X \setminus \bigcup_{\{h \in I(e)^+ : L(h)=0\}} \{x \in X : h(x) > 0\}$ as seen in the preliminaries. Recall that this is the support of the measure m corresponding to L.

LEMMA 3.1. *$L(f) = \int f \, dm$ for any f in E^+*

PROOF. Consider any positive function f in E that that is not in $C(X)$. Then, define functions $f_n = f \wedge ne$. So $\vee f_n = f$ and $f_n \nearrow f$. Since L is continuous, it follows that $L(f_n) \nearrow L(f)$. Furthermore, each f_n is in $C(X)$. Hence, $L(f_n) = \int f_n \, dm$. Additionally, since the f_n are countable and $f_n \nearrow f$, the Monotone Convergence Theorem gives $\int f_n \, dm \nearrow \int f \, dm$. Combining the above information, it follows that $L(f) = \int f \, dm$ for any f in E^+. □

Now a Dedekind complete Banach lattice E, which has an order continuous norm and quasi-interior point, e is considered. E will be identified with a subspace of $C^\infty(X)$, where X is compact and extremally disconnected. Collections of functions of the form $\mathcal{F}_p = \{f_\alpha \geq 0 : f_\alpha(p) = 1\}$ will be explored and the measure of a point related to the operator on these sets will be discussed.

Recall the definition of an atom. A vector $u > 0$ in a Riesz space is said to be an atom whenever $0 \leq x \leq u, 0 \leq y \leq u$, and $x \wedge y = 0$ imply that either $x = 0$ or $y = 0$. Also, note that χ_p denotes the characteristic function of the point p.

LEMMA 3.2. *If E is a Banach lattice as above, then u is an atom if and only if $u = \lambda \chi_p$ for $\lambda > 0$.*

PROOF. Suppose u is an atom and $u \neq \lambda \chi_p$. Then u is not zero at more than one point. Without loss of generality, let $u(x_1) \neq 0$ and $u(x_2) \neq 0$. Then there exists two continuous functions u_1 and u_2 such that $u_1(x_1) = 0, u_2(x_1) \neq 0, u_1(x_2) \neq 0$, and $u_2(x_2) = 0$. So, $u_1 \leq u$ and $u_2 \leq u$. Furthermore, $u_1 \wedge u_2 = 0$. But neither u_1 nor u_2 are identically zero. Hence, u is not an atom. This contradiction gives u can only be zero at exactly one point.

By the Representation Theorem, $C(X) \subset E \subset C^\infty(X)$ and $\overline{C(X)} = E$. So there is no function u such that $u(p) = \infty$ and $u(x) = 0$ for every $x \neq p$. Hence, $0 < u(p) < \infty$ and $u(x) = 0$ for every $x \neq p$. So, $u = \lambda \chi_p$, for some real number λ.

Now suppose $u = \lambda\chi_p$ for some real λ. So if $0 \leq u_1 \leq u$ and $0 \leq u_2 \leq u$, then $u_1 = 0$ or $u_1 = \lambda\chi_p$ and $u_2 = 0$ or $u_2 = \lambda\chi_p$. Thus if $u_1 \wedge u_2 = 0$, then at least one of u_1 or u_2 must equal zero. Therefore, u is on atom. □

LEMMA 3.3. *Let $\mathcal{F} = \{f_\alpha \geq 0 : f_\alpha(p) = 1\}$. Then for this downward directed set, $\wedge f_\alpha \neq 0$ if and only if $\wedge f_\alpha$ is an atom.*

PROOF. Suppose $\wedge f_\alpha = h \neq 0$. Notice that for every $y \neq p$ there exists f_α such that $f_\alpha(y) = 0$. So, $h(y) \leq f_\alpha(y)$ for every α and hence $h(y) = 0$ for every $y \neq p$. Thus, $h = \chi_p$. Thus by Lemma 3.2 $h = \wedge f_\alpha$ is an atom.

Now suppose $\wedge f_\alpha = h$ is an atom. Then by Lemma 3.2, $h = \chi_p \neq 0$. So $\wedge f_\alpha \neq 0$ if and only if $\wedge f_\alpha$ is an atom. □

Recall in the reals, a decreasing sequence $\{x_\alpha\}$ of positive elements has $\wedge x_\alpha = 0$ if and only if $x_\alpha \searrow 0$. Thus, for any positive, continuous functional, T, this results in $\wedge T(\mathcal{G}) = 0$ if and only if $T(g_\alpha) \searrow 0$ for any downward directed set $\mathcal{G}$ and $g_\alpha \geq 0$.

Let E be a Banach lattice as previously described and $L : E \to \mathbb{R}$ be positive, continuous, and linear. By the Riesz Representation Theorem, there exists a Baire measure m such that $L(f) = \int f dm$ for any $f \in C(X)$. The following theorem utilizes this Baire measure.

THEOREM 3.4. *Let E be a Banach lattice with an order continuous norm and a quasi-interior point, e. Furthermore, let $L : E \to \mathbb{R}$ be a positive, continuous, linear functional. Define $\mathcal{F}_p = \{f_\alpha : f_\alpha(p) = 1\}$. Then the following hold true:*

1. *If either p is measurable, then $m(p) = 0$ if and only if $\wedge L(\mathcal{F}_p) = 0$*
2. *If m is complete, then $m(p) = 0$ if and only if $\wedge L(\mathcal{F}_p) = 0$*
3. *$m(p) > 0$ if and only if for every f there exists $\lambda > 0$ such that $L(f) \geq \lambda f(p)$*

PROOF. (1) Consider the case where p is measurable and open. Suppose $m(p) = 0$. Since p is open, χ_p is in E. Therefore, $L(\chi_p) = \int \chi_p dm = m(p) = 0$. But, $\chi_p(p) = 1$, so $\chi_p \in \mathcal{F}_p$. Hence, $\wedge L(\mathcal{F}_p) = 0$.

Now consider the case where p is measurable and not open. Since p is not open $\wedge f_\alpha = 0$. Otherwise, by Lemma 3.3 and Lemma 3.2, $\wedge f_\alpha = \chi_p$ and p would have to be clopen. Additionally, since E has an order continuous norm, $\wedge f_\alpha = 0$ implies $f_\alpha \searrow 0$. Since L is continuous, $L(f_\alpha) \searrow L(0) = 0$. Thus, $L(f_\alpha) \searrow 0$ (regardless of $m(p)$ in this setting). Thus, $\wedge L(\mathcal{F}_p) = 0$.

Now suppose $\wedge L(\mathcal{F}_p) = 0$. Then, $L(f_\alpha) \searrow 0$. If p is measurable, then χ_p is a measurable function and so $\int \chi_p dm$ can be examined. Furthermore, $\chi_p \leq f_\alpha$ for all α. Hence, $0 \leq m(p) = \int \chi_p dm \leq \int f_\alpha\, dm = L(f_\alpha)$. But, $L(f_\alpha) \searrow 0$. Thus, $m(p) = 0$.

Thus, if p is measurable, then $m(p) = 0$ if and only if $\wedge L(\mathcal{F}_p) = 0$

(2) Next consider the case where m is complete. Then sets $A_n = \bigcap_\alpha \{x : f_\alpha(x) \geq 1 - \dfrac{1}{n}\}$ can be examined. Notice $\{x : f_\alpha \geq 1 - \frac{1}{n}\}$ is closed. Thus, $A_n = \bigcap_\alpha \{x : f_\alpha(x) \geq 1 - \dfrac{1}{n}\}$ is also closed. Therefore, $\{A_n\}$ are Borel sets, and hence are measurable.

Now notice for every $\varepsilon = \frac{n-1}{n^2}$ there exists α_0 such that $L(f_\alpha) < \varepsilon$ for every $\alpha > \alpha_0$. That is, $\int f_\alpha\, dm < \frac{n-1}{n^2}$ for $\alpha > \alpha_0$. Thus, for $\alpha > \alpha_0$, $\frac{n-1}{n^2} >$

$\int f_\alpha \, dm > \displaystyle\int_{A_n} (1-\frac{1}{n}) \, dm = (1-\frac{1}{n})m(A_n) = (\frac{n-1}{n})m(A_n)$. Therefore, $m(A_n) < (\frac{n-1}{n^2})(\frac{n}{n-1}) = \frac{1}{n}$.

Now, since the collection, $\mathcal{B}$, of Borel sets is a σ-algebra, and σ-algebras are closed under countable intersections, it follows that $\bigcap_{n=1}^{\infty} A_n \in \mathcal{B}$. Thus, $\bigcap_{n=1}^{\infty} A_n$ is measurable. Furthermore, notice that $m(A_1) < 1 < \infty$ and $A_{n+1} \subset A_n$. It follows that $m(\bigcap_{n=1}^{\infty} A_n) = \lim_{n\to\infty} m(A_n) = \lim_{n\to\infty} \frac{1}{n} = 0$.

Since $f_\alpha(p) = 1$ for every α, $p \in A_n$ for every n. Therefore, $p \in \bigcap_{n=1}^{\infty} A_n$. So, p is a subset of a set of measure zero, and since m is complete, it follows that p is measurable. Furthermore, $m(p) < m(\bigcap_{n=1}^{\infty} A_n) = 0$. So, $m(p) = 0$.

Thus, if m is complete, then p is measurable. So by part 1, $m(p) = 0$ if and only if $\wedge L(\mathcal{F}_p) = 0$

(3) Suppose $m(p) > 0$. Notice if $f(p) = 0$ the statement is trivial. So consider f such that $f(p) \neq 0$. Without loss of generality, consider $f(p) = 1$. Now $f \geq \chi_p$. Thus, $L(f) = \int f \, dm \geq \int \chi_p \, dm = m(p)$. Define $\lambda = m(p) > 0$. So, $L(f) \geq m(p) = \lambda = \lambda f(p)$. Thus, if $m(p) > 0$, then for every f there exists a $\lambda > 0$ such that $L(f) \geq \lambda f(p)$.

Now suppose for every f there exists $\lambda > 0$ such that $L(f) \geq \lambda f(p)$. Note that $m(p) \geq 0$. Since m is a Baire measure and X is compact, m is regular, and hence outer regular. So $m(p) = inf\{m(W_\varepsilon) : p \in W_\varepsilon \text{ and } W_\varepsilon \text{ is open}\}$. If $m(p) = 0$, then for every $\varepsilon > 0$ there exists an open set W_ε containing p such that $m(W_\varepsilon) < \varepsilon$. Then, for every W_ε there exists f_{W_ε} such that $f_{W_\varepsilon}(p) = 1$ and the support of f_{W_ε} is a subset of W_ε. Thus, $0 < \lambda = \lambda f_{W_\varepsilon}(p) \leq L(f_{W_\varepsilon}) = \int f_{W_\varepsilon} \, dm \leq \int \chi_{W_\varepsilon} \, dm = m(W_\varepsilon) < \varepsilon$ for every ε. It follows that $\lambda = 0$; a contradiction. Thus $m(p) \neq 0$, and so $m(p) > 0$. □

In [**1**], a nonlinear operator $T : E \to \mathbb{R}$, which was orthogonally additive, continuous, monotonic, and subhomogeneous was explored. A positive, continuous, linear functional, L, that was associated with T through a measure $\overline{\mu}$ was then found.

This assoicated linear operator L and the measure $\overline{\mu}$ that was found will now be used to explore results related to the support sets of nonlinear operators.

Notice that by construction, $\overline{\mu}$ is a complete measure.

THEOREM 3.5. *Let E be a Banach lattice with an order continuous norm and a quasi-interior point, e. Furthermore, let $T : E \to \mathbb{R}$ be an orthogonally additive, continuous, monotonic, and subhomogeneous nonlinear operator. Additionally, let L be a positive, continuous, linear functional associated with T through the measure $\overline{\mu}$. Define $\mathcal{F}_p = \{f_\alpha : f_\alpha(p) = 1\}$. Then the following hold true:*

(1) $\overline{\mu}(p) = 0$ *if and only if* $\wedge T(\mathcal{F}_p) = 0$
(2) $\overline{\mu}(p) > 0$ *if and only if for every f there exists $\lambda > 0$ such that $T(f) \geq \lambda f(p)$*

PROOF. (1) Suppose $\overline{\mu}(p) = 0$. Then by Theorem 3.4, $\wedge L(\mathcal{F}_p) = 0$. So $L(f_\alpha) \searrow 0$. If $\wedge f_\alpha = 0$, then $f_\alpha \searrow 0$ because E has an order continuous norm. Furthermore, since T is continuous, $T(f_\alpha) \searrow T(0) = 0$. Thus, $\wedge T(\mathcal{F}_p) = 0$.

Now if $\wedge f_\alpha \neq 0$, then $\wedge f_\alpha = \chi_p$ by Lemma 3.3. Again, E has an order continuous norm, so $f_\alpha \searrow \chi_p$. L is continuous, so $L(f_\alpha) \searrow L(\chi_p)$. Since $L(f_\alpha) \searrow 0$, $L(\chi_p) = 0$. By construction, $T(\chi_p) = L(\chi_p)$, so $T(\chi_p) = 0$. Since χ_p is in the collection $\mathcal{F}_p$, $\wedge T(\mathcal{F}_p) = 0$.

Next, suppose $\wedge T(\mathcal{F}_p) = 0$. Then $T(f_\alpha) \searrow 0$. Consider the case where $\wedge f_\alpha = 0$, then $f_\alpha \searrow 0$ by the order continuous norm. Thus, by the continuity of L, $L(f_\alpha) \searrow L(0) = 0$, and thus $\wedge L(\mathcal{F}_p) = 0$. So by Theorem 3.4, $\overline{\mu}(p) = 0$.

Now consider the case where $\wedge f_\alpha \neq 0$, then $\wedge f_\alpha = \chi_p$ by Lemma 3.3. Again, E has an order continuous norm, so $f_\alpha \searrow \chi_p$. L is continuous, so $L(f_\alpha) \searrow L(\chi_p)$. Likewise, T is continuous, so $T(f_\alpha) \searrow T(\chi_p)$. But, $T(f_\alpha) \searrow 0$, so $T(\chi_p) = 0$. L and T agree on the characteristic functions, so it follows that $L(\chi_p) = 0$. Thus, $L(f_\alpha) \searrow 0$, and $\wedge L(\mathcal{F}_p) = 0$. Hence, by Theorem 3.4, $\overline{\mu}(p) = 0$.

Thus, $\overline{\mu}(p) = 0$ if and only if $\wedge T(\mathcal{F}_p) = 0$.

(2) Suppose $\overline{\mu}(p) > 0$. By Theorem 3.4, for every f there exists a $\lambda_1 > 0$ such that $L(f) \geq \lambda_1 f(p)$. Notice if $f(p) = 0$, then $T(f) \geq \lambda f(p)$ is trivial. So, assume $f(p) = c \neq 0$. Since f is continuous and $f(p) = c$, there exists a clopen set H and a constant $0 < \alpha < c$ such that p is in H, $\alpha\chi_H(p) \neq 0$, and $\alpha\chi_H \leq f$. But T is monotonic, and so $T(f) > T(\alpha\chi_H)$. Furthermore, since T is subhomogeneous, there exists a positive constant $m(\alpha)$ such that $T(\alpha\chi_H) > m(\alpha)T(\chi_H)$. But L and T agree on the characteristic functions, so $m(\alpha)T(\chi_H) = m(\alpha)L(\chi_H) \geq m(\alpha)\lambda_1\chi_H(p) = m(\alpha)\lambda_1$. Now define $\lambda_2 = m(\alpha)\lambda_1$. Since $\frac{1}{c}f(p) = 1$, then $T(f) > \lambda_2 = \lambda_2(\frac{1}{c}f(p))$. Define $\lambda = \frac{\lambda_2}{c}$. It follows that $T(f) > \lambda f(p)$.

Now suppose that for every f there exists a $\lambda > 0$ such that $T(f) \geq \lambda f(p)$. Let H be a clopen set containing p. Then, $T(\chi_H) = \overline{\mu}(H) \geq \overline{\mu}(p)$. Furthermore, $T(\chi_H) \geq \lambda\chi_H(p) = \lambda$. It follows that $\overline{\mu}(H) \geq \lambda$ for each such set H and thus $\wedge\overline{\mu}(H) \geq \lambda$. Since $\overline{\mu}$ is a Baire measure and X is a Hausdorff space, $\overline{\mu}$ is regular. Thus $\overline{\mu}(p) = inf\{\overline{\mu}(H) : p \in H \text{ and } H \text{ is open }\} = inf\{\overline{\mu}(H) : p \in H \text{ and } H \text{ is clopen }\}$. It follows that $\overline{\mu}(p) \geq \lambda > 0$.

So $\overline{\mu}(p) > 0$ if and only if for every f there exists $\lambda > 0$ such that $T(f) \geq \lambda f(p)$. □

4. Sets With Empty Interior

For this section, the relationship between sets with empty interior and countable, downward directed sets with infimum zero is considered. The Banach lattices investigated are of the type from Section 2.

LEMMA 4.1. *Let E be a Banach lattice with an order continuous norm and a quasi-interior point, e. Furthermore, let $L : E \to \mathbb{R}$ be a positive, continuous, linear functional and define $Q = \{\mathcal{F} : \wedge\mathcal{F} = 0, \mathcal{F}$ is directed downward and countable$\}$. Then, for every $\mathcal{F}$ in Q, $\mathcal{F} \searrow 0$ pointwise except possibly on a set K with $int(K) = \emptyset$.*

PROOF. Assume there exists a $K \subset X$ such that $int(K) \neq \emptyset$ and $\mathcal{F} \not\searrow 0$ pointwise at any point in K. Since $int(K) \neq \emptyset$, there exists an open set $\mathcal{O}$ contained in K. Furthermore, since X is a compact Hausdorff space, X is normal. Thus, by the separation axioms, there exists a nonempty open set P such that $\overline{P} \subset \mathcal{O}$. Notice that $\overline{P}$ is a closed subset of a compact Hausdorff space, and thus is also a compact Hausdorff space. Hence, $\overline{P}$ is a Baire space.

Now since $\mathcal{F} \not\searrow 0$ pointwise at any point in K, given $x \in \overline{P}$, there exists n such that for every $f_\alpha \in \mathcal{F}$, $f_\alpha(x) \geq \frac{1}{n}$. Define $A_n = \bigcap_{f_\alpha \in \mathcal{F}} \{x \in \overline{P} : f_\alpha(x) \geq \frac{1}{n}\}$. Note that each A_n is an intersection of closed sets and is therefore closed. Furthermore, $\overline{P} = \bigcup_{n=1}^{\infty} A_n$.

Since $\overline{P}$ is a Baire space, $\overline{P}$ is second category. Thus, $\overline{P}$ cannot be written as a union of nowhere dense sets. It follows that at least one of the A_n is not nowhere dense in $\overline{P}$. That is, there exist a n such that $int(\overline{A_n}) \neq \emptyset$. But, each A_n is closed, so $\overline{A_n} = A_n$. Thus, $int(A_n) \neq \emptyset$. So, there exists an open set U contained in A_n.

Again, since X is normal, by the separation axioms there exists a nonempty open subset W such that $\overline{W} \subset U$, and also a nonempty open subset H such that $\overline{H} \subset W$. Since W is open, W^C is closed. Furthermore, $\overline{H}$ and W^C are disjoint sets in X. Thus, Urysohn's Lemma gives a continuous real-valued function g_1 on X such that $0 \leq g_1 \leq 1$ on X and $g_1(\overline{H}) = 1$ and $g_1(W^C) = 0$.

Now, define $g = (\frac{1}{n+1})g_1$. So, $0 \leq g \leq \frac{1}{n+1}$, $g(W^C) = 0$, and $g(\overline{H}) = \frac{1}{n+1}$. Since $\overline{H} \subset A_n$, $f_\alpha(\overline{H}) \geq \frac{1}{n}$. Therefore, $0 \neq g < f_\alpha$ for every α. Thus, g is a lower bound of $\mathcal{F}$. So, $\wedge\mathcal{F} \neq 0$; a contradiction. It follows that $\mathcal{F} \searrow 0$ pointwise on K. Thus, for every $\mathcal{F}$ in Q, $\mathcal{F} \searrow 0$ pointwise except possibly on a set K with $int(K) = \emptyset$. □

Now, the measure of sets with empty interior is considered.

LEMMA 4.2. *Let E be a Banach lattice with an order continuous norm and a quasi-interior point, e. Furthermore, let $L : E \to \mathbb{R}$ be a positive, continuous, linear functional and define $Q = \{\mathcal{F} : \wedge\mathcal{F} = 0, \mathcal{F}$ is directed downward and countable$\}$. Then, for every $\mathcal{F} \in Q$ and for every K where $\mathcal{F} \not\searrow 0$ pointwise on K, $m(K) = 0$ if and only if $\wedge L(\mathcal{F}) = 0$.*

PROOF. From Lemma 4.1, $\mathcal{F} \searrow 0$ pointwise, except possibly on sets K where $int(K) = \emptyset$. So, $\mathcal{F} \not\searrow 0$ pointwise at any point in K gives $int(K) = \emptyset$. Suppose $m(K) = 0$. Then, $\mathcal{F} \searrow 0$ pointwise on K^C. So, $L(f_\alpha) = \int f_\alpha \, dm = \int_K f_\alpha \, dm + \int_{K^C} f_\alpha \, dm$. But, $f_\alpha \to 0$ on K^C. Thus, $\int_{K^C} f_\alpha \, dm \to \int_{K^C} 0 \, dm = 0$ by the Monotone Convergence Theorem.

Additionally, for any f_α in E, define functions $f_{\alpha_n} = f_\alpha \wedge ne$. Now the f_{α_n} are bounded and directed upward. So, $f_{\alpha_n} \leq M$ for all α_n. Thus, $\int_K f_{\alpha_n} \, dm \leq \int_K M \, dm = M \int \chi_K \, dm = M(m(K)) = M(0) = 0$. Furthermore, $f_{\alpha_n} \nearrow f_\alpha$. Thus by the Monotone Convergence Theorem, $\int_K f_{\alpha_n} \, dm \nearrow \int_K f_\alpha \, dm$. But, $\int_K f_{\alpha_n} \, dm = 0$. Thus, $\int_K f_\alpha \, dm = 0$. It follows that $L(f_\alpha) \searrow 0$, and $\wedge L(\mathcal{F}) = 0$

Now suppose $\wedge L(\mathcal{F}) = 0$. Let $K = \{x : f_\alpha(x) \not\searrow 0\}$. Thus, for every $x \in K$, there exists n such that $f_\alpha(x) \geq \frac{1}{n}$ for all α. So let, $A_n = \bigcap_\alpha \{x : f_\alpha(x) \geq \frac{1}{n}\}$.

Notice that A_n is an intersection of closed sets, which means A_n is closed. It follows that A_n is Borel, and hence measurable. Since $\wedge L(\mathcal{F}) = 0$, $L(f_\alpha) \searrow 0$. So, for every $\varepsilon = \frac{1}{n^2}$ there exists α_0 such that $L(f_\alpha) < \varepsilon$ for any $\alpha > \alpha_0$. It follows that for every $\alpha > \alpha_0$, $\frac{1}{n^2} > L(f_\alpha) = \int f_\alpha \, dm > \displaystyle\int_{A_n} \frac{1}{n} \, dm = \frac{1}{n} m(A_n)$. Thus, $m(A_n) < \frac{1}{n^2}(n) = \frac{1}{n}$.

Furthermore, notice that $\{A_n\}$ is an increasing collection and $K = \displaystyle\bigcup_{n=1}^{\infty} A_n$. So, K is a countable union of closed sets. So, K is $\mathcal{F}_\sigma$, and thus measurable. So, $0 \leq m(K) = m(\displaystyle\bigcup_{n=1}^{\infty} A_n) = \lim_{w \to \infty} m(\bigcup_{n=1}^{w} A_n) = \lim_{w \to \infty} (A_w) < \lim_{w \to \infty} \frac{1}{w} = 0$. It follows that $m(K) = 0$. □

Next, an operator L with the interior of its support is empty is discussed. Recall $K_L = X \setminus \bigcup_{\{h \in I(e)^+ : L(h) = 0\}} \{x \in X : h(x) > 0\}$ is the support of L.

Lemma 4.3. *Let E be a Banach lattice with an order continuous norm and a quasi-interior point, e. Furthermore, let $L : E \to \mathbb{R}$ be a positive, continuous, linear functional. Then $int(K_L) = \emptyset$ if and only if $L \equiv 0$.*

Proof. Suppose $int(K_L) = \emptyset$. Since the support of L is closed, ${K_L}^C$ is open. Furthermore, since $int(K_L) = \emptyset$, ${K_T}^C$ is dense in X. Let $\{x_\alpha\} \in {K_L}^C$ and define $A_\beta = \displaystyle\bigcup_{\alpha=1}^{n} x_\alpha$. Since A_β is a finite union of closed sets, it is closed. Furthermore, A_β and K_L are disjoint sets. Thus by Urysohn's Lemma, for every $A_\beta \in {K_L}^C$, there exists a continuous real-valued function, f_β, where $0 \leq f_\beta \leq 1$, $f_\beta(A_\beta) = 1$, and $f_\beta(K_L) = 0$.

Now consider the set $D = \{f_\beta : f_\beta(K_L) = 0, 0 \leq f_\beta \leq 1\}$. Notice that if f_1 is in D and if f_2 is in D, then their pointwise supremum $f_1 \vee f_2$ is also in D. Since $f_1 \vee f_2 \geq f_1$ and $f_1 \vee f_2 \geq f_2$, it follows that D is directed upward.

Define $g = \vee f_\beta$. So, g is continuous and real-valued. Since g is the least upper bound of $\{f_\beta\}, 0 \leq f_\beta \leq 1$, and ${K_L}^C$ is dense in X, $g \equiv 1$. Furthermore, since the $\{f_\beta\}$ are directed upward, and E has an order continuous norm, $f_\beta \nearrow g$. Since L is continuous, $L(f_\beta) \nearrow L(g) = L(1)$. But, $L(f_\beta) = 0$ for every β. It follows that $L(1) = 0$.

Now, for every $f > 0$, define $f_n = f \wedge n(1)$. Notice $f_n \leq n(1)$ for all n. Since L is positive and linear, it follows that $0 \leq L(f_n) \leq L(n(1)) = nL(1) = 0$. Thus, $L(f_n) = 0$ for all n. Furthermore, since $f_n \nearrow f$ and L is continuous, $L(f_n) \nearrow L(f)$. Hence, $L(f) = 0$ for all $f > 0$. That is, $L \equiv 0$.

Now suppose $L \equiv 0$. Then $K_L = \emptyset$. Therefore, $int(K_L) = int(\emptyset) = \emptyset$. □

Next, the closure of the interior of the support of an operator in relation to the original support set is explored.

Theorem 4.4. *Let E be a Banach lattice with an order continuous norm and a quasi-interior point, e. Furthermore, let $L : E \to \mathbb{R}$ be a positive, continuous, linear functional. Then $K_L = \overline{int(K_L)}$.*

Proof. If $K_L = \emptyset$, then $int(K_L) = \emptyset$ and thus $K_L = \overline{int(K_L)}$. Now suppose $K_L \neq \emptyset$ and suppose $K_L \neq \overline{int(K_L)}$. Then, there exists a p in K_L such that p is not in $\overline{int(K_L)}$. Since X is a compact, Hausdorff space, X is normal. So, there exists an open set H such that p is not in $\overline{H}$. Thus, by Urysohn's Lemma, there exists a continuous real-valued function g such that $0 \leq g \leq 1$, $g(p) = 1$, and $g(\overline{H}) = 0$. Notice that this means $g(\overline{int(K_L)}) = 0$.

Now $\overline{H} \subset \{x : g(x) = 0\}$. Thus, $\{x : g(x) = 0\}^C \subset \overline{H}^C$. But, $\overline{\{x : g(x) > 0\}} = \overline{\{x : g(x) = 0\}^C} \subset \overline{\overline{H}^C}$ and $\overline{int(K_L)} \subset H$. Furthermore, $\overline{\overline{H}^C} \cap H = \emptyset$. Hence, $\overline{\{x : g(x) > 0\}} \cap \overline{int(K_L)} = \emptyset$.

Now, define $S(f) = L(fg)$. Notice that S is linear. For any f such that $f(p) > 0$, $L(fg) > 0$. Thus $S(f) = L(fg) > 0$. Hence, S is positive and $S \not\equiv 0$. Define K_S to be the support of S. Furthermore, since $fg \leq f$ and L is positive, it follows that $L(fg) \leq L(f)$. Thus, if $L(f) = 0, S(f) = L(fg) = 0$. Therefore, ${K_L}^C \subset {K_S}^C$, and $K_S \subset K_L$.

Now, choose $x_0 \notin \overline{\{x : g(x) > 0\}}$. By Urysohn's Lemma, there is a continuous real-valued function f such that $f(x_0) = 1$ and $f(\overline{\{x : g(x) > 0\}}) = 0$. Notice that $g(\overline{\{x : g(x) > 0\}}^C) = 0$, so $fg \equiv 0$. Hence, $S(f) = 0$. Therefore, $x_0 \in {K_S}^C$. Since this holds true for any $x_0 \notin \overline{\{x : g(x) > 0\}}$, then $\overline{\{x : g(x) > 0\}}^C \subset {K_S}^C$. It follows that $K_S \subset \overline{\{x : g(x) > 0\}}$.

Since $K_S \subset \overline{\{x : g(x) > 0\}}$ and $\overline{\{x : g(x) > 0\}} \cap \overline{int(K_L)} = \emptyset$, it follows that $K_S \cap \overline{int(K_L)} = \emptyset$. Furthermore, $int(K_S) \subset K_S$. Thus, $\overline{int(K_S)} \subset \overline{K_S} = K_S$, since the support is closed. It follows that $\overline{int(K_S)} \cap \overline{int(K_L)} = \emptyset$. Moreover, $K_S \subset K_L$, and thus $\overline{int(K_S)} \subset \overline{int(K_L)}$. Therefore, $int(K_S) = \emptyset$. Thus, by Lemma 4.3, $S \equiv 0$, which is a contradiction.

Therefore, $K_L = \overline{int(K_L)}$. □

Now Theorem 4.4 is extended to nonlinear functionals.

THEOREM 4.5. *Let E be a Banach lattice with an order continuous norm and a quasi-interior point, e, and let $T : E \to \mathbb{R}$ be an orthogonally additive, continuous, monotonic, and subhomogeneous nonlinear operator. Then $K_T = \overline{int(K_T)}$.*

PROOF. Let L be a positive, continuous, linear functional associated with T through the measure μ and let K_L denote the support of L. Then by Theorem 4.4, $K_L = \overline{int(K_L)}$. But, by construction $K_T = K_L$, so it follows that $K_T = \overline{int(K_T)}$. □

This result can now be used to extend Lemma 4.3 to nonlinear functionals.

LEMMA 4.6. *Let E be a Banach lattice with an order continuous norm and a quasi-interior point, e, and let $T : E \to \mathbb{R}$ be an orthogonally additive, continuous, monotonic, and subhomogeneous nonlinear operator. Then, $int(K_T) = \emptyset$ if and only if $T \equiv 0$.*

PROOF. Suppose $int(K_T) = \emptyset$. Then by Theorem 4.5, $K_T = \overline{int(K_T)}$. But, since $int(K_T) = \emptyset$, it follows that $\overline{int(K_T)} = \overline{\emptyset} = \emptyset$. So $K_T = \emptyset$. Thus, by the definition of the support, it follows that $T \equiv 0$.

Now suppose $T \equiv 0$. Then, $K_T = \emptyset$. Therefore, $int(K_T) = int(\emptyset) = \emptyset$.

□

5. Banach Lattices Without Order Continuous Norms

In this section, Banach lattices that do not have order continuous norms are examined. Without the order continuous norm, there is no guaranteed that the space is Dedekind complete. Therefore, E will now be assumed to be a Dedekind complete space, and thus X is extremally disconnected. What it means for the interior of the support of a functional to be empty will be explored.

THEOREM 5.1. *Let E be a Dedekind complete Banach lattice with a quasi-interior point, e. Furthermore, let $L : E \to \mathbb{R}$ be a positive, continuous, linear functional. Then, $int(K_L) = \emptyset$ is equivalent to for every f, there exists g, where $0 < g \leq f$ and $L(g) = 0$.*

PROOF. Suppose $int(K_L) = \emptyset$. This implies that K_L does not contain any open sets. Thus, for any $f \in E, \{x : f(x) > 0\} \not\subset K_L$. Thus, there exists $p \in \{x : f(x) > 0\}$ such that $p \cap K_L = \emptyset$. By Urysohn's Lemma, there exists a continuous, real-valued function h, where $0 \leq h \leq 1$, $h(p) = 1$, and $h(K_L) = 0$. Define $g = h \wedge f$. Now, $0 < g \leq f$. Furthermore, $g(K_L) = 0$. Hence, $L(g) = 0$.

Now suppose $int(K_L) \neq \emptyset$. Then, K_L contains an open set. Furthermore, since X is normal, the separation axioms once again guarantee that there exists an open set B such that $x \in B \subset K_L$. Notice that x and B^C are disjoint closed sets. Thus, by Urysohn's Lemma there is a continuous real-valued function f such that $0 \leq f \leq 1$, $f(x) = 1$, and $f(B^C) = 0$. It follows that $\{x : f(x) > 0\} \subset B \subset K_L$. Then consider any function $g \not\equiv 0$ such that $g \leq f$. Since $g \leq f$, if $f(x) = 0$, then $g(x) = 0$. Therefore, $\{x : f(x) = 0\} \subset \{x : g(x) = 0\}$. Thus, $\{x : g(x) = 0\}^C \subset \{x : f(x) = 0\}^C$. Hence, $\{x : g(x) > 0\} \subset \{x : f(x) > 0\} \subset K_L$. It follows that $L(g) \neq 0$. Hence, if $int(K_L) \neq \emptyset$, then $L(g) \neq 0$. Therefore, if for every f, there exists g, where $0 < g \leq f$ and $L(g) = 0$, then $int(K_L) = \emptyset$. □

LEMMA 5.2. *If L is associated with T, then for any $h \in E^+$, $L(h) = 0$ if and only if $T(h) = 0$.*

PROOF. Let K_L denote the support of L. Since $K_T = K_L$, then ${K_T}^C = {K_L}^C$. As a result of the definition of the support, $\bigcup_{\{h \in I(e)^+ : T(h) = 0\}} \{x \in X : h(x) > 0\} = \bigcup_{\{h \in I(e)^+ : L(h) = 0\}} \{x \in X : h(x) > 0\}$. Thus, $\{h : T(h) = 0\} = \{h : L(h) = 0\}$. It follows that $L(h) = 0$ if and only if $T(h) = 0$. □

Finally, Theorem 5.1 is extended to nonlinear functionals.

THEOREM 5.3. *Let E be a Dedekind complete Banach lattice with a quasi-interior point, e, and let $T : E \to \mathbb{R}$ be an orthogonally additive, continuous, monotonic, and subhomogeneous nonlinear operator. Then, $int(K_T) = \emptyset$ is equivalent to for every f, there exists g, where $0 < g \leq f$ and $T(g) = 0$.*

PROOF. For this proof, let L be a positive, continuous, linear functional associated with T through the measure $\overline{\mu}$ and let K_L denote the support of L. Suppose $int(K_T) = \emptyset$. Since $K_T = K_L$, it follows that $int(K_L) = \emptyset$. Hence by Theorem 5.1, for every f, there exists g, where $0 < g \leq f$ and $L(g) = 0$. By Lemma 5.2, this gives $T(g) = 0$.

Similarly, suppose for every f, there exists g, where $0 < g \leq f$ and $T(g) = 0$. By Lemma 5.2, $L(g) = 0$. Thus, by Theorem 5.1, $int(K_L) = \emptyset$. But, $K_L = K_T$ and so $int(K_L) = int(K_T)$. So, $int(K_T) = \emptyset$.

Thus $int(K_T) = \emptyset$ if and only if for every f, there exists g, where $0 < g \leq f$ and $T(g) = 0$. $\square$

References

[1] J. E. Stovall and W. A. Feldman, *Associating linear and nonlinear operators*, Problems and recent methods in operator theory, Contemp. Math., vol. 687, Amer. Math. Soc., Providence, RI, 2017, pp. 225–230. MR3635766

[2] W. Feldman, *A factorization for orthogonally additive operators on Banach lattices*, J. Math. Anal. Appl. **472** (2019), no. 1, 238–245, DOI 10.1016/j.jmaa.2018.11.021. MR3906371

[3] W. A. Feldman and P. Singh, *A characterization of positively decomposable non-linear maps between Banach lattices*, Positivity **12** (2008), no. 3, 495–502, DOI 10.1007/s11117-007-2115-5. MR2421147

[4] W. Feldman, *Nonlinear Carleman operators on Banach lattices*, Proc. Amer. Math. Soc. **127** (1999), no. 7, 2109–2115, DOI 10.1090/S0002-9939-99-04729-2. MR1485472

[5] W. Feldman, *Separation properties for Carleman operators on Banach lattices*, Positivity **7** (2003), no. 1-2, 41–45, DOI 10.1023/A:1025818931354. Positivity and its applications (Nijmegen, 2001). MR2028365

[6] J. L. Kelley and I. Namioka, *Linear topological spaces*, Springer-Verlag, New York-Heidelberg, 1976. With the collaboration of W. F. Donoghue, Jr., Kenneth R. Lucas, B. J. Pettis, Ebbe Thue Poulsen, G. Baley Price, Wendy Robertson, W. R. Scott, and Kennan T. Smith; Second corrected printing; Graduate Texts in Mathematics, No. 36. MR0394084

[7] H. H. Schaefer, *Banach lattices and positive operators*, Springer-Verlag, New York-Heidelberg, 1974. Die Grundlehren der mathematischen Wissenschaften, Band 215. MR0423039

[8] J. E. Stovall, *Nonlinear functionals on banach lattices and their support sets*, Ph.D. thesis, The University of Arkansas, 2011.

Department of Mathematics, University of North Alabama, Florence, Alabama 35632

Email address: `jstovall@una.edu`

Department of Mathematics, University of Arkansas, Fayetteville, Arkansas 72701

Email address: `wfeldman@uark.edu`

Contemporary Mathematics
Volume **737**, 2019
https://doi.org/10.1090/conm/737/14867

Statistical approximation by generalized unitary discrete operators

Merve Kester Thomas

ABSTRACT. In this paper, after providing some useful estimates, we show that it is possible to approximate a function with generalized unitary discrete Picard and Gauss-Weierstrass operators in statistical sense. These operators are not positive in general. Additionally, we present an example which shows that our statistical approximation results are stronger than the classical uniform approximations.

1. Introduction

This article is mainly motivated by [**4**], in which the authors studided the uniform approximation properties with rates of generalized unitary discrete Picard and Gauss-Weierstrass singular operators. We are also motivated by [**1**] and [**2**] in which the authors studied the statiscial approximation properties of double Picard and Gauss-Weierstrass singular integral operators.

In this paper, we examine the the statistical approxmation properties of generalized unitary discrete Picard and Gauss-Weierstrass singular operators which do not need to be positive in general. We also show that our statistical approximation results more powerful than the classical uniform approximation.

2. Background

In [**4**], the authors defined the following operators :
For $r \in \mathbb{N}$ and $m \in \mathbb{N}_0 = \mathbb{N} \cup \{0\}$, they introduced

$$\alpha_{j,r}^{[m]} := \begin{cases} (-1)^{r-j}\binom{r}{j} j^{-m}, & j = 1, ..., r, \\ 1 - \sum\limits_{k=1}^{r} (-1)^{r-k}\binom{r}{k} k^{-m}, & j = 0 \end{cases} \tag{2.1}$$

and

$$\delta_{k,r}^{[m]} = \sum_{j=1}^{r} \alpha_{j,r}^{[m]} j^k; \text{ where } k = 1, 2, ..., m \in \mathbb{N}. \tag{2.2}$$

Now let $f \in C^m(\mathbb{R})$, $x \in \mathbb{R}$, $n, r \in \mathbb{N}$, $m \in \mathbb{N}_0$, and $(\xi_n)_{n\in\mathbb{N}}$ be any sequence of real numbers such that $0 < \xi_n \leq 1$,

---

2010 *Mathematics Subject Classification.* 26A15, 40G15, 41A17, 41A25.

Key words and phrases. A-statistical convergence , statistical approximation, discrete singular operator, modulus of smoothness.

(1) They defined the generalized unitary discrete Picard operators as:

$$P_{r,n}^{*[m]}(f;x) := \frac{\sum\limits_{\nu=-\infty}^{\infty}\left(\sum\limits_{j=0}^{r}\alpha_{j,r}^{[m]}f(x+j\nu)\right)e^{-\frac{|\nu|}{\xi_n}}}{\sum\limits_{\nu=-\infty}^{\infty}e^{-\frac{|\nu|}{\xi_n}}}. \tag{2.3}$$

(2) They defined the generalized unitary discrete Gauss-Weierstrass operators as:

$$W_{r,n}^{*[m]}(f;x) := \frac{\sum\limits_{\nu=-\infty}^{\infty}\left(\sum\limits_{j=0}^{r}\alpha_{j,r}^{[m]}f(x+j\nu)\right)e^{-\frac{\nu^2}{\xi_n}}}{\sum\limits_{\nu=-\infty}^{\infty}e^{-\frac{\nu^2}{\xi_n}}}. \tag{2.4}$$

.

In [**4**], the authors stated that, for c constant

$$P_{r,n}^{*[m]}(c;x) = W_{r,n}^{*[m]}(c;x) = c.$$

They assume also that the operators $P_{r,n}^{*[m]}$, $W_{r,n}^{*[m]} \in \mathbb{R}$ for $x \in \mathbb{R}$. This is the case when $\|f\|_{\infty,\mathbb{R}} < \infty$.

Additionally, in [**4**], the authors observed that the operators $P_{r,n}^{*[m]}$ and $W_{r,n}^{*[m]}$ are not positive in general.

Next, [**4**], they proved the following propositions:

PROPOSITION 2.1. *Let $r \in \mathbb{N}$, $k = 1, 2, ..., m \in \mathbb{N}$,*

$$\lambda_m := \sum_{\nu=1}^{2m} \nu^m e^{-\frac{\nu}{2}},$$

and

$$K_{r,m} := 2^{2r+1}r!\left[\lambda_m + (2m+1)^{m+1}e^{-\frac{(2m+1)}{2}} + m!2^{m+1}\right]. \tag{2.5}$$

Then

$$\frac{\sum\limits_{\nu=-\infty}^{\infty}|\nu|^m\left(1+\frac{|\nu|}{\xi_n}\right)^r e^{-\frac{|\nu|}{\xi_n}}}{\sum\limits_{\nu=-\infty}^{\infty}e^{-\frac{|\nu|}{\xi_n}}} \le K_{r,m} < \infty, \tag{2.6}$$

and

$$c_{k,\xi_n}^{*} := \frac{\sum\limits_{\nu=-\infty}^{\infty}\nu^k e^{-\frac{|\nu|}{\xi_n}}}{\sum\limits_{\nu=-\infty}^{\infty}e^{-\frac{|\nu|}{\xi_n}}} < \infty, \tag{2.7}$$

for all $\xi_n \in (0,1]$, $n \in \mathbb{N}$.

PROPOSITION 2.2. *Let $r \in \mathbb{N}$, $k = 1, 2, ..., m \in \mathbb{N}$, and $K_{r,m}$ be given as* (2.5). *Then*

$$\frac{\sum\limits_{\nu=-\infty}^{\infty} |\nu|^m \left(1 + \frac{|\nu|}{\xi_n}\right)^r e^{-\frac{\nu^2}{\xi_n}}}{\sum\limits_{\nu=-\infty}^{\infty} e^{-\frac{\nu^2}{\xi_n}}} \leq K_{r,m} < \infty, \tag{2.8}$$

and

$$p^*_{k,\xi_n} := \frac{\sum\limits_{\nu=-\infty}^{\infty} \nu^k e^{-\frac{\nu^2}{\xi_n}}}{\sum\limits_{\nu=-\infty}^{\infty} e^{-\frac{\nu^2}{\xi_n}}} < \infty, \tag{2.9}$$

for all $\xi_n \in (0, 1]$, $n \in \mathbb{N}$.

Next, in [**4**], the *rth* modulus of smoothness finite given as

$$\omega_r(f^{(m)}, h) := \sup_{|u| \leq h} \|\Delta^r_u f^{(m)}(x)\|_{\infty,x} < \infty, \quad h > 0, \tag{2.10}$$

where$\|.\|_{\infty,x}$ is the supremum norm with respect to x, $f \in C^m(\mathbb{R})$, $m \in \mathbb{N}_0$, and

$$\Delta^r_u f^{(m)}(x) := \sum_{j=0}^{r} (-1)^{r-j} \binom{r}{j} f^{(m)}(x + ju). \tag{2.11}$$

In [**4**], the authors proved the following estimate for the operators $P^{*[m]}_{r,n}$. For the case of $m \in \mathbb{N}$, they had

THEOREM 2.3. *Let $f \in C^m(\mathbb{R})$ with $f^{(m)} \in C_u(\mathbb{R})$ (uniformly continuous functions), $m, r \in \mathbb{N}$, and $(\xi_n)_{n \in \mathbb{N}}$ be any sequence of real numbers such that $0 < \xi_n \leq 1$. Then*

$$\left\| P^{*[m]}_{r,n}(f; x) - f(x) - \sum_{k=1}^{m} \frac{f^{(k)}(x)}{k!} \delta^{[m]}_{k,r} c^*_{k,\xi_n} \right\|_{\infty,x} \leq \frac{K_{r,m}}{m!} \omega_r(f^{(m)}, \xi_n), \tag{2.12}$$

where $\delta^{[m]}_{k,r}$, c^*_{k,ξ_n}, and $K_{r,m}$ given as in (2.2), (2.7), *and* (2.5); *respectively.*

For the case of $m = 0$, they showed

THEOREM 2.4. *Let $f \in C_u(\mathbb{R})$, $r \in \mathbb{N}$, and $(\xi_n)_{n \in \mathbb{N}}$ be any sequence of real numbers such that $0 < \xi_n \leq 1$. Then*

$$\left\| P^{*[0]}_{r,n}(f; x) - f(x) \right\|_{\infty,x} \leq \omega_r(f, \xi_n) \left(\frac{\sum\limits_{\nu=-\infty}^{\infty} \left(1 + \frac{|\nu|}{\xi_n}\right)^r e^{-\frac{|\nu|}{\xi_n}}}{\sum\limits_{\nu=-\infty}^{\infty} e^{-\frac{|\nu|}{\xi_n}}} \right). \tag{2.13}$$

In [**4**], the authors proved following results for the operators $W^{*[m]}_{r,n}$. For the case of $m \in \mathbb{N}$, they demonstrated

THEOREM 2.5. *Let $f \in C^m(\mathbb{R})$ with $f^{(m)} \in C_u(\mathbb{R})$, $m, r \in \mathbb{N}$, and $(\xi_n)_{n\in\mathbb{N}}$ be any sequence of real numbers such that $0 < \xi_n \leq 1$. Then*

$$(2.14) \qquad \left\| W_{r,n}^{*[m]}(f;x) - f(x) - \sum_{k=1}^{m} \frac{f^{(k)}(x)}{k!} \delta_{k,r}^{[m]} p_{k,\xi_n}^{*} \right\|_{\infty,x} \leq \frac{K_{r,m}}{m!} \omega_r(f^{(m)}, \xi_n),$$

where $\delta_{k,r}^{[m]}$, p_{k,ξ_n}^{*}, and $K_{r,m}$ given as in (2.2) , (2.9), *and* (2.5) ; *respectively.*

Finally, for the case of $m = 0$, they showed

THEOREM 2.6. *Let $f \in C_u(\mathbb{R})$, $r \in \mathbb{N}$, and $(\xi_n)_{n\in\mathbb{N}}$ be any sequence of real numbers such that $0 < \xi_n \leq 1$. Then*

$$(2.15) \qquad \left\| W_{r,n}^{*[0]}(f;x) - f(x) \right\|_{\infty,x} \leq \omega_r(f, \xi_n) \left(\frac{\sum\limits_{\nu=-\infty}^{\infty} \left(1 + \frac{|\nu|}{\xi_n}\right)^r e^{-\frac{\nu^2}{\xi_n}}}{\sum\limits_{\nu=-\infty}^{\infty} e^{-\frac{\nu^2}{\xi_n}}} \right).$$

Now, we recall the definition of $A-$statistical convergence (see [**6**]).

DEFINITION 2.7. Let $A := [a_{jn}]$, $j, n \in \mathbb{N}$ be a non-negative regular summability matrix and (x_n) be a sequence. (x_n) is called $A-$statistically convergent to L, denoted by

$$st_A \quad \lim_n x_n = L,$$

if

$$\lim_{j\to\infty} \sum_{n:|x_n - L| \geq \epsilon} a_{jn} = 0,$$

for every $\epsilon > 0$.

It is easy to see that if $A = C_1 = [c_{jn}]$, the Cesáro matrix of order one defined by

$$c_{jn} = \begin{cases} \frac{1}{j}, & \text{if } 1 \leq n \leq j \\ 0, & \text{otherwise} \end{cases},$$

then C_1-statistical convergence becomes statistical convergence which was introduced by Fast (see [**5**]) . Moreover, if we take A as the identity matrix I, then $I-$statistical convergence coincides with the ordinary convergence. Observe that every ordinary convergent sequence is $A-$statistical convergent; however, not every $A-$statistical convergent sequence is ordinary convergent.

3. Main Results

We start with the following proposition.

PROPOSITION 3.1. *For each fixed $m \in \mathbb{N}$ and for every $(\xi_n)_{n\in\mathbb{N}}$ such that $0 < \xi_n \leq 1$,*

$$(3.1) \qquad \sum_{k=1}^{m} c_{k,\xi_n}^{*} \leq S_m \left(e^{-\frac{1}{\xi_n}} + \xi_n \right)$$

holds. Here

$$(3.2) \qquad S_m = 4m(m+1)^{m+1}$$

is a constant depending on m and c_{k,ξ_n}^{*} is given as (2.7) .

PROOF. Observe that

$$\sum_{\nu=-\infty}^{\infty} \nu^k e^{-\frac{|\nu|}{\xi_n}} = \begin{cases} 0, & k \text{ is odd} \\ 2\sum\limits_{\nu=1}^{\infty} \nu^k e^{-\frac{\nu}{\xi_n}}, & k \text{ is even} \end{cases}$$

Therefore,

$$c_{k,\xi_n}^{*} = \frac{\sum\limits_{\nu=-\infty}^{\infty} \nu^k e^{-\frac{|\nu|}{\xi_n}}}{\sum\limits_{\nu=-\infty}^{\infty} e^{-\frac{|\nu|}{\xi_n}}} \leq \frac{2\sum\limits_{\nu=1}^{\infty} \nu^k e^{-\frac{\nu}{\xi_n}}}{\sum\limits_{\nu=-\infty}^{\infty} e^{-\frac{|\nu|}{\xi_n}}}. \tag{3.3}$$

Additionally, since

$$\sum_{\nu=-\infty}^{\infty} e^{-\frac{|\nu|}{\xi_n}} > 1,$$

we have

$$\frac{1}{\sum\limits_{\nu=-\infty}^{\infty} e^{-\frac{|\nu|}{\xi_n}}} < 1. \tag{3.4}$$

Thus, by (3.3) and (3.4), we obtain

$$c_{k,\xi_n}^{*} \leq 2\sum_{\nu=1}^{\infty} \nu^k e^{-\frac{\nu}{\xi_n}}. \tag{3.5}$$

Now, let $f(\nu) = \nu^k e^{-\frac{\nu}{\xi_n}}$. Observe that f is continuous, positive-valued, and decreasing for $\nu > k\xi_n$. Hence by shifted triple inequality similar to [**7**] we get

$$\begin{aligned}
2\sum_{\nu=1}^{\infty} \nu^k e^{-\frac{\nu}{\xi_n}} &= 2\left[\sum_{\nu=1}^{\lceil k\xi_n\rceil} \nu^k e^{-\frac{\nu}{\xi_n}} + \sum_{\nu=\lceil k\xi_n\rceil+1}^{\infty} \nu^k e^{-\frac{\nu}{\xi_n}}\right] \\
&\leq 2\sum_{\nu=1}^{\lceil k\xi_n\rceil} \nu^k e^{-\frac{\nu}{\xi_n}} + 2\int_{\lceil k\xi_n\rceil+1}^{\infty} \nu^k e^{-\frac{\nu}{\xi_n}}\,dv + 2f\left(\lceil k\xi_n\rceil + 1\right) \\
&= 2\sum_{\nu=1}^{\lceil k\xi_n\rceil} \nu^k e^{-\frac{\nu}{\xi_n}} + 2\int_{\lceil k\xi_n\rceil+1}^{\infty} \nu^k e^{-\frac{\nu}{\xi_n}}\,dv + 2\left(\lceil k\xi_n\rceil + 1\right)^k e^{-\frac{\lceil k\xi_n\rceil+1}{\xi_n}} \\
&\leq 2\sum_{\nu=1}^{\lceil k\xi_n\rceil} \nu^k e^{-\frac{\nu}{\xi_n}} + 2\int_{0}^{\infty} \nu^k e^{-\frac{\nu}{\xi_n}}\,dv + 2\left(\lceil k\xi_n\rceil + 1\right)^k e^{-\frac{\lceil k\xi_n\rceil+1}{\xi_n}}
\end{aligned} \tag{3.6}$$

Next, since for all $\nu \in [1, \lceil k\xi_n\rceil]$,

$$0 < e^{-\frac{\nu}{\xi_n}} \leq e^{-\frac{1}{\xi_n}}$$

and

$$0 < \nu^k \leq \left(\lceil k\xi_n\rceil\right)^k,$$

we have

$$0 < \nu^k e^{-\frac{\nu}{\xi_n}} \leq \left(\lceil k\xi_n\rceil\right)^k e^{-\frac{1}{\xi_n}}. \tag{3.7}$$

Thus, by (3.7), we obtain

$$
\begin{aligned}
\sum_{\nu=1}^{\lceil k\xi_n\rceil} \nu^k e^{-\frac{\nu}{\xi_n}} &\leq \sum_{\nu=1}^{\lceil k\xi_n\rceil} \left(\lceil k\xi_n\rceil\right)^k e^{-\frac{1}{\xi_n}} \\
&= \left(\lceil k\xi_n\rceil\right)^{k+1} e^{-\frac{1}{\xi_n}} \\
&\leq (k+1)^{k+1} e^{-\frac{1}{\xi_n}}.
\end{aligned} \tag{3.8}
$$

Additionally, by the integral calculation in [**3**], p.86, we get

$$
\int_0^\infty \nu^k e^{-\frac{\nu}{\xi_n}} dv = k! \left(\xi_n\right)^{k+1} \leq (k+1)^{k+1} \left(\xi_n\right)^{k+1} \leq (k+1)^{k+1} \xi_n. \tag{3.9}
$$

By, combining the results in (3.6), (3.8), and (3.9), for all $k = 1, 2, ..., m \in \mathbb{N}$ and $\xi_n \in (0, 1]$, we have

$$
\begin{aligned}
2\sum_{\nu=1}^{\infty} \nu^k e^{-\frac{\nu}{\xi_n}} &\leq 2(k+1)^{k+1} e^{-\frac{1}{\xi_n}} + 2(k+1)^{k+1} \xi_n + 2\left(\lceil k\xi_n\rceil + 1\right)^k e^{-\frac{\lceil k\xi_n\rceil+1}{\xi_n}} \\
&\leq 2(k+1)^{k+1} e^{-\frac{1}{\xi_n}} + 2(k+1)^{k+1} \xi_n + 2(k+1)^{k+1} e^{-\frac{1}{\xi_n}} \\
&\leq 4(k+1)^{k+1} e^{-\frac{1}{\xi_n}} + 4(k+1)^{k+1} \xi_n \\
&= 4(k+1)^{k+1} \left(e^{-\frac{1}{\xi_n}} + \xi_n\right) \\
&\leq 4(m+1)^{m+1} \left(e^{-\frac{1}{\xi_n}} + \xi_n\right).
\end{aligned} \tag{3.10}
$$

Thus, by (3.5) and (3.10), we conclude that

$$
\begin{aligned}
\sum_{k=1}^{m} c^*_{k,\xi_n} &\leq \sum_{k=1}^{m} 4(m+1)^{m+1} \left(e^{-\frac{1}{\xi_n}} + \xi_n\right) \\
&= 4m(m+1)^{m+1} \left(e^{-\frac{1}{\xi_n}} + \xi_n\right) \\
&= S_m \left(e^{-\frac{1}{\xi_n}} + \xi_n\right)
\end{aligned} \tag{3.11}
$$

where

$$
S_m := 4m(m+1)^{m+1}. \tag{3.12}
$$

□

Now, with the help of the previous proposition, we give the following estimate for the operators $P^{*[m]}_{r,n}$ defined as in (2.3) for $m \in \mathbb{N}$.

THEOREM 3.2. *Let $(\xi_n)_{n\in\mathbb{N}}$, $0 < \xi_n \leq 1$, be a sequence of positive real numbers. Then, for each fixed $m, r \in \mathbb{N}$ and for every $f \in C^m(\mathbb{R})$ with $f^{(m)} \in C_u(\mathbb{R})$ such that*

$$
\left\| f^{(m)}(x) \right\|_{\infty,x} < \infty,
$$

we have

$$
\left\| P^{*[m]}_{r,n}(f;x) - f(x) \right\|_{\infty,x} \leq C_{r,m}\omega_r(f^{(m)}, \xi_n) + T_{r,m}\left(e^{-\frac{1}{\xi_n}} + \xi_n\right), \tag{3.13}
$$

where $C_{r,m}$ and $T_{r,m}$ are constants depending on r and m.

PROOF. By (2.12) and the sub-additivity of the norm, we get

$$\left\|P_{r,n}^{*[m]}(f;x)-f(x)\right\|_{\infty,x}\leq\frac{K_{r,m}}{m!}\omega_r(f^{(m)},\xi_n)+\left\|\sum_{k=1}^{m}\frac{f^{(k)}(x)}{k!}\delta_{k,r}^{[m]}c_{k,\xi_n}^{*}\right\|_{\infty,x}. \tag{3.14}$$

In order to estimate

$$\left\|\sum_{k=1}^{m}\frac{f^{(k)}(x)}{k!}\delta_{k,r}^{[m]}c_{k,\xi_n}^{*}\right\|_{\infty,x},$$

we observe that

$$\left\|\sum_{k=1}^{m}\frac{f^{(k)}(x)}{k!}\delta_{k,r}^{[m]}c_{k,\xi_n}^{*}\right\|_{\infty,x}\leq\sum_{k=1}^{m}\frac{\left\|f^{(k)}(x)\right\|_{\infty,x}}{k!}\delta_{k,r}^{[m]}c_{k,\xi_n}^{*}. \tag{3.15}$$

By combining the results (3.1) and (3.15), we have

$$\begin{aligned}&\left\|\sum_{k=1}^{m}\frac{f^{(k)}(x)}{k!}\delta_{k,r}^{[m]}c_{k,\xi_n}^{*}\right\|_{\infty,x}\\ &\leq T_{r,m}\left(e^{-\frac{1}{\xi_n}}+\xi_n\right)\end{aligned} \tag{3.16}$$

where

$$T_{r,m}:=\max_{1\leq k\leq m}\left\{\frac{\left\|f^{(k)}(x)\right\|_{\infty,x}}{k!}\delta_{k,r}^{[m]}\right\}S_m. \tag{3.17}$$

Finally, by (3.14) and (3.16), we conclude that

$$\left\|P_{r,n}^{*[m]}(f;x)-f(x)\right\|_{\infty,x}\leq C_{r,m}\omega_r(f^{(m)},\xi_n)+T_{r,m}\left(e^{-\frac{1}{\xi_n}}+\xi_n\right)$$

where

$$C_{r,m}:=\frac{K_{r,m}}{m!}. \tag{3.18}$$

□

Next, we need the following proposition.

PROPOSITION 3.3. *Let $r\in\mathbb{N}$. For every $(\xi_n)_{n\in\mathbb{N}}$ such that $0<\xi_n\leq 1$,*

$$\frac{\sum\limits_{\nu=-\infty}^{\infty}\left(1+\frac{|\nu|}{\xi_n}\right)^r e^{-\frac{|\nu|}{\xi_n}}}{\sum\limits_{\nu=-\infty}^{\infty}e^{-\frac{|\nu|}{\xi_n}}}<H_r \tag{3.19}$$

holds. Here, H_r is a constant depending on r.

PROOF. Notice that

$$\sum_{\nu=-\infty}^{\infty}\left(1+\frac{|\nu|}{\xi_n}\right)^r e^{-\frac{|\nu|}{\xi_n}}=1+2\sum_{\nu=1}^{\infty}\left(1+\frac{\nu}{\xi_n}\right)^r e^{-\frac{\nu}{\xi_n}} \tag{3.20}$$

and

$$\frac{1}{\sum\limits_{\nu=-\infty}^{\infty}e^{-\frac{|\nu|}{\xi_n}}}<1. \tag{3.21}$$

Therefore, by (3.20) and (3.21) , we have

$$\frac{\sum\limits_{\nu=-\infty}^{\infty}\left(1+\frac{|\nu|}{\xi_n}\right)^r e^{-\frac{|\nu|}{\xi_n}}}{\sum\limits_{\nu=-\infty}^{\infty} e^{-\frac{|\nu|}{\xi_n}}} \tag{3.22}$$

$$< 1+2\sum_{\nu=1}^{\infty}\left(1+\frac{\nu}{\xi_n}\right)^r e^{-\frac{\nu}{\xi_n}}$$

$$= 1+2\sum_{\nu=1}^{\infty}\left[\left(1+\frac{\nu}{\xi_n}\right)^r e^{-\frac{\nu}{2\xi_n}}\right] e^{-\frac{\nu}{2\xi_n}}.$$

Since for all $\nu \geq 1$, we get $\frac{\nu}{\xi_n} \geq 1$, we can say that

$$\left(1+\frac{\nu}{\xi_n}\right)^r e^{-\frac{\nu}{2\xi_n}} \leq \frac{2^r\left(\frac{\nu}{\xi_n}\right)^r}{e^{\frac{\nu}{2\xi_n}}}. \tag{3.23}$$

Moreover, since

$$e^{\frac{\nu}{2\xi_n}} = \sum_{i=0}^{\infty}\frac{\left(\frac{\nu}{2\xi_n}\right)^i}{i!} \tag{3.24}$$

$$\geq \frac{\left(\frac{\nu}{2\xi_n}\right)^r}{r!}$$

$$= \frac{\left(\frac{\nu}{\xi_n}\right)^r}{2^r r!},$$

we have

$$\left(\frac{\nu}{\xi_n}\right)^r e^{-\frac{\nu}{2\xi_n}} \leq 2^r r!. \tag{3.25}$$

Thus, by (3.22) , (3.23) , and (3.25) , we get

$$\frac{\sum\limits_{\nu=-\infty}^{\infty}\left(1+\frac{|\nu|}{\xi_n}\right)^r e^{-\frac{|\nu|}{\xi_n}}}{\sum\limits_{\nu=-\infty}^{\infty} e^{-\frac{|\nu|}{\xi_n}}} < 1+2^{2r+1}r!\sum_{\nu=1}^{\infty} e^{-\frac{\nu}{2\xi_n}} \tag{3.26}$$

$$\leq 1+2^{2r+1}r!\sum_{\nu=1}^{\infty} e^{-\frac{\nu}{2}}.$$

Now, we define the function $f(\nu) = e^{-\frac{\nu}{2}}$. Observe that f is positive, continuous, and decreasing for all $\nu \geq 1$. Hence, by the shifted triple inequality similar to [**7**], we have

$$\sum_{\nu=1}^{\infty} e^{-\frac{\nu}{2}} \leq \int_{1}^{\infty} e^{-\frac{\nu}{2}} d\nu + f(1) \tag{3.27}$$

$$\leq \int_{0}^{\infty} e^{-\frac{\nu}{2}} d\nu + e^{-\frac{1}{2}}$$

$$= 2 + e^{-\frac{1}{2}}.$$

Thus, by (3.26) and (3.27), we get the desired result

$$\frac{\sum\limits_{\nu=-\infty}^{\infty} \left(1 + \frac{|\nu|}{\xi_n}\right)^r e^{-\frac{|\nu|}{\xi_n}}}{\sum\limits_{\nu=-\infty}^{\infty} e^{-\frac{|\nu|}{\xi_n}}} < H_r \tag{3.28}$$

where

$$H_r := 1 + 2^{2r+1} r! \left(2 + e^{-\frac{1}{2}}\right). \tag{3.29}$$

□

Next, we give our estimate for the operators $P_{r,n}^{*[m]}$ for $m = 0$.

THEOREM 3.4. *Let $(\xi_n)_{n\in\mathbb{N}}$, $0 < \xi_n \leq 1$, be a sequence of positive real numbers. Then, for each fixed $r \in \mathbb{N}$ and for every $f \in C_u(\mathbb{R})$ such that*

$$\|f(x)\|_{\infty,x} < \infty,$$

we have

$$\left\| P_{r,n}^{*[0]}(f;x) - f(x) \right\|_{\infty,x} \leq H_r \omega_r(f, \xi_n). \tag{3.30}$$

Here H_r is a constant, depending on r, given as (3.29).

PROOF. By Theorem 2.4 and Proposition 3.3. □

Before representing our statistical approximation results, by the right continuity of rth modulus of smoothness, we give the following result.

LEMMA 3.5. *Let $A = [a_{jn}]$ be a non-negative regular summability matrix, and let $(\xi_n)_{n\in\mathbb{N}}$, $0 < \xi_n \leq 1$, be a sequence of positive real numbers for which*

$$st_A - \lim_n \xi_n = 0.$$

Then , for every $m \in \mathbb{N}_0$ and for every $f \in C^m(\mathbb{R})$ with $f^{(m)} \in C_u(\mathbb{R})$ such that

$$\left\| f^{(m)}(x) \right\|_{\infty,x} < \infty,$$

we have

$$st_A - \lim_n \omega_r(f^{(m)}, \xi_n) = 0.$$

Now, we show our statistical approximation result for the operators $P_{r,n}^{*[m]}$ for the case of $m \in \mathbb{N}$.

THEOREM 3.6. *Let $m, r \in \mathbb{N}$ and $A = [a_{jn}]$ be a non-negative regular summability matrix, and let $(\xi_n)_{n\in\mathbb{N}}$, $0 < \xi_n \leq 1$, be a sequence of positive real numbers for which*

$$st_A - \lim_n \xi_n = 0. \tag{3.31}$$

Then, for each fixed $m, r \in \mathbb{N}$ and for every $f \in C^m(\mathbb{R})$ with $f^{(m)} \in C_u(\mathbb{R})$ such that

$$\left\| f^{(m)}(x) \right\|_{\infty,x} < \infty,$$

we have

$$st_A - \lim_n \left\| P^{*[m]}_{r,n}(f;x) - f(x) \right\|_{\infty,x} = 0. \tag{3.32}$$

PROOF. By Lemma 3.5 and (3.31), we can say that

$$st_A - \lim_n \omega_r(f^{(m)}, \xi_n) = 0 \tag{3.33}$$

and

$$st_A - \lim_n e^{-\frac{1}{\xi_n}} = 0. \tag{3.34}$$

Now, for given $\epsilon > 0$, we define the following sets

$$\begin{aligned} D &: = \left\{ n \in \mathbb{N} : \left\| P^{*[m]}_{r,n}(f;x) - f(x) \right\|_{\infty,x} \geq \epsilon \right\}, \\ D_1 &: = \left\{ n \in \mathbb{N} : \omega_r(f^{(m)}, \xi_n) \geq \frac{\epsilon}{2C_{r,m}} \right\}, \\ D_2 &: = \left\{ n \in \mathbb{N} : e^{-\frac{1}{\xi_n}} \geq \frac{\epsilon}{4T_{r,m}} \right\}, \\ D_3 &: = \left\{ n \in \mathbb{N} : \xi_n \geq \frac{\epsilon}{4T_{r,m}} \right\}, \end{aligned} \tag{3.35}$$

where $T_{r,m}$ and $C_{r,m}$ defined as (3.17) and (3.18); respectively. Then, it follows from (3.13) that

$$D \subseteq \bigcup_{i=1}^{3} D_i$$

which yields, for every $j \in \mathbb{N}$, that

$$\sum_{n\in D} a_{jn} \leq \sum_{i=1}^{3} \sum_{n\in D_i} a_{jn}. \tag{3.36}$$

If we take the limit of the both sides of (3.36) as $j \to \infty$, by (3.31),(3.33), and (3.34), we have

$$\lim_{j\to\infty} \sum_{n\in D} a_{jn} = 0$$

which completes the proof. □

When $m = 0$, we have the following statistical approximation result.

THEOREM 3.7. *Let $r \in \mathbb{N}$ and $A = [a_{jn}]$ be a non-negative regular summability matrix, and let $(\xi_n)_{n\in\mathbb{N}}$, $0 < \xi_n \leq 1$, be a sequence of positive real numbers for which*

$$st_A - \lim_n \xi_n = 0. \tag{3.37}$$

Then, for each fixed $r \in N$ and for every $f \in C_u(\mathbb{R})$ such that

$$\|f(x)\|_{\infty,x} < \infty,$$

we have

$$st_A - \lim_n \left\| P_{r,n}^{*[0]}(f;x) - f(x) \right\|_{\infty,x} = 0. \tag{3.38}$$

PROOF. By Lemma 3.5 and (3.37), we can say that

$$st_A - \lim_n \omega_r(f, \xi_n) = 0 \tag{3.39}$$

Now, for given $\epsilon > 0$, we define the following sets

$$\begin{aligned} D \quad &: = \left\{ n \in \mathbb{N} : \left\| P_{r,n}^{*[0]}(f;x) - f(x) \right\|_{\infty,x} \geq \epsilon \right\}, \\ D_1 \quad &: = \left\{ n \in \mathbb{N} : \omega_r(f, \xi_n) \geq \frac{\epsilon}{H_r} \right\}, \end{aligned} \tag{3.40}$$

where H_r is defined as (3.29). Then, it follows from (3.30) that

$$D \subseteq D_1$$

which yields, for every $j \in \mathbb{N}$, that

$$\sum_{n\in D} a_{jn} \leq \sum_{n\in D_1} a_{jn}. \tag{3.41}$$

If we take the limit of the both sides of (3.41) as $j \to \infty$, by (3.39), we have

$$\lim_{j\to\infty} \sum_{n\in D} a_{jn} = 0,$$

whence the result. □

Before we demonstrate our estimate for the operators $W_{r,n}^{*[m]}$ defined as in (2.4) for $m \in \mathbb{N}$, we need the following proposition.

PROPOSITION 3.8. *For each fixed $m \in \mathbb{N}$ and for every $(\xi_n)_{n\in\mathbb{N}}$ such that $0 < \xi_n \leq 1$,*

$$\sum_{k=1}^{m} p_{k,\xi_n}^* \leq S_m \left(e^{-\frac{1}{\xi_n}} + \xi_n \right) \tag{3.42}$$

holds.
Here, S_m is a constant depending on m defined as (3.2) and p_{k,ξ_n}^* is given as (2.9).

PROOF. Observe that

$$\sum_{\nu=-\infty}^{\infty} \nu^k e^{-\frac{\nu^2}{\xi_n}} = \begin{cases} 0, & k \text{ is odd} \\ 2\sum_{\nu=1}^{\infty} \nu^k e^{-\frac{\nu^2}{\xi_n}}, & k \text{ is even} \end{cases}$$

Thus,

$$p^{*}_{k,\xi_n}=\frac{\sum\limits_{\nu=-\infty}^{\infty}\nu^{k}e^{-\frac{\nu^2}{\xi_n}}}{\sum\limits_{\nu=-\infty}^{\infty}e^{-\frac{\nu^2}{\xi_n}}}\leq\frac{2\sum\limits_{\nu=1}^{\infty}\nu^{k}e^{-\frac{\nu2}{\xi_n}}}{\sum\limits_{\nu=-\infty}^{\infty}e^{-\frac{\nu2}{\xi_n}}}. \tag{3.43}$$

Moreover, since

$$\sum_{\nu=-\infty}^{\infty}e^{-\frac{\nu^2}{\xi_n}}>1,$$

we get

$$\frac{1}{\sum\limits_{\nu=-\infty}^{\infty}e^{-\frac{\nu^2}{\xi_n}}}<1. \tag{3.44}$$

Therefore, by (3.43) and (3.44) , we obtain

$$p^{*}_{k,\xi_n}\leq 2\sum_{\nu=1}^{\infty}\nu^{k}e^{-\frac{\nu^2}{\xi_n}}. \tag{3.45}$$

Additionally, for all $\nu\in\mathbb{N}$, and $(\xi_n)_{n\in\mathbb{N}}$ such that $0<\xi_n\leq 1$,

$$e^{-\frac{\nu}{\xi_n}}\geq e^{-\frac{\nu^2}{\xi_n}}$$

holds. Hence, for all $k=1,2,...,m\in\mathbb{N}$, we have

$$\nu^{k}e^{-\frac{\nu}{\xi_n}}\geq \nu^{k}e^{-\frac{\nu^2}{\xi_n}}$$

which implies

$$\sum_{\nu=1}^{\infty}\nu^{k}e^{-\frac{\nu}{\xi_n}}\geq\sum_{\nu=1}^{\infty}\nu^{k}e^{-\frac{\nu^2}{\xi_n}}. \tag{3.46}$$

Hence, by (3.10) , (3.45) , and (3.46) , for all $k=1,2,...,m\in\mathbb{N}$ and $\xi_n\in(0,1]$, we get

$$p^{*}_{k,\xi_n}\leq 4\,(m+1)^{m+1}\left(e^{-\frac{1}{\xi_n}}+\xi_n\right). \tag{3.47}$$

Thus, by (3.47) , we have

$$\sum_{k=1}^{m}p^{*}_{k,\xi_n}\leq 4m\,(m+1)^{m+1}\left(e^{-\frac{1}{\xi_n}}+\xi_n\right)=S_m\left(e^{-\frac{1}{\xi_n}}+\xi_n\right)$$

which completes the proof. □

For $m\in\mathbb{N}$, we give the following estimate.

THEOREM 3.9. *Let $(\xi_n)_{n\in\mathbb{N}}$, $0<\xi_n\leq 1$, be a sequence of positive real numbers. Then, for each fixed $m,r\in\mathbb{N}$ and for every $f\in C^{m}(\mathbb{R})$ with $f^{(m)}\in C_u(\mathbb{R})$ such that*

$$\left\|f^{(m)}(x)\right\|_{\infty,x}<\infty,$$

we have

$$\left\|W^{*[m]}_{r,n}(f;x)-f(x)\right\|_{\infty,x}\leq C_{r,m}\omega_r(f^{(m)},\xi_n)+T_{r,m}\left(e^{-\frac{1}{\xi_n}}+\xi_n\right), \tag{3.48}$$

where $T_{r,m}$ *and* $C_{r,m}$ *are constants depending on* r *and* m *defined as in* (3.17) *and* (3.18) ; *respectively.*

PROOF. By (2.14) and the sub-additivity of the norm, we get

$$\left\| W_{r,n}^{*[m]}(f;x) - f(x) \right\|_{\infty,x} \leq \frac{K_{r,m}}{m!}\omega_r(f^{(m)},\xi_n) + \left\| \sum_{k=1}^{m} \frac{f^{(k)}(x)}{k!}\delta_{k,r}^{[m]} p_{k,\xi_n}^* \right\|_{\infty,x}. \tag{3.49}$$

In order to estimate

$$\left\| \sum_{k=1}^{m} \frac{f^{(k)}(x)}{k!}\delta_{k,r}^{[m]} p_{k,\xi_n}^* \right\|_{\infty,x},$$

we observe that

$$\left\| \sum_{k=1}^{m} \frac{f^{(k)}(x)}{k!}\delta_{k,r}^{[m]} p_{k,\xi_n}^* \right\|_{\infty,x} \leq \sum_{k=1}^{m} \frac{\left\| f^{(k)}(x) \right\|_{\infty,x}}{k!}\delta_{k,r}^{[m]} p_{k,\xi_n}^*. \tag{3.50}$$

By combining the results (3.42) and (3.50) , we have

$$\left\| \sum_{k=1}^{m} \frac{f^{(k)}(x)}{k!}\delta_{k,r}^{[m]} p_{k,\xi_n}^* \right\|_{\infty,x} \leq T_{r,m}\left(e^{-\frac{1}{\xi_n}} + \xi_n\right). \tag{3.51}$$

By (3.49) and (3.51) , we get the desired result. □

Next, we have

PROPOSITION 3.10. *Let* $r \in \mathbb{N}$. *For every* $(\xi_n)_{n\in\mathbb{N}}$ *such that* $0 < \xi_n \leq 1$,

$$\frac{\sum\limits_{\nu=-\infty}^{\infty}\left(1+\frac{|\nu|}{\xi_n}\right)^r e^{-\frac{\nu^2}{\xi_n}}}{\sum\limits_{\nu=-\infty}^{\infty} e^{-\frac{\nu^2}{\xi_n}}} < H_r \tag{3.52}$$

holds. Here, H_r *is a constant depending on* r, *given as* (3.29) .

PROOF. Since

$$\sum_{\nu=-\infty}^{\infty} e^{-\frac{\nu^2}{\xi_n}} > 1,$$

we get

$$\frac{1}{\sum\limits_{\nu=-\infty}^{\infty} e^{-\frac{\nu^2}{\xi_n}}} < 1. \tag{3.53}$$

Therefore, by (3.53), we obtain

$$\frac{\sum_{\nu=-\infty}^{\infty}\left(1+\frac{|\nu|}{\xi_n}\right)^r e^{-\frac{\nu^2}{\xi_n}}}{\sum_{\nu=-\infty}^{\infty} e^{-\frac{\nu^2}{\xi_n}}} < \sum_{\nu=-\infty}^{\infty}\left(1+\frac{|\nu|}{\xi_n}\right)^r e^{-\frac{\nu^2}{\xi_n}} = 1+2\sum_{\nu=1}^{\infty}\left(1+\frac{\nu}{\xi_n}\right)^r e^{-\frac{\nu^2}{\xi_n}}. \tag{3.54}$$

Additionally, for all $\nu \in \mathbb{N}$, and $(\xi_n)_{n\in\mathbb{N}}$ such that $0 < \xi_n \leq 1$,

$$e^{-\frac{\nu}{\xi_n}} \geq e^{-\frac{\nu^2}{\xi_n}}$$

holds. Hence, for all $r \in \mathbb{N}$, we have

$$\left(1+\frac{\nu}{\xi_n}\right)^r e^{-\frac{\nu}{\xi_n}} \geq \left(1+\frac{\nu}{\xi_n}\right)^r e^{-\frac{\nu^2}{\xi_n}}$$

which implies

$$\sum_{\nu=1}^{\infty}\left(1+\frac{\nu}{\xi_n}\right)^r e^{-\frac{\nu}{\xi_n}} \geq \sum_{\nu=1}^{\infty}\left(1+\frac{\nu}{\xi_n}\right)^r e^{-\frac{\nu^2}{\xi_n}}. \tag{3.55}$$

Thus, by (3.54) and (3.55), we obtain

$$\frac{\sum_{\nu=-\infty}^{\infty}\left(1+\frac{|\nu|}{\xi_n}\right)^r e^{-\frac{\nu^2}{\xi_n}}}{\sum_{\nu=-\infty}^{\infty} e^{-\frac{\nu^2}{\xi_n}}} < 1+2\sum_{\nu=1}^{\infty}\left(1+\frac{\nu}{\xi_n}\right)^r e^{-\frac{\nu}{\xi_n}}. \tag{3.56}$$

Hence, by (3.56) and the proof of Proposition 3.3, we get the desired result. □

For $m = 0$, we give the following estimate.

THEOREM 3.11. *Let $(\xi_n)_{n\in\mathbb{N}}$, $0 < \xi_n \leq 1$, be a sequence of positive real numbers. Then, for each fixed $r \in \mathbb{N}$ and for every $f \in C_u(\mathbb{R})$ such that*

$$\|f(x)\|_{\infty,x} < \infty,$$

we have

$$\left\| W_{r,n}^{*[0]}(f;x) - f(x) \right\|_{\infty,x} \leq H_r \omega_r(f,\xi_n). \tag{3.57}$$

Here H_r is a constant, depending on r, given as (3.29).

PROOF. By Theorem 2.6 and Proposition 3.10. □

Now, we demonstrate our statistical approximation result for the operators $W_{r,n}^{*[m]}$ for the case of $m \in \mathbb{N}$.

THEOREM 3.12. *Let $m, r \in \mathbb{N}$ and $A = [a_{jn}]$ be a non-negative regular summability matrix, and let $(\xi_n)_{n\in\mathbb{N}}$, $0 < \xi_n \leq 1$, be a sequence of positive real numbers for which*

$$st_A - \lim_n \xi_n = 0. \tag{3.58}$$

Then, for each fixed $m, r \in \mathbb{N}$ and for every $f \in C^m(\mathbb{R})$ with $f^{(m)} \in C_u(\mathbb{R})$ such that

$$\left\| f^{(m)}(x) \right\|_{\infty,x} < \infty,$$

we have

$$st_A - \lim_n \left\| W_{r,n}^{*[m]}(f;x) - f(x) \right\|_{\infty,x} = 0. \tag{3.59}$$

PROOF. By Lemma 3.5 and (3.58), we can say that

$$st_A - \lim_n \omega_r(f^{(m)}, \xi_n) = 0 \tag{3.60}$$

and

$$st_A - \lim_n e^{-\frac{1}{\xi_n}} = 0. \tag{3.61}$$

Now, for given $\epsilon > 0$, we define the following sets

$$\begin{aligned} R \; &: = \left\{ n \in \mathbb{N} : \left\| W_{r,n}^{*[m]}(f;x) - f(x) \right\|_{\infty,x} \geq \epsilon \right\}, \\ R_1 \; &: = \left\{ n \in \mathbb{N} : \omega_r(f^{(m)}, \xi_n) \geq \frac{\epsilon}{2C_{r,m}} \right\}, \\ R_2 \; &: = \left\{ n \in \mathbb{N} : e^{-\frac{1}{\xi_n}} \geq \frac{\epsilon}{4T_{r,m}} \right\}, \\ R_3 \; &: = \left\{ n \in \mathbb{N} : \xi_n \geq \frac{\epsilon}{4T_{r,m}} \right\}, \end{aligned} \tag{3.62}$$

where $T_{r,m}$ and $C_{r,m}$ defined as (3.17) and (3.18); respectively. Then, it follows from (3.48) that

$$R \subseteq \bigcup_{i=1}^{3} R_i$$

which yields, for every $j \in \mathbb{N}$, that

$$\sum_{n \in R} a_{jn} \leq \sum_{i=1}^{3} \sum_{n \in R_i} a_{jn}. \tag{3.63}$$

If we take the limit of the both sides of (3.36) as $j \to \infty$, by (3.58), (3.60), and (3.61), we have

$$\lim_{j \to \infty} \sum_{n \in R} a_{jn} = 0,$$

whence the result. □

When $m = 0$, we have the following statistical approximation result for the operators $W_{r,n}^{*[m]}$.

THEOREM 3.13. *Let $r \in \mathbb{N}$ and $A = [a_{jn}]$ be a non-negative regular summability matrix, and let $(\xi_n)_{n \in \mathbb{N}}$, $0 < \xi_n \leq 1$, be a sequence of positive real numbers for which*

$$st_A - \lim_n \xi_n = 0. \tag{3.64}$$

Then, for each fixed $r \in N$ and for every $f \in C_u(\mathbb{R})$ such that

$$\left\| f(x) \right\|_{\infty,x} < \infty,$$

we have

$$st_A - \lim_n \left\| W_{r,n}^{*[0]}(f;x) - f(x) \right\|_{\infty,x} = 0. \tag{3.65}$$

PROOF. By Lemma 3.5 and (3.64), we can say that

$$st_A - \lim_n \omega_r(f, \xi_n) = 0 \tag{3.66}$$

Now, for given $\epsilon > 0$, we define the following sets

$$\begin{aligned} R &: = \left\{ n \in \mathbb{N} : \left\| W_{r,n}^{*[0]}(f;x) - f(x) \right\|_{\infty,x} \geq \epsilon \right\}, \\ R_1 &: = \left\{ n \in \mathbb{N} : \omega_r(f, \xi_n) \geq \frac{\epsilon}{H_r} \right\}, \end{aligned} \tag{3.67}$$

where H_r is given as (3.29). Then, it follows from (3.57) that

$$R \subseteq R_1$$

which yields, for every $j \in \mathbb{N}$, that

$$\sum_{n \in R} a_{jn} \leq \sum_{n \in R_1} a_{jn}. \tag{3.68}$$

If we take the limit of the both sides of (3.68) as $j \to \infty$, by (3.66), we have

$$\lim_{j \to \infty} \sum_{n \in R} a_{jn} = 0,$$

which completes the proof. □

Finally, we present an example which shows that our statistical approximation results in Theorems 3.6, 3.7, 3.12, and 3.13 are stronger than the classical uniform approximation. Let $A = C_1$, the the Cesáro matrix of order one, and define

$$\xi_n = \begin{cases} \frac{n}{n+1}, & \text{if } n = k^2,\ k = 1, 2, ... \\ \frac{1}{n}, & \text{otherwise.} \end{cases}$$

Observe that $(\xi_n)_{n \in \mathbb{N}}$ is not convergent in ordinary sense. However,

$$st_{C_1} - \lim_n \xi_n = st - \lim_n \xi_n = 0.$$

Then the Theorems 3.6, 3.7, 3.12, and 3.13 are valid for the operators $P_{r,n}^{*[m]}$ and $W_{r,n}^{*[m]}$ constructed by $(\xi_n)_{n \in \mathbb{N}}$; but, in the usual sense, it is impossible to approximate f by these operators.

References

[1] G. A. Anastassiou and O. Duman, *Statistical approximation by double Picard singular integral operators*, Stud. Univ. Babeş-Bolyai Math. **55** (2010), no. 1, 3–20. MR2645955

[2] G. A. Anastassiou and O. Duman, *Uniform approximation in statistical sense by double Gauss-Weierstrass singular integral operators*, Nonlinear Funct. Anal. Appl. **15** (2010), no. 3, 487–502. MR2906655

[3] G. A. Anastassiou and R. A. Mezei, *Approximation by singular integrals*, Cambridge Scientific Publishers, Cambridge, UK, 2012.

[4] G. A. Anastassiou and M. Kester, *Quantitative uniform approximation by generalized discrete singular operators*, Stud. Univ. Babeş-Bolyai Math. **60** (2015), no. 1, 39–60. MR3335783

[5] H. Fast, *Sur la convergence statistique* (French), Colloquium Math. **2** (1951), 241–244 (1952), DOI 10.4064/cm-2-3-4-241-244. MR0048548

[6] A. R. Freedman and J. J. Sember, *Densities and summability*, Pacific J. Math. **95** (1981), no. 2, 293–305. MR632187

[7] F. Smarandache, *A triple inequality with series and improper integrals*, Bull. Pure Appl. Sci. Sect. E Math. Stat. **25** (2006), no. 1, 215–217. MR2513941

DEPARTMENT OF MATHEMATICS AND COMPUTER SCIENCE, TUSCULUM UNIVERSITY, TUSCULUM, TENNESSEE 37745

Email address: `mkester@tusculum.edu`

SELECTED PUBLISHED TITLES IN THIS SERIES

737 **Fernanda Botelho, Editor,** Recent Trends in Operator Theory and Applications, 2019

734 **Peter Kuchment and Evgeny Semenov, Editors,** Differential Equations, Mathematical Physics, and Applications, 2019

733 **Peter Kuchment and Evgeny Semenov, Editors,** Functional Analysis and Geometry, 2019

732 **Samuele Anni, Jay Jorgenson, Lejla Smajlović, and Lynne Walling, Editors,** Automorphic Forms and Related Topics, 2019

731 **Robert G. Niemeyer, Erin P. J. Pearse, John A. Rock, and Tony Samuel, Editors,** Horizons of Fractal Geometry and Complex Dimensions, 2019

730 **Alberto Facchini, Lorna Gregory, Sonia L'Innocente, and Marcus Tressl, Editors,** Model Theory of Modules, Algebras and Categories, 2019

729 **Daniel G. Davis, Hans-Werner Henn, J. F. Jardine, Mark W. Johnson, and Charles Rezk, Editors,** Homotopy Theory: Tools and Applications, 2019

728 **Nicolás Andruskiewitsch and Dmitri Nikshych, Editors,** Tensor Categories and Hopf Algebras, 2019

727 **André Leroy, Christian Lomp, Sergio López-Permouth, and Frédérique Oggier, Editors,** Rings, Modules and Codes, 2019

726 **Eugene Plotkin, Editor,** Groups, Algebras and Identities, 2019

725 **Shijun Zheng, Marius Beceanu, Jerry Bona, Geng Chen, Tuoc Van Phan, and Avy Soffer, Editors,** Nonlinear Dispersive Waves and Fluids, 2019

724 **Lubjana Beshaj and Tony Shaska, Editors,** Algebraic Curves and Their Applications, 2019

723 **Donatella Danielli, Arshak Petrosyan, and Camelia A. Pop, Editors,** New Developments in the Analysis of Nonlocal Operators, 2019

722 **Yves Aubry, Everett W. Howe, and Christophe Ritzenthaler, Editors,** Arithmetic Geometry: Computation and Applications, 2019

721 **Petr Vojtěchovský, Murray R. Bremner, J. Scott Carter, Anthony B. Evans, John Huerta, Michael K. Kinyon, G. Eric Moorhouse, and Jonathan D. H. Smith, Editors,** Nonassociative Mathematics and its Applications, 2019

720 **Alexandre Girouard, Editor,** Spectral Theory and Applications, 2018

719 **Florian Sobieczky, Editor,** Unimodularity in Randomly Generated Graphs, 2018

718 **David Ayala, Daniel S. Freed, and Ryan E. Grady, Editors,** Topology and Quantum Theory in Interaction, 2018

717 **Federico Bonetto, David Borthwick, Evans Harrell, and Michael Loss, Editors,** Mathematical Problems in Quantum Physics, 2018

716 **Alex Martsinkovsky, Kiyoshi Igusa, and Gordana Todorov, Editors,** Surveys in Representation Theory of Algebras, 2018

715 **Sergio R. López-Permouth, Jae Keol Park, S. Tariq Rizvi, and Cosmin S. Roman, Editors,** Advances in Rings and Modules, 2018

714 **Jens Gerlach Christensen, Susanna Dann, and Matthew Dawson, Editors,** Representation Theory and Harmonic Analysis on Symmetric Spaces, 2018

713 **Naihuan Jing and Kailash C. Misra, Editors,** Representations of Lie Algebras, Quantum Groups and Related Topics, 2018

712 **Nero Budur, Tommaso de Fernex, Roi Docampo, and Kevin Tucker, Editors,** Local and Global Methods in Algebraic Geometry, 2018

711 **Thomas Creutzig and Andrew R. Linshaw, Editors,** Vertex Algebras and Geometry, 2018

For a complete list of titles in this series, visit the
AMS Bookstore at **www.ams.org/bookstore/conmseries/**.